NOBEL LECTURES

PHYSIOLOGY
OR
MEDICINE

2001 – 2005

Nobel Lectures

Including Presentation Speeches and Laureates' Biographies

Physics

Chemistry

Physiology or Medicine

Literature

Peace

Economic Sciences

NOBEL LECTURES

INCLUDING PRESENTATION SPEECHES
AND LAUREATES' BIOGRAPHIES

PHYSIOLOGY OR MEDICINE

2001–2005

EDITOR

Hans Jörnvall

Karolinska Institutet
Sweden

World Scientific

NEW JERSEY • LONDON • SINGAPORE • BEIJING • SHANGHAI • HONG KONG • TAIPEI • CHENNAI

Published by

World Scientific Publishing Co. Pte. Ltd.
5 Toh Tuck Link, Singapore 596224
USA office: 27 Warren Street, Suite 401-402, Hackensack, NJ 07601
UK office: 57 Shelton Street, Covent Garden, London WC2H 9HE

NOBEL LECTURES IN PHYSIOLOGY OR MEDICINE (2001–2005)

ISBN-13 978-981-279-441-3
ISBN-10 981-279-441-7
ISBN-13 978-981-279-442-0 (pbk)
ISBN-10 981-279-442-5 (pbk)

Printed in Singapore by Mainland Press Pte Ltd

FOREWORD

Since 1901 the Nobel Foundation has published annually *Les Prix Nobel* with reports from the Nobel award ceremonies in Stockholm and Oslo as well as the biographies and Nobel lectures of the Nobel laureates. In order to make the lectures available for people with special interests in the different fields the Nobel Foundation gave Elsevier Publishing Company the right to publish in English the lectures for 1901–1970, which were published in 1964–1972 through the following volumes:

Physics 1901–1970	4 vols.
Chemistry 1901–1970	4 vols.
Physiology or Medicine 1901–1970	4 vols.
Literature 1901–1967	1 vol.
Peace 1901–1970	3 vols.

Thereafter, and until the year 2000, the Nobel Foundation has given World Scientific Publishing Company the right to bring the series up to date and also publish the Prize Lectures in Economics from the year 1969. Between the years 2001–2005, the Nobel Foundation has given World Scientific Publishing Company a publishing right through an agreement between Nobel Media AB and World Scientific Publishing Company.

The Nobel Foundation is very pleased that the intellectual and spiritual message to the world laid down in the laureates' lectures, thanks to the efforts of World Scientific, will reach new readers all over the world.

Michael Sohlman
Executive Director
Stockholm, April 2008

PREFACE

This volume presents great discoveries in Physiology or Medicine as reflected by the Nobel Prizes in this subject given for the years 2001–2005. It also details the backgrounds of the Laureates as told by themselves. The discoveries illustrate the progress of natural science in relation to life, resulting in new possibilities for mankind. Alfred Nobel gave the conditions for the Prize: "to those who shall have conferred the greatest benefit on mankind" within the "domain of physiology or medicine". Indeed, the scientists and fields recognized by the Nobel Assembly at Karolinska Institutet during these five years cover central discoveries in the medical Prize domain:

* "key regulators of the cell cycle" — in 2001 to Leland Hartwell, Tim Hunt and Sir Paul Nurse — showed the components and mechanisms behind cell divisions, and hence growth. Obviously, those discoveries are important for understanding both normal physiology and many disease developments, including those in tumor control.
* "genetic regulation of organ development and programmed cell death" — in 2002 to Sydney Brenner, Robert Horvitz and John Sulston — showed the components and mechanisms behind formation and removal of cellular patterns in organs and tissues. Again, discoveries of importance to understand normal physiology and disease states.
* "magnetic resonance imaging" — in 2003 to Paul Lauterbur and Sir Peter Mansfield — opened a new mode of imaging. This method is now in routine use clinically to detect and follow internal alterations, thus facilitating diagnosis, monitoring of disease, and selection of treatments in many patients.
* "odorant receptors and the organization of the olfactory system" — in 2004 to Richard Axel and Linda Buck — clarified the mode of action of the sense of smell and of its pathway in signaling to the brain. A rare event, to define molecular mechanisms in a human sense organ, and to trace relationships explaining complex functions.
* "the bacterium Helicobacter pylori and its role in gastritis and peptic ulcer disease" — in 2005 to Barry Marshall and Robin Warren — a discovery that led to immediate benefit for patients and hospitals by introducing effective treatment and changing the whole progress for a common human disease previously giving much suffering and considered to be chronic.

This 5-year period, with all its Awardees and Awards, excellently illustrates the progress of research to give molecular and cellular understanding of diseases and normal physiology. It introduces novel and effective tools in diagnosis and treatment, for the benefit on mankind as desired by Alfred Nobel.

Finally, the Awards illustrate that even today, when research often requires large teams, heavy equipment, and big grants, it is still possible to discern individual achievements, the role of personal creativity and the importance of simple observations in experimental designs and clinical practice. During this 5-year period, the Award was given three times to two and just twice to three scientists. We are grateful to Alfred Nobel for his vision of promoting science through recognition of discoveries, and to the Laureates for their creativity and insight that have opened novel prospects to human well-being in health and disease.

Hans Jörnvall

CONTENTS

Physiology or Medicine 2001

LELAND H. HARTWELL, TIM HUNT and Sir PAUL M. NURSE

"for their discoveries of key regulators of the cell cycle"

THE NOBEL PRIZE IN PHYSIOLOGY OR MEDICINE

Speech by Professor Anders Zetterberg of the Nobel Committee at Karolinska Institutet.
Translation of the Swedish text.

Your Majesties, Your Royal Highnesses, Honoured Nobel Laureates, Ladies and Gentlemen,

Cell division is a fundamental process of life. All living organisms on earth are descended from an ancestral cell that appeared about 3 billion years ago, and which has undergone an unbroken series of cell divisions since then. Each human being also began life as one single cell – a cell that divided repeatedly to give rise to all one hundred thousand billion cells that we consist of. Every second millions of cells divide in our body.

The cycle of events that a cell completes from one division to the next is called the cell cycle. During the cell cycle the cell grows in size, it duplicates its hereditary material – that is, it copies the DNA molecules in the chromosomes – and it divides into two daughter cells.

This year's Nobel Laureates have discovered the key regulators of the cell cycle, cyclin dependent kinase (CDK) and cyclin. Together these two components form an enzyme, in which CDK is comparable to a "molecular engine" that drives the cell through the cell cycle by altering the structure and function of other proteins in the cell. Cyclin is the main switch that turns the "CDK engine" on and off. This cell cycle engine operates in the same way in such widely disparate organisms as yeast cells, plants, animals and humans.

How were the key regulators CDK and cyclin discovered?

Lee Hartwell realized the great potential of genetic methods for cell cycle studies. He chose baker's yeast as a model organism. In the microscope he could identify genetically altered cells – mutated cells – that stopped in the cell cycle when they were cultured at an elevated temperature. Using this method Hartwell discovered, in the early 1970s, dozens of genes specific to the cell division cycle, which he named CDC genes. One of these genes, CDC28, controls the initiation of each cell cycle, the "start" function. Hartwell also formulated the concept of "checkpoints," which ensure that cell cycle events occur in the correct order. Checkpoints are comparable to the program in a washing machine that checks if one step has been properly completed before the next can start. Checkpoint defects are considered to be one of the reasons behind the transformation of normal cells into cancer cells.

Paul Nurse also used the genetic approach in his cell cycle studies, but in a different kind of yeast. In the late 1970s and early 1980s he discovered the gene cdc2, which could be mutated in two different ways. Either the cells did not divide, or they divided too early. From this he correctly concluded that cdc2 controls cell division. He later discovered that cdc2 not only controls cell division, the final event of the cell cycle, but has a key regulatory function for the whole cell cycle, including that described for CDC28 in baker's yeast. This key function was shown to be that of CDK in the cell cycle engine. By moving human genes into yeast cells, in 1987 Nurse isolated a human cdc2 gene. This human cdc2 gene functioned perfectly in yeast cells. Thus, the CDK function in the cell cycle engine had been conserved through more than one billion years of evolution – from yeast to man.

Tim Hunt discovered the other key component of the cell cycle engine, the protein cyclin, which regulates the function of the CDK molecule. Working with sea urchin eggs as a model organism, in 1982 he discovered a specific protein that increased in amount before cell division but disappeared abruptly when the cells divided. Because of these cyclical variations, he named the protein cyclin. These experiments not only led to the discovery of cyclin, but also demonstrated the existence of periodic protein degradation in the cell cycle – a fundamental control mechanism. Hunt also showed the existence of cyclins in other, unrelated species. Thus cyclins, like CDK, had been conserved during evolution.

It is now almost fifty years since the structure of the DNA molecule – the double helix – was discovered, leading to a molecular explanation of how a gene can make a copy of itself. With the discoveries of CDK and cyclin we are now beginning to understand, at the molecular level, how the cell can make a copy of itself.

Dr. Hartwell, Dr. Hunt and Dr. Nurse,

Your fundamental discoveries have profoundly increased our understanding of how the cell cycle is controlled. This new knowledge has a huge impact on cell biology with broad applications in many fields of biology and medicine.

On behalf of the Nobel Assembly at Karolinska Institutet I wish to convey our warmest congratulations, and I now ask you to step forward to receive the Nobel Prize from the hands of His Majesty the King.

LELAND H. HARTWELL

I don't remember much about my early childhood. I am told that before I began school, my older cousin would return from her school and teach me what she had learned. By the time I was 10 or so, I was displaying a curiosity that might have suggested an academic bent. I was an avid collector of bugs, butterflies, lizards, snakes, and spiders. I remember on one of these adventures learning to be skeptical of everything one reads in books. I had read that lizards do not have teeth. I had grabbed a very big lizard and stared in disbelief as it turned its head, displayed a fine set of teeth and sank them into my thumb. From age 13 to 17 I was preoccupied with sports, girls and cars. But midway through high school a fortunate event happened. I became unhappy with the football coach and bored of carousing with my car club friends. A close friend of mine suggested we transfer to the cross town high school and my accommodating mother moved apartments so I could do so. At the new school I was fortunate to find some teachers that challenged me, particularly a physics teacher and my grades improved dramatically. Since no one in my family had ever been to college, I had little in the way of career guidance. However, I was good at math and physics and I liked mechanical drawing, so I thought I might become an engineer. I went for one year to a two-year junior college, Glendale Junior College, in preparation, taking math, chemistry and physics. I did well, and my counselor encouraged me to meet with a recruiter from the California Institute of Technology. This was my next major break. I took the entrance exams and was admitted into the sophomore class.

For the first time, I was in an environment of real science and it was magical. In hindsight, I can see that I was destined for science. I have always been curious and wanted to understand things at a fundamental level. For example, I used to get annoyed at my father, a neon sign man, because he would not explain to me, how neon signs work. I only realized later, that I wanted to know about electricity and atomic emissions, topics that he knew nothing about. At Cal Tech, I enrolled as a physics major. During my sophomore year, I took a lecture course in Biology, taught by James Bonner. He was incredibly enthusiastic. This must have been around 1958, only five years after the discovery of the DNA structure, and the course was all about DNA, RNA and protein. The clincher for me was an optional evening laboratory in bacteriophage genetics taught by Bob Edgar and Charlie Steinberg. Here, I learned that biology could be quantitative and it was little time before I changed my major to biology. At Cal Tech I was able to do research nearly every quarter and summer, in chemistry, biochemistry, and genetics. I worked with Hildegard Lamfrom, Howard Temin and Bob Edgar. Another feature of the

Cal Tech undergraduate education was the opportunity to take credits as reading which I did most quarters, reading the original literature in phage genetics and gene regulation.

I graduated in 1961 and went to MIT for graduate school. I had decided that I wanted to work on gene regulation and I knew of only two prominent labs in the US, Arthur Pardee at Berkeley, and Boris Magasanik, at MIT. I visited Arthur Pardee at Berkeley and he was not encouraging so I went to MIT. Being a graduate student at MIT was great fun and very intense. Boris Magasanik would come through the lab each afternoon and stop at each desk to discuss the previous day's results. I wanted a new result each day so that set quite a pace! I was there only about two and a half years when Boris came to me and said that I had essentially finished my thesis but since no one could leave that soon I would have to stay another year.

By this time Francois Jacob and Jacques Monod had published their famous work on the lac operon and I figured that gene regulation was pretty well understood. I thought it appropriate to do my postdoctoral work on something much less well understood and I decided to work on the control of cell growth. Going to the literature again, I was most impressed with the work coming out of Renato Dulbecco's laboratory, so I applied to Renato and he accepted me. My timing was excellent because Renato was moving to the Salk Institute but the buildings had not been built. We were quartered in temporary buildings that kept the lab groups very small. Most of the original fellows at the Salk had only one or two postdocs so it was very intimate. I spent a lot of time talking with Marguerite Vogt, Renato's long time collaborator. Marguerite's enthusiasm for cancer research and her adeptness at cell culture were a joy to behold and try to learn from. During that period I spent about half my time working on polyoma virus infected cells, the main research program of the laboratory, and about half the time trying to get something of my own started. The polyoma work went very well. DNA hybridization was a relatively new technique and Nada Ledinko had come to Renato's lab with experience with this technique. We used it to show that polyoma infected cells induced cellular DNA synthesis and a complement of cellular DNA replication enzymes. This seemed like it might be fundamentally interesting to the cancer problem. However, none of the three projects that I worked on as possible starting points for my own entry into research worked. I left the lab after only about 18 months as a postdoc, taking a position as assistant professor at the new University of California, Irvine, in 1965.

I had applied for and gotten a grant to study the control of cellular DNA synthesis but I was not enthusiastic about the prospects for reaching fundamental insights with mammalian cells. It was again a fortunate circumstance. While I was waiting for my newly ordered equipment to arrive, I spent considerable time in the library trying to figure out what to do. Dan Wulff, another new recruit, suggested that I find a model eukaryotic cell that could be used for genetic studies of cell growth. I thought this was a great idea and went to the library to discover that *Saccharomyces cerevisiae*, baker's yeast, was one of the only singled-celled eukaryotic organisms with facile genetics. I visited Bob

Mortimer at Berkeley and Herschel Roman and Don Hawthorne at Seattle to find out how one worked with yeast. They were both incredibly kind and generous. Hersch and Don loaned me a micromanipulator to start my work.

I began by isolating temperature-sensitive mutants and analyzing them for macromolecule synthesis and cell division at the restrictive temperature. This approach came directly from my earlier experience with Bob Edgar at Cal Tech where I had seen the power of conditional mutants to analyze phage morphogenesis. Three years later, Herschel Roman invited me to join the faculty at the genetics department in Seattle which I did. Soon after that move, we discovered the power of time-lapse photomicroscopy to study the cell cycle.

Looking back over my life, I feel in some sense destined because of my native interest in understanding things at a fundamental level. And I also feel extremely fortunate for a great number of people who intervened or mentored me on my path. I sought a rigorous approach to mysterious problems and found people at each step that pointed the way. I feel very fortunate to have become a biologist at a time when curiosity was paramount and to have been trained by people who had high standards.

YEAST AND CANCER

Nobel Lecture, December 9, 2001

by

LELAND H. HARTWELL

Fred Hutchinson Cancer Research Center, 100 Fairview Avenue North, D1-060 Seattle, WA 98109-2024, USA.

My research career has been motivated by a desire to understand cancer. Each time I have identified an intriguing aspect of the cancer problem, I have found that it could be approached more effectively in the simpler eukaryotic cell, *Saccharomyces cerevisiae,* than the human cell. Each time the yeast cell has revealed some of its secrets. I will relate four vignettes involving studies on the genetics of cell division, the control of genome fidelity, therapeutics for cancer and the role of natural genetic variation in disease susceptibility. In the first two instances, yeast has told us something that is relevant to mankind. For the other two, it is too soon to tell.

When I was finishing my graduate studies and thinking about what area of science to pick for my postdoctoral work, I wanted to study a field that was still a mystery and would offer ample opportunity for a career. I chose to study the control of growth and proliferation in relation to cancer cells with Renato Dulbecco and Marguerite Vogt. Even in plastic petri dishes, cancer cells grew when normal cells did not and I became thoroughly engrossed with the problem. However, while in Renato's laboratory I had not been able to establish an experimental system that I thought could lead to fundamental insights. When I set up my own laboratory at the University of California at Irvine I was not sure what to work on.

I had applied for and received a grant to work on the control of DNA synthesis in mammalian cells. However, after much pondering and an influential conversation with Dan Wulff, I decided to work on a model system that permitted genetic analysis. I had been imprinted as an undergraduate with the insights that Bob Edgar and Bill Wood had attained on the pathway of viral morphogenesis using genetics. The processes of the cell cycle, chromosome replication and segregation, seemed like morphogenetic problems of the same nature. Yeast was the obvious choice because it grew as single cells, an important property for studies of cell division. Moreover, Don Williamson had shown that *S. cerevisiae* had a eukaryotic cell cycle with G1, S, G2 and M periods (1) and C.F. Robinow had demonstrated the presence of an intranuclear spindle (2). These were important facts since yeast has, at various times, been accused of not being a proper eukaryote. There was even a time when people thought that yeast lacked DNA.

GENES THAT CONTROL CELL DIVISION

Since cell division was an essential process, I set out isolating temperature-sensitive mutants that could grow at room temperature but not at 36°C. We isolated about a thousand mutants and characterized each for protein, RNA, and DNA synthesis, cell division, and cell morphology following a shift from the permissive to the restrictive temperature. Although most of the mutants had unremarkable phenotypes, a significant number behaved as if defective in a specific one of these processes (3).

I thought it important to demonstrate that we could identify the defective protein in some mutants. Calvin McLaughlin (Fig. 1) who was an expert in protein synthesis generously agreed to collaborate and we began studying the protein synthesis mutants. We did identify mutations in genes that coded for two aminoacyl-tRNA synthetases and another mutant defective in the initiation of protein synthesis (4).

CDC genes. Three years into the project, I moved to the Genetics department at the University of Washington. I have had a number of very bright undergraduates in my lab over the years and one, Brian Reid (Fig. 2), set us on the path to the cell cycle. Although we had a few mutants that behaved like DNA synthesis or cell division mutants from the original survey, we didn't have any clever ideas about how to analyze them. Brian, through a very amusing bit of serendipity that I have described elsewhere (5), discovered how useful photomicroscopy was for analyzing the *S. cerevisiae* cell cycle. This was because the cell forms a small bud on its surface at the beginning of a cell cycle and the bud grows in size as the cycle progresses. Thus one can easily tell where a cell is in the cycle merely from its morphology. Mutants with specific defects in the cell cycle were recognized because the asynchronous population of cells grown at the permissive temperature became synchronously arrested in the cell cycle at the restrictive temperature (Fig.3).

Figure 1. Calvin McLaughlin.

Figure 2. Brian Reid.

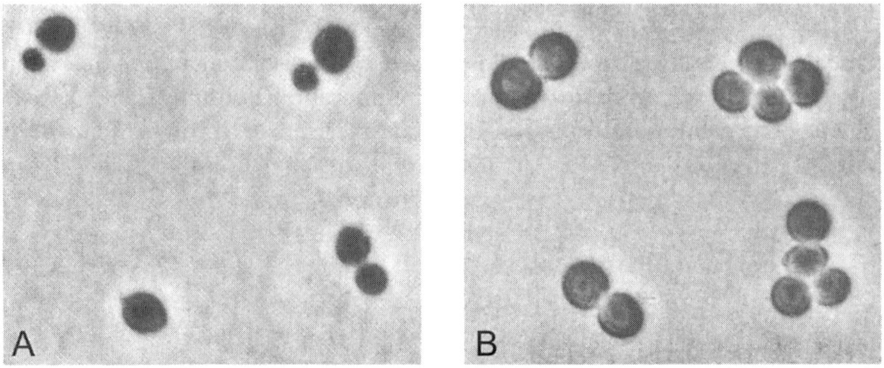

Figure 3. Time-lapse photomicroscopy of *cdc* mutant cells growing at the permissive temperature (A) and several hours after a shift to the restrictive temperature (B).

When we realized this, we were able to find hundreds of cell cycle mutants in a few months. C.F. Robinow agreed to visit the lab and teach us how to stain the nuclei of cells, and Joe Culotti (Fig. 4), a new graduate student, was able to show that the mutants that arrested with a uniform cell morphology also arrested with a uniform nuclear morphology. At this point we were excited that we had some very interesting mutants (Fig. 5) and we were confident that they would reveal new insights into the cell cycle (6).

Figure 4. Joe Culotti.

Pathways. We ordered the mutants in a number of ways. First, by how far they traversed the cycle before arresting. We analyzed the following cell cycle events: budding, DNA synthesis, nuclear division cytokinesis and cell division. The event that stopped first after a shift from the permissive to the restrictive temperature was considered the primary defect. After the primary defect, other cell cycle events would occur or not depending upon the particular mutant. Eventually, the cell would arrest development and generate a terminal phenotype, depending on which events occurred and which did not. Mutants were found with primary defects in each of the cell cycle events that we monitored. The phenotypes of the mutants suggested a relatively simple pathway of dependent events leading to cell division. The first event,

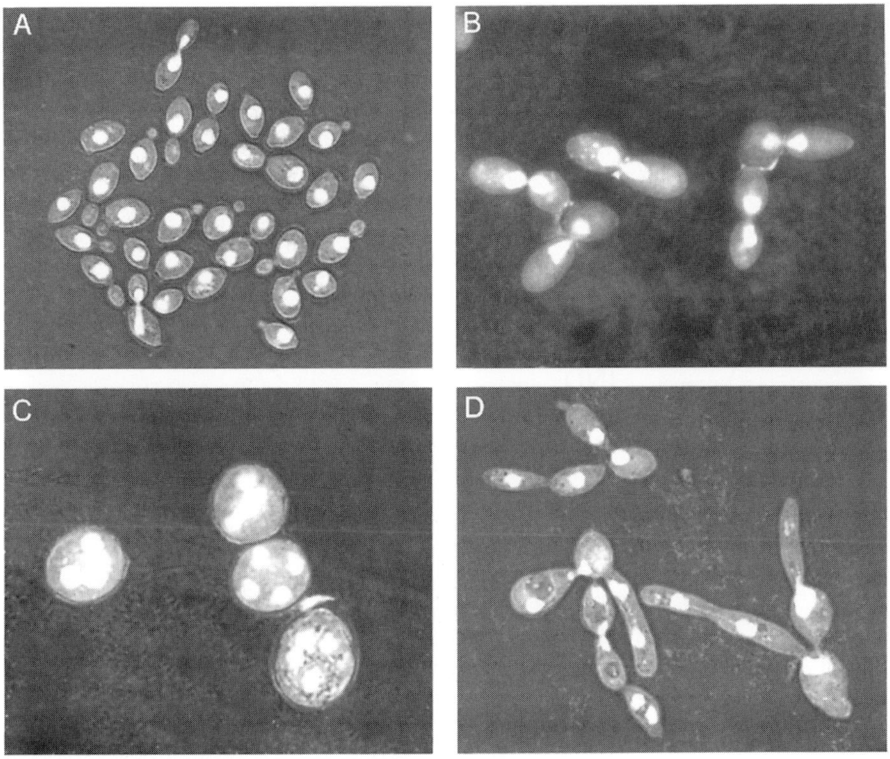

Figure 5. Normal cells and *cdc* mutant cells several hours after incubation at the restrictive temperature. (A) wild type, (B) *cdc8* (C) *cdc24* (D) *cdc10.*

controlled by the *CDC28* gene, was required for initiating two pathways, one of which led to budding, nuclear migration, cytokinesis and cell division (Fig. 6). The second led to DNA replication, nuclear division and joined the first prior to cytokinesis and cell division. We tested our conclusions by constructing all possible double mutants and, indeed, the phenotype of double mutants was precisely that expected for the pathway (7).

It is interesting, in hindsight, that the pathway of dependent events was not the least surprising to me. Although the results could have been very different, the pathway we found was sensible with late events requiring the completion of earlier events in the cycle. This is just the way that bacteriophage morphogenesis was organized.

I thought we might learn more about the mutants by looking at them at higher resolution in the electron microscope. When I approached Breck Byers about this, he immediately responded that there was only one structure worth looking at and that was the spindle and its poles. He and Loretta Goetsch examined the spindle and were able to resolve additional events of spindle pole duplication, separation of the two poles, and elongation of the spindle (Fig. 7). This analysis resolved three mutants, all required for the initiation of DNA synthesis into successive steps involving duplication of the spindle pole body (*CDC28*), separation of the poles (*CDC4*) and the initiation of DNA synthesis (*CDC7*) (8).

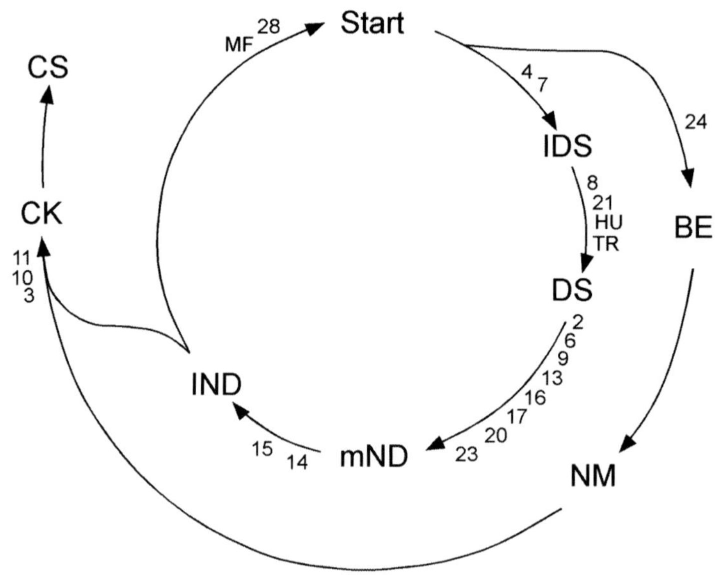

Figure 6. A pathway of gene controlled events in the *S. cerevisiae* cell cycle. Numbers refer to *cdc* genes. Abbreviations are: iDS, initiation of DNA synthesis, DS, DNA synthesis, mND, medial nuclear division; lND, late nuclear division; BE, bud emergence; NM, nuclear migration; CK, cytokinesis; CS cell separation; MF mating factor. Reprinted from ref 7 with permission.

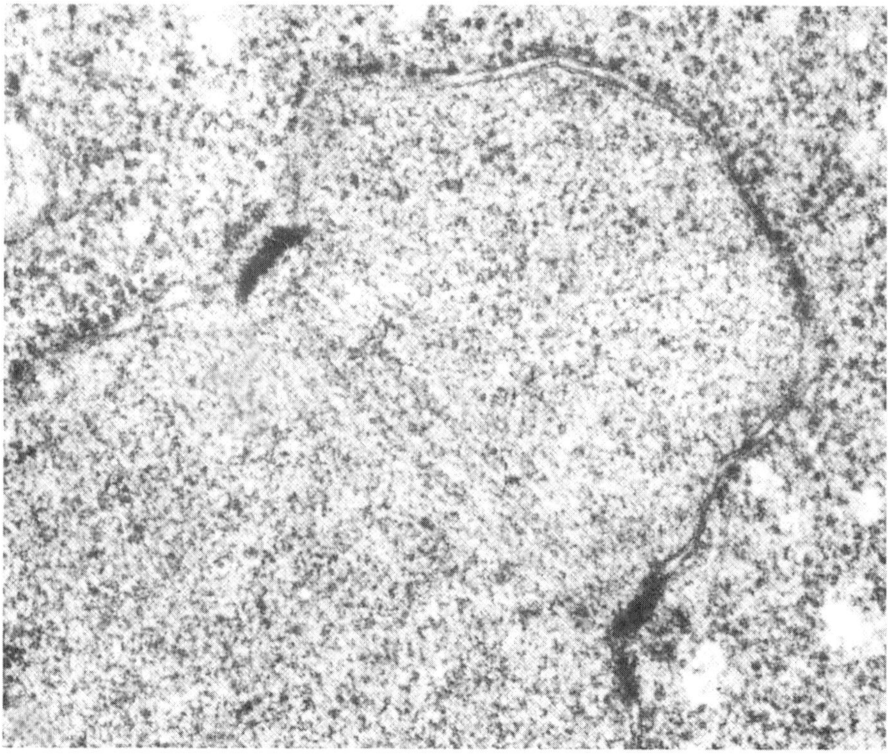

Figure 7. Electron microscopic image of the yeast nucleus revealing the electron dense spindle pole bodies imbedded in the nuclear membrane with microtubules emanating from them. Reprinted from ref 8 with permission.

The methods that we had used to order mutants into pathways utilized the phenotypes of single and double mutants. Lynna Hereford and Joe Culotti thought of a method we called "reciprocal shifts" that utilized two conditional agents that arrested the cell cycle independently. Jarvik and Botstein (9) independently introduced the same rationale to study bacteriophage morphogenesis. Lynna applied this method to mating factor (see below) and the *cdc* mutants showing that mating factor arrested cells at the *CDC28* step prior to the *CDC4* and *CDC7* steps (10). These results were in agreement with the spindle pole phenotypes determined by Byers & Goetsch.

Mating and cell division. Wolfgang Duntze and Tom Maney had discovered the presence of a mating pheromone made by *Matα* cells and Duntze had found that it inhibited DNA synthesis of *Matα* cells (11). We collaborated with them to analyze the cell cycle response to mating pheromone and found that it arrested cells at the *CDC28* step (12). Over the next few years, we investigated the relationship between the mating pathway and the division pathway. I found that when asynchronous cultures of *Mata* and *Matα* cells were mixed, they arrested one another in G1, and at the time of cell fusion, both cells were at the G1 stage (13; Fig. 8). John Pringle and Linda Wilkenson found evidence for a *Mata* pheromone that arrested *Matα* cells at the *CDC28* step (14). These experiments demonstrated that cell division was integrated with cell mating because both haploid cell types produced pheromones that arrested the other cell at the beginning of the cell cycle.

Breck Byers studied mating cells by electron microscopy and found that nuclear fusion in the zygote occurred by fusion of the two unduplicated spindle pole bodies resulting in a single nucleus with a single large spindle pole body (8). This assures that the newly formed diploid nucleus begins the cell cycle in G1 at the unduplicated spindle pole body stage. Brian Reid challenged cells to mate at other stages of the cell cycle by first arresting them with different cell cycle mutations. He found that only cells arrested at the *CDC28* controlled step were capable of mating (15).

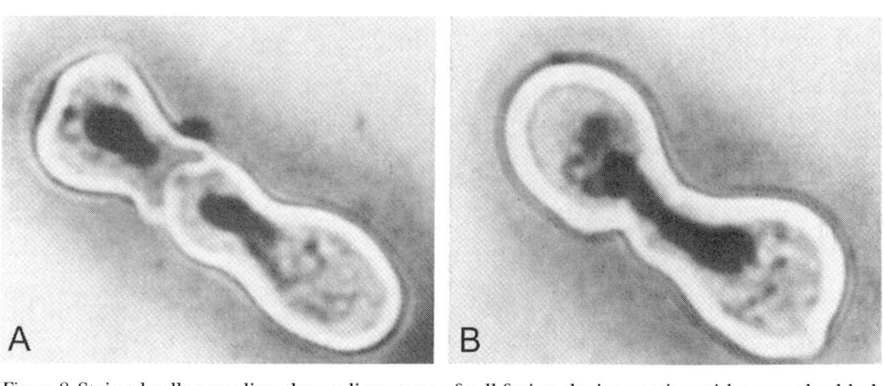

Figure 8. Stained cells revealing the earliest stage of cell fusion during mating with two unbudded, G1 cells undergoing cell fusion (A) followed by nuclear fusion (B). Reprinted from ref 13 with permission.

Figure 9. John Pringle.

Growth and division. John Pringle (Fig. 9) became interested in the regulation of growth in the cell cycle. Williamson and Scopes had found that stationary cultures of yeast cells were unbudded (16). John Pringle and Gerry Johnston followed up on this and showed that growth is integrated with the division cycle prior to budding (17), once again at the *CDC28* step. They found that cells were able to complete the cell cycle after abrupt nitrogen starvation in the absence of any net growth, but did not start new cell cycles (Fig. 10). However, if division was interrupted with a cell cycle mutation, the growth of the cell continued unabated for several hours.

Yeast monitored other essential nutrients prior to budding as well. Michael Unger used a series of mutations in the sulfur assimilation pathway to ask whether an intermediate in sulfur assimilation acted as a signal to arrest cells. The idea was that interruption of the pathway after the signal would fail to produce arrest as unbudded

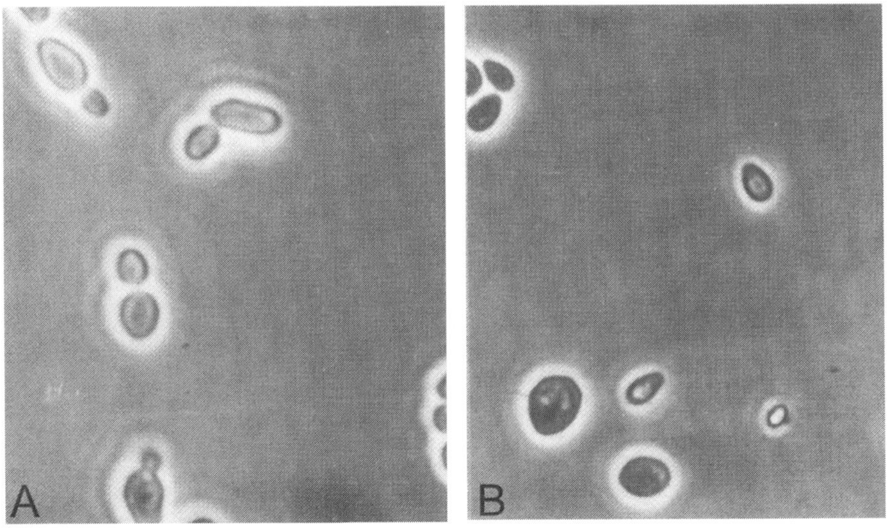

Figure 10. Growing (A) and nitrogen starved (B) yeast cells illustrating that starved cells complete cycles without growth and generate very tiny cells from new buds; all cells end up arrested in G1 as unbudded cells. Reprinted from ref 17 with permission.

cells. He found that limitation for sulfur assimilation at any step, including the charging of methionyl-tRNA resulted in arrest at the *CDC28* step, suggesting that the cell was monitoring protein synthesis or the accumulation of protein (18).

Start. The *CDC28* step was the first step in the cell cycle. From it branched the two pathways of gene activity leading to the cell surface events, budding and cytokinesis, and the nuclear events, spindle pole duplication, separation and elongation, and DNA replication and nuclear division. It was also the step where growth was integrated with division. *CDC28* is not executed until the cell reaches a specific size but once the *CDC28* event was completed the cell could finish the rest of the cycle without appreciable growth. *CDC28* was also the step at which mating was coordinated with the cell cycle. Both pheromones arrested cells at a position, interdependent with the *CDC28* step so that when the two cells fused in G1 they were synchronized at the unduplicated spindle pole body stage. Thus *CDC28* was of central importance. We called the event that was performed by *CDC28* "Start" (Fig. 11).

Given the fundamental importance of the *CDC28* gene, it is amusing how close we came to missing it in our mutant screen. We obtained only one allele in the original mutant screen and that allele was present in a mutant strain with another temperature-sensitive mutation that limited growth at the high temperature. The double mutant was a clear cell cycle mutant. However, when the second mutation was crossed away from the *cdc28* allele, the strain grew extensively and became very oddly shaped after prolonged incubation at the restrictive temperature. It is quite possible that we would have missed it all together without the second mutation in the background. As with most areas of science, the cell cycle is replete with serendipity. Paul Nurse discovered the very important wee mutants at the wrong end of a size gradient and Tim Hunt discovered cyclin while looking for something quite different.

Cyclin Dependent Kinase. My initial work on temperature-sensitive mutants had been motivated by an interest in cancer with the hope of learning about the genes that controlled cell division. The genes we discovered, especially *CDC28* seemed like they might eventually shed some light on why cancer cells divided. When Steve Reed (Fig. 12) joined the lab as a postdoc, he decided to

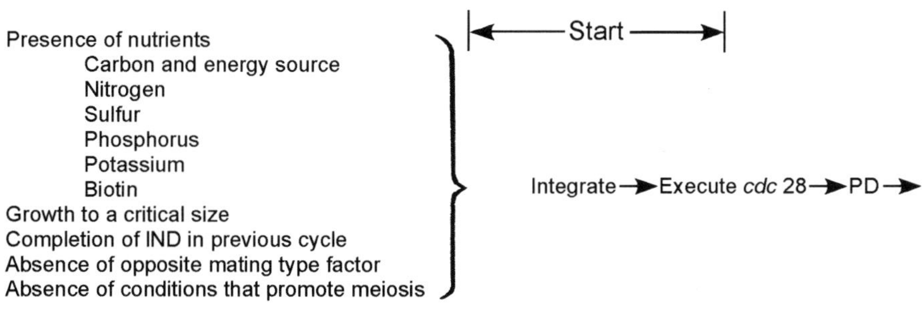

Figure 11. Summary of the events that are integrated at "Start". Reprinted from ref 49 with permission.

Figure 12. Steve Reed.

focus on the *CDC28* gene, setting up selections that uncovered new alleles as well as new genes that arrested at the same step. In 1980 he and Kim Nasmyth used recently developed techniques to clone the *CDC28* gene (19) and after Steve set up his own laboratory, he found that it had protein kinase activity (20). The ultimate unification of the cell cycle field with the discoveries relating *CDC28* of *S. cerevisiae*, *cdc2* of *S. pombe*, and MPF of Xenopus embryos to cyclin dependent kinases centered largely in the laboratory of Paul Nurse with important contributions by many others and has been elegantly described recently by Kim Nasmyth (21).

Expectations fulfilled. I find it quite interesting philosophically that two different groups approached the cell cycle with different expectations and each found exactly what they were looking for. At first, the results looked very different, perhaps because we were imposing our expectations. I approached the study of the cell cycle with the paradigm of bacteriophage morphogenesis in mind. Indeed what we found looked a lot like phage morphogenesis – a series of gene controlled, dependent events leading to progressive stages in division. Paul Nurse, on the other hand, was most interested in rate limiting steps. He found a rate-limiting step at mitosis controlled by the *cdc2* gene of *S. pombe* (22). From this point of view, the most important gene of fission yeast, *cdc2*, seemed very different than what we viewed as the most important gene in budding yeast, *CDC28*, because the former acted in G2 while the latter acted in G1. Once it became clear that the two genes could substitute for one another (23), the apparent differences were resolved. In fact, both proteins acted in each organism at G1 and G2. Under rapid growth conditions the G1 activation of *CDC28* in *S. cerevisiae* is dependent on cell growth while in *S. pombe* it is the G2 activation of *cdc2* that is dependent on growth.

GENOME FIDELITY

Cancer cells not only divide when they shouldn't but they also reproduce with far poorer fidelity than normal cells. Chromosome aberrations and chromosome losses are very common in cancer cells. This loss of fidelity seemed central to cancer because it would permit the cancer cells to evolve more quickly. We thought we might be able to learn more about how chromosome fidelity is maintained in the normal cell cycle by studying the fidelity of chromosome transmission in yeast cells.

Essential machinery. We found that normal yeast cells lose a chromosome or undergo mitotic recombination at rates of about one event in 10^5 cells. We wondered whether limitation of any of the essential cell cycle functions would alter that fidelity. An undergraduate, David Smith and I studied chromosome loss, recombination and mutation in temperature-sensitive cell cycle mutants when they were grown at their maximum permissive temperature. We found that most mutants had greatly elevated rates of chromosome loss, recombination or mutation (Fig. 13), implying that defects in these functions could be important in the fidelity of cell division (24).

Other essential components were also important for fidelity. Doug Koshland examined the role of centromere structure and dicentrics on chromosome fidelity (25); Meeks-Wagner defined the importance of correct ratios of histones (26); and Megan Brown studied a new gene involved in centromere function, MIF2 (27).

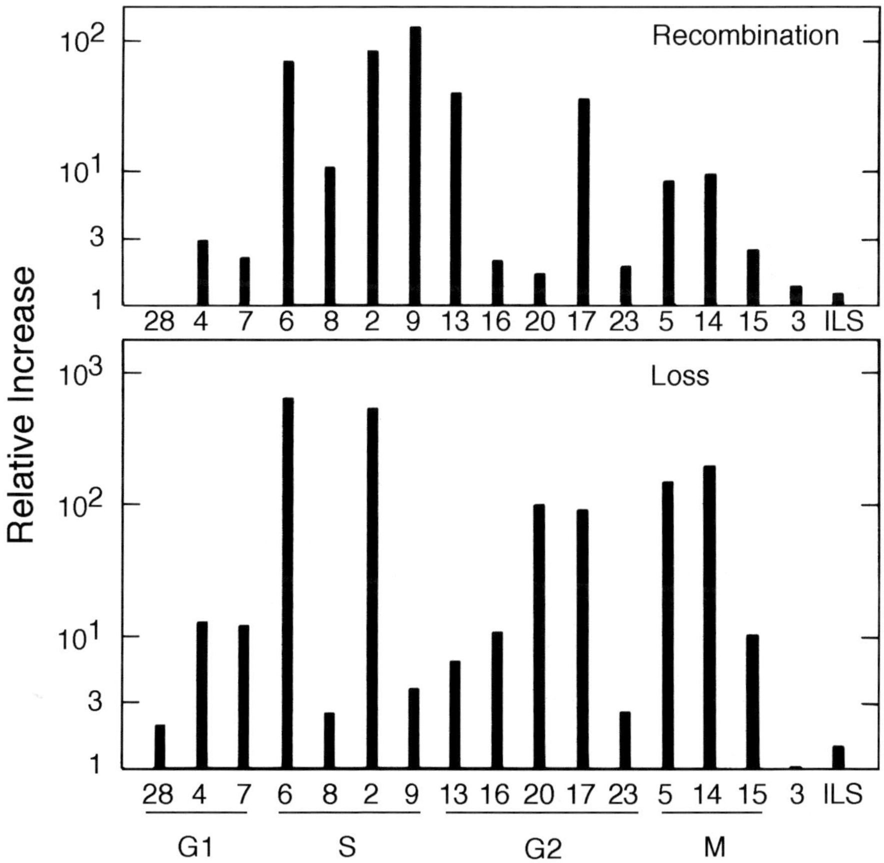

Figure 13. Rates of mitotic recombination and chromosome loss in *cdc* mutants grown at their maximum permissive temperature. Reproduced from ref 24 with permission.

These results suggested that mutations in essential components of the cell cycle machinery might be responsible for the genetic instability of tumor cells in addition to the known role of DNA repair defects in certain forms of cancer susceptibility (e.g. DNA excision repair in skin cancer and mismatch repair in colon cancer).

Figure 14. Ted Weinert.

DNA damage checkpoint. Fortunately, my interest in genomic instability coincided with Ted Weinert's (Fig. 14) interest in studying the regulation of cell division. He thought it likely that all of our cell cycle mutants were identifying genes that contributed to the machinery of cell division and was interested in studying something that was more clearly an example of cell cycle regulation. I had remembered noticing that yeast cells became arrested synchronously in the cell cycle by radiation and mutagens and he began looking at radiation sensitive mutants to see if any were altered for their cell cycle response. He quickly found that some radiation sensitive mutants failed to arrest the cell cycle in response to radiation (28). He demonstrated that deletion of the *RAD9* gene eliminated the regulation of the cell cycle by radiation, demonstrating a regulatory role for this gene (Fig. 15) and discovered a number of additional genes involved in the DNA damage checkpoint (29).

Ted's discovery led to an appreciation of the role of checkpoints in the fidelity of chromosome transmission. The *rad9* mutant exhibited a 20-fold increase in the rate of chromosome loss in the absence of any extrinsic DNA damage. An explanation for the increase in chromosome loss in checkpoint defective cells was provided by Ted's discovery that the same mutations that rendered cells insensitive to arrest of mitosis by radiation also rendered cells insensitive to arrest by defects in DNA replication (29; Fig. 16). This suggested that intrinsic errors in DNA replication occurred stochastically in rare cells of the population and that the checkpoint function was needed in those cells to assure correct repair of the damage. The same checkpoint genes were later found by Mandy Paulovitch to control the rate of replication over damaged DNA (30) and by Siede, Friedberg and Friedberg to control the rate at which cells enter S phase when they experience damage in G1 (31).

The role of the DNA damage checkpoint in assuring genomic fidelity was further elucidated by Daivd Toczyski when he was a postdoc in the lab. San-

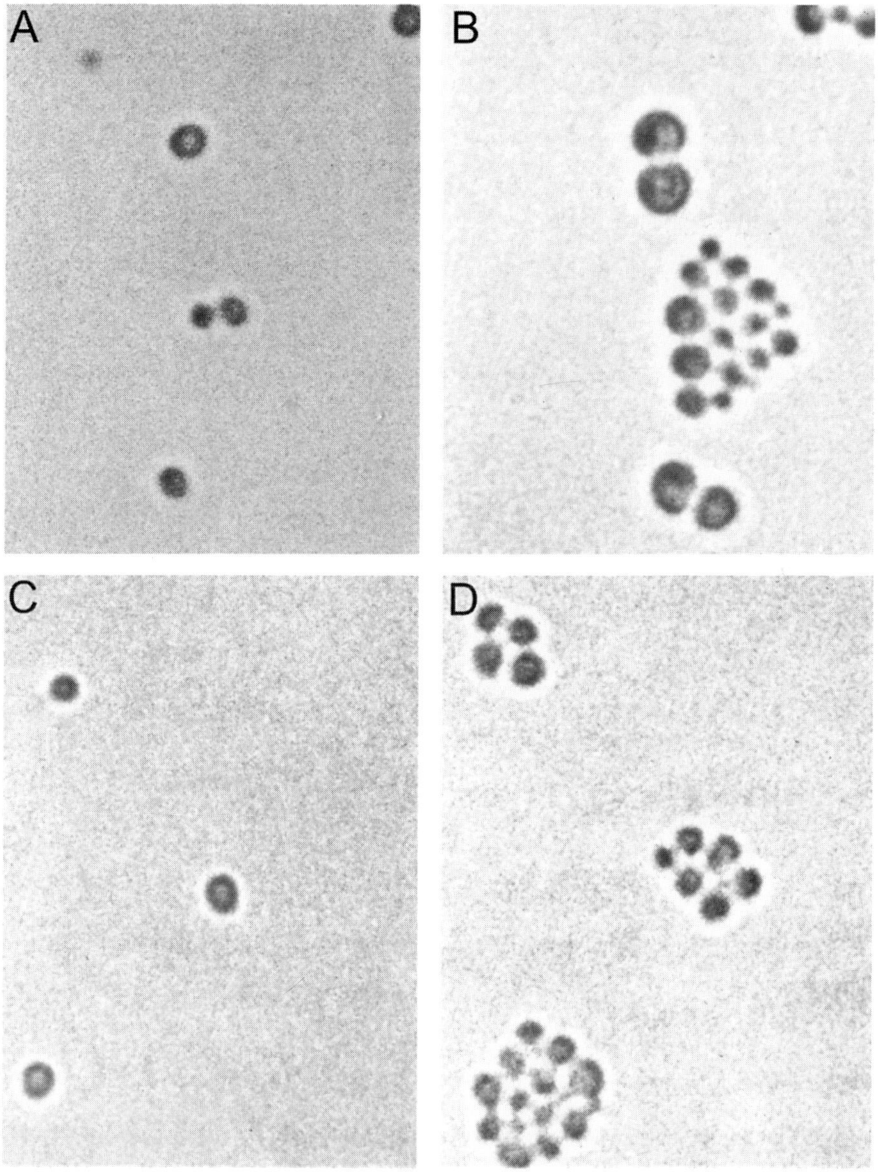

Figure 15. Time-lapse photomicroscopy of cells irradiated with X-rays. Wild type cells at the time of irradiation (A) and several hours later (B). Note that the originally unbudded G1 (A) cells have remained arrested as large budded cells (B) while the budded G2 cell (A) has resumed cell division (B). G1 haploid cells are very inefficient at repairing double strand breaks because of the lack of a template for homologous recombinational repair. *rad9* mutants cells at the time of irradiation (C) and several hours later (D). Note that the G1 unbudded *rad9* cells (C) do not arrest division but continue dividing producing dead microcolonies (D).

dell and Zakian had demonstrated, very elegantly, that cells which contain a single double strand break arrest at mitosis for a number of hours and then begin dividing again even if they cannot repair the break (32). This result indicated that the checkpoint pathway has an intrinsic adaptation component.

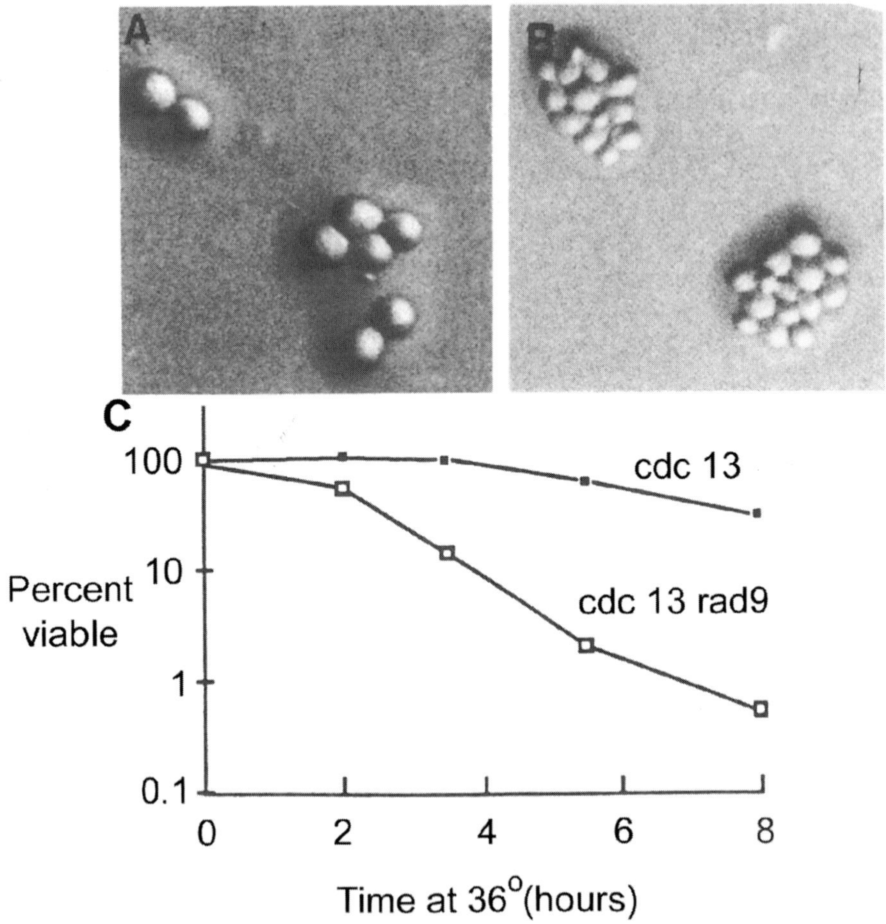

Figure 16. Cells defective in telomere replication fail to arrest cell division at the restrictive temperature. *cdc13 RAD9* cells incubated at the restrictive temperature for several hours have arrested cell division and retain viability (A) but *cdc13 rad9* cells failed to arrest and die (B). Reproduced from ref 29 with permission.

David discovered mutants defective in adaptation of the DNA damage checkpoint and showed that wild type cells actually transmit more damage to offspring than adaptation defective cells (33). The p53 tumor suppressor gene (34) acts as a cell cycle checkpoint gene and its function is also regulated by adaptation. Several other cell cycle checkpoints have been discovered. One controls mitosis if chromosomes are not attached to the mitotic spindle (35). A recent review of the DNA damage checkpoint has appeared (36).

Paradox resolved. After the discovery of cyclin dependant kinases, there remained an important paradox in our understanding of the cell cycle. By 1988 it was clear that the CDK played a controlling role in the cell cycle of frogs as well as yeast. Gerhart, Wu and Kirschner had demonstrated that the maturation promoting activity (CDK) of Xenopus eggs displayed a periodic activation at the onset of mitosis (37). Murray, Solomon and and Kirschner had shown that accumulation of the cyclin mRNA was the limiting component for activating mitosis (38). Thus it seemed that the Xenopus cell cycle was driven by periodic activation of the CDK through cyclin synthesis. Moreover, this CDK "clock" was autonomous from the events of the cell cycle. Inhibition of DNA replication does not block mitosis during the earliest divisions of the egg.

How could we reconcile this picture with that derived from the two yeasts? The yeast cell cycle did not behave as though it were being run by a clock undergoing periodic activation. Instead, it had the appearance of a series of dependent events where each event provided the substrate or activated the subsequent event. Growth was the activator of the CDK each cycle and there was no need for a clock. To phrase the problem more precisely, if periodic activation of the CDK was driving cell cycle events, why did the cell cycle stop in yeast when the machinery for DNA replication or mitosis was defective? Why wouldn't the clock keep ticking and activate mitosis in the absence of DNA replication or cytokinesis in the absence of mitosis? The discovery of mutants that no longer arrested division in response to radiation provided the answer – a series of regulatory signaling pathways, checkpoints, that keep the cell informed of each event's progress. If an event stops, checkpoint signaling arrests the clock. The frog embryo is different because the checkpoint only becomes active after about the twelfth embryonic division (39).

A unified view. The genetic control of cell division provided two important lessons that have been repeated over and over in molecular, cellular and developmental biology. The first is the conservation of proteins and their functions throughout evolution. This was not a surprising conclusion because all living organisms share a common ancestor. However, we did not know how conserved the machinery would be. We did not know that a newly sequenced gene from nearly any organism would be seen to fit into a family of proteins like kinases, myosins, or anthranylate synthetases. Indeed, the molecular conservation of many of the proteins involved in the cell cycle and in cell cycle checkpoints is not very evident across the prokaryotic / eukaryotic boundary but within eukaryotes the conservation is striking. A second, and of course related issue is how easily biology uses the same parts to make different things. Here again, the cell cycle is a dramatic example. Of the three organisms most studied, Xenopus, fission yeast and budding yeast there appeared to be little in common. The frog embryonic cell cycle was controlled by a clock, the fission yeast by a rate controlling step in G2 and the budding yeast by a gene that initiated spindle pole duplication in G1. Yet all three activities turned out to encode the same cyclin dependent protein kinase, and the differences were only superficial.

CANCER THERAPEUTICS

The genetic instability of cancer cells are due to genetic defects that affect the cell cycle machinery, DNA repair, or cell cycle checkpoints. Examples of the latter two are well known. Since we know a lot about the biochemistry of each of these processes it has been possible to define the genetic changes that lead to loss of fidelity in some cancers and it should be possible in many cancers. These defects should create a vulnerability for the cancer cell relative to the normal cell that could provide a powerful therapeutic advantage if the appropriate vulnerability were targeted for therapeutic intervention.

This brings me to a third application of the yeast model to cancer research. A new adventure was in the making when I had the good fortune to be invited to give a seminar at Massachusetts General Hospital at about the time that Stephen Friend was exploring his interest in drug discovery. Stephen had spent some time investigating the drug discovery programs at some of the major pharmaceutical companies. Mark Groudine, Jim Roberts, and Bob Eisenman had gotten me interested in the potential for drug discovery in the area of cell cycle checkpoints. Stephen got interested in this idea as well and was recruited to the FHCRC to head up a program in molecular pharmacology. Stephen and I set up a drug discovery program that was originally supported by Merck and the NCI to use yeast genetics to discover and validate cancer targets.

Our goal was to identify drugs and drug targets that would present a therapeutic advantage. That is, where the cancer cells would be more sensitive to the drug or to inhibition of the target than the normal cells. The idea was to construct yeast cells that contained mutations characteristic of specific tumors (altered in mismatch repair, cyclin, activated telomerase, etc.). The mutant and normal cells were then screened for drugs that killed the mutant yeast more effectively than a wild type yeast (40). The results have been encouraging. A test of 23 commonly used chemotherapeutic agents revealed that some had a high "therapeutic advantage" (41). Cis-platin preferentially kills cells with defects in post-replication repair genes, *rad6* and *rad18* (Fig. 17); the nucleotide analogue, cytarabine is preferential for an *sgs1* mutant, a homologue of the Bloom's syndrome and Werner's Syndrome helicases; camptothecin, a topoisomerase I inhibitor, and idarubicin and mitoxantone, two topoisomerase II inhibitors, are specific for cells with defects in double strand break repair, *rad50, rad51* and *rad52.* Other agents were selective for a broader range of DNA damage repair mutants and some agents were nonspecific.

This approach identifies drugs with therapeutic advantage. We were also interested in identifying protein targets which if inhibited by drugs would provide therapeutic advantage. These would be "validated" targets. Yeast lends itself to this goal as well. There are methods for identifying mutations in an unknown gene that are synthetic-lethal with mutations in a second known gene (42). So, we could start with a mutation characteristic of certain types of cancer and identify mutations in other genes that would be synthetic-lethal with the first. These genes would be potential drug targets. Targets were identified that would kill mutants defective in the DNA damage checkpoint. These targets were enzymes of deoxynucleotide biosynthesis and components of the DNA replication apparatus (Eric Foss and Patrick Paddison, unpublished).

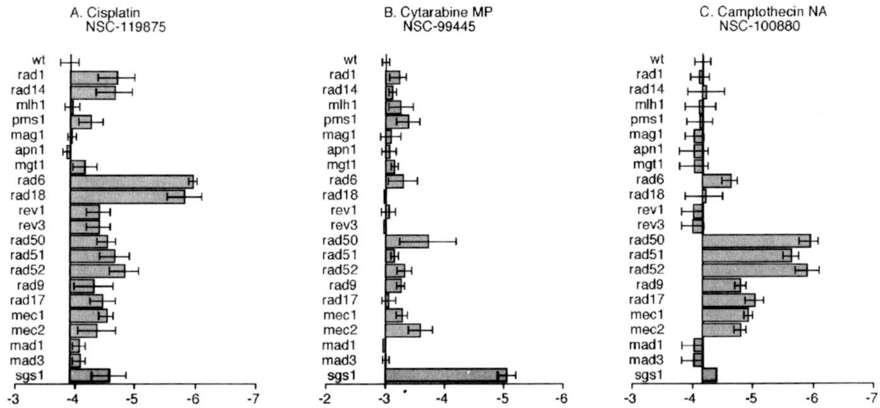

Figure 17. Sensitivities of different DNA repair and checkpoint mutants to common chemotherapeutic agents. (A) Cisplatin, (B) Cytarabine, (C) Camptothecin. The horizontal axis is a log scale. Reproduced from 41 with permission.

NATURAL GENETIC VARIATION

I am currently interested in the individual differences in susceptibility to cancer. How does the genetic variation that exists in the human population manifest in differences in disease susceptibility? Will we be able to assess disease predisposition for most people at an early age so they can choose life styles or frequent screening that will lead to better health? Will there be an individualized medicine?

The current preoccupation in molecular genetics is in the use of inbred organisms as models of human disease. When a gene is assigned a function in a mouse we reason that it is likely to play a similar role in the human. Yet the model systems that we study are genetically homogeneous except for the single gene perturbation that we study, whereas the human population is genetically outbred such that all individuals are very different (except for identical twins). It is clear that this genetic individuality has a very important role in the transformation of genotype into phenotype because the same mutation in different inbred strains of mice or yeast often have different phenotypes. Are we missing something by studying inbred model organisms?

I seem to have extremely good luck in finding colleagues who point the way. Just as I was considering these issues, I had the good fortune of meeting Stan Leibler who taught me about the robustness of biological circuits (43). I have come to think that this concept of robustness is critical to understanding the phenotypic consequences of natural genetic variation. Because of buffering, some genetic variation in protein activity or amount will have no phenotypic consequence.

There are several seminal contributions to our understanding of robustness. Redundant pathways, feed back control, and certain metabolic pathways have this robust behavior. One is the insight of Kczar and Burns to the robustness of metabolic pathways (44). They demonstrated that the flux of metabolic pathways (in which the enzymes followed Michaelis-Menton kinetics and none of the enzymes are saturated for their substrates) would be relatively insensitive to changes in the amount or activity of individual enzymes in the pathway. Stan Leibler and his colleagues have come to the same conclusion for the more complicated bacterial chemotaxis behavioral response because of its intrinsic feed-back regulation (45). An investigation of the genetic circuitry that controls Drosophila segmentation by vonDassow, Meir, Munroe and Odell has come to the surprising conclusion that robustness may be a property of almost any circuit with a significant complexity (46).

Since this buffering capacity is a product of the circuit and its components there will be natural genetic variants that compromise the buffering capacity itself. The consequence will be that many polymorphisms in a population will present no phenotypic consequence in the most prevalent genotypes. However, in rare genotypes that happen to compromise the buffering capacity for the particular polymorphism in question, a phenotype may result. If this reasoning is correct, then many disease susceptibilities will not be assignable to single polymorphisms but only to combinations.

How can we identify alleles that interact to produce phenotypes in combination that neither produce alone? Here again, yeast is an ideal organism in which to investigate this issue at its most fundamental level. As described above, there are methods in yeast for identifying synthetic-lethal interactions or synthetic-phenotypes (42).

For my purpose, which is to understand the genetic basis of phenotypic variation in an outbred population, it is important to understand the number of synthetic relationships that each gene is subject to. Obviously, the more interactions that can create synthetic phenotypes, the greater the genetic complexity potentially underlying a phenotype. Recent work in yeast from Tong, *et al.* (47) and unpublished work by John Hartman in my own laboratory suggest that genes interact with 20 or more other genes in the genome to create synthetic phenotypes. Or, to put it another way, variations in the activity of any single protein are buffered by interactions with as many as 20 other proteins in the genome.

It is clear that the prospects for genetic complexity underlying disease phenotypes in outbred populations is very great. So great, in fact, that if even a small percentage of the potential genetic variation were present in the population, then nearly all phenotypes would be the result of highly complex genetic interactions and most individuals with the same phenotype would have different combinations of alleles generating that phenotype. Attempts to identify natural polymorphisms that account for the differences in cancer susceptibility between two inbred mouse strains often does uncover great complexity (48). What is not clear, is how much functional variation is actually present in the human population. This depends on the history of the population, its size, its age, etc. For the human population, we know that about 1 nucleotide in 600 is polymorphic but we do not have any idea how many of these polymorphisms effect function or expression of gene products.

In order to learn more about the principles underlying the genetic complexity of human phenotypes we can hope to learn again from yeast. Important questions are the number and identity of genes that can interact synergistically to generate a phenotype, the patterns in which they interact, and the fraction of natural polymorphisms that have functional consequences. An investigation of these questions in yeast might help identify the set of candidate genes that could be associated with a particular human disease.

We still have a lot to learn from yeast and other model organisms about the nature of human disease.

REFERENCES

1. Williamson, D.H. **1965**. The timing of deoxyribonucleic acid synthesis in the cell cycle of *Saccharomyces cerevisiae. J Cell Biol* 25(3); 517–28.
2. Robinow, C.F., Marak, J. **1966**. A fiber apparatus in the nucleus of the yeast cell. *J Cell Biol* 29(1): 129–51.
3. Hartwell, L.H. **1967**. Macromolecule synthesis in temperature-sensitive mutants of yeast. *J. Bact.* 93: 1662–1670.
4. Hartwell, L.H. and McLaughlin, C.S. **1968**. Mutants of yeast with temperature-sensitive isoleucyl-tRNA systhetases. *Proc. Nat. Acad. Sci* 59: 422–428.
5. Hartwell, L.H. **1993**. Getting started in the cell cycle. In the early days of yeast genetics, 307–314. M.N. Hall and P. Linder. Cold Spring Harbor Laboratory Press, Cold Spring Harbor, NY.
6. Hartwell, L.H., Culotti, J. and Reid, B. **1970**. Genetic control of the cell-division cycle in yeast, I. Detection of mutants. *Proc. Nat. Acad. Sci* 66:352–359.
7. Hartwell, L.H., Culotti, J., Pringle, J.R. and Reid, B.J. **1974**. Genetic control of the cell division cycle in yeast. *Science* 183: 46–51.
8. Byers, B. and Goetsch, L. **1975**. Behaviors of spindles and spindle plaques in the cell cycle and conjugation in *Saccharomyces cerevisiae. J. Bacteriol.* 124: 511–523.
9. Jarvik, J. and Botstein, D. **1973**. A genetic method for determining the order of events in a biological pathway. *Proc. Natl. Acad. Sci.* 70: 2046–2050.
10. Hereford, L.M. and Hartwell, L.H. **1974**. Sequential gene function in the initiation of *Saccharomyces cerevisiae* DNA synthesis. *J. Mol. Biol.* 84: 445–461.
11. Duntze, W., MacKay, V., and Manney, T. **1970**. *Saccharomyces cerevisiae*: A diffusible sex factor. *Science* 168: 1472–1743.
12. Bucking-Throm, E., Duntze, W., Hartwell, L.H. and Manney, T.R. **1973**. Reversible arrest of haploid yeast cells at the initiation of DNA synthesis by a diffusible sex factor. *Exp. Cell Res.* 76:99–110.
13. Hartwell, L.H. **1973**. Synthronization of haploid yeast cell cycles, a prelude to conjugation. *Exp. Cell Res.* 76:111–117.
14. Wilkinson, L.E. and Pringle, J.R. **1974**. Transient GI arrest of *S. cerevisiae* cells of mating type α by a factor produced by cells of mating type α *Exp. Cell Res.* 89: 175–187.
15. Reid, Brian J. and Hartwell, L.H. **1977**. Regulation of mating in the cell cycle of *Saccharomyces cerevisiae. J. Cell Biol.* 75: 355–365.
16. Okasis.
17. Johnston, G.C., Pringle, J.R. and Hartwell, L.H. **1977**. Coordination of growth with cell division in the yeast *Saccharomyces cerevisiae. Exp. Cell Res.* 105: 79–98.
18. Unger, M.W. and Hartwell, L.H. **1976**. Control of cell division in *Saccharomyces cerevisiae* by methionyl-tRNA. *Proc. Natl. Acad. Sci.* 73: 1664–1668.
19. Nasmyth, K.A. and Reed, S.I. **1980**. Isolation of genes by complementation in yeast: Molecular cloning of a cell-cycle gene. *Proc. Nat. Acad. Sci.* 77: 2119–2123.

20. Reed, S.I., Hadwiger, J.a., and Lorinez, A.T. **1985**. Protein kinase activity associated with the product of the yeast cell division cycle gene CDC28. *Proc. Natl. Acad. Sci* 82(12): 4055–9.
21. Nasmyth, K. **2001**. A Prize for Proliferation. *Cell* 107:689–701.
22. Nurse, P. and Thuriaux, P. **1980**. Regulatory genes controlling mitosis in the fission yeast *Schizosaccharomyces pombe. Genetics* 96: 627–637.
23. Beach, D., Durkacz, B., and Nurse, P. **1982**. Functionally homologous cell cycle control genes in budding and fission yeast. *Nature* 300: 706–709.
24. Hartwell, L.H. and Smith, D. **1985**. Altered fidelity of mitotic chromosome transmission in cell cycle mutants of *S. cerevisiae. Genetics* 110: 381–395.
25. Koshland, D., Kent., J.C. and Hartwell, L. **1985**. Genetic analysis of mitotic transmission of minichromosomes. *Cell* 40: 393–403.
26. Meeks–Wagner, D. and Hartwell, L.H. **1986**. Normal stoichiometry of histone dimer sets is necessary for the high fidelity of mitotic chromosome transmission. *Cell* 44: 43–52.
27. Brown, M.T., Goetsch, L. and Hartwell, L.H. **1993**. *MIF2* is required for mitotic spindle integrity during anaphase spindle elongation in *Saccharomyces cerevisiae. J. Cell Biol.* 123: 387–403.
28. Weinert, T.A., and Hartwell, L.H. **1988**. The *RAD9* gene controls the cell cycle response to DNA damage in *Saccharomyces cerevisiae. Science* 241: 317–322.
29. Weinert, T.A., Kiser, G.L. and Hartwell, L.H. **1994**. Mitotic checkpoint genes in budding yeast and the dependence of mitosis on DNA replication and repair. *Genes Dev.* 8: 652–665.
30. Paulovich, A.G. and Hartwell, L.H. **1995**. A checkpoint regulates the rate of progression through S phase in *S. cerevisiae* in response to DNA damage. *Cell* 82: 841–7.
31. Siede, W., Friedberg, A.S., and Friedberg, E.C. **1993**. RAD9-dependent G1 arrest defines a second checkpoint for damaged DNA in the cell cycle of *Saccharomyces cerevisiae. Proc. Natl. Acad. Sci.* Sep 1 90(17): 7985–9.
32. Sandell, L.L. and Zakian, V.a. **1993**. Loss of a yeast telomere: arrest, recovery, and chromosome loss. *Cell* 75: 729–739.
33. Toczyski, D.P., Galgoczy, D.J., and Hartwell, L.H. **1997**. CDC5 and CKII control adaptation to the yeast DNA damage checkpoint. *Cell* 90:1097–1106.
34. Kastan, M.B., Zhan, Q., El-Deiry, W.S., Carrier, F., Jacks, T., Walsh, W.V., Plunkett, B.S., Vogelstein, B., and Fornace, A.J. Jr. **1992**. A mammalian cell cycle checkpoint pathway utilizingi p53 and GADD45 is defective in ataxia-telangiectasia. *Cell* 71(4): 587–97.
35. Rudner, A.D. and Murray, A.W. **1996**. The spindle assembly checkpoint. *Curr. Opin. Cell Biol.* 8(6): 773–80.
36. Zhou, B.S. and Elledge, S.J. **2000**. The DNA damage response: putting checkpoints in perspective. *Nature* 408:433–9.
37. Gerhart, J., Wu, M., Kirschner, M. **1984**. Cell cycle dynamics of an M-phase-specific cytoplasmic factor in Xenopus laevis oocytes and eggs. *J. Cell Biol.* 98(4): 1247–55.
38. Murray, A.W., Solomon, M.J., and Kirschner, M.S. **1989**. The role of cyclin synthesis and degradation in the control of maturation promoting factor activity. *Nature* 339: 280–286.
39. Newport, J. and Dasso, M. **1989**. On the coupling between DNA replication and mitosis. *J. Cell. Sci. Suppl.* 12:149–60.
40. Hartwell, L.H., Szankasi, P., Roberts, C.J., Murray, A.W., and Friend, S.H. **1997**. Integrating Genetic Approaches into the discovery of anti-cancer drugs. *Science* 278(5340): 1064–1068.
41. Simon, J.A., Szankasi, P., Nguyen, D.K., Ludlow, C., Dunstan, H.M., Roberts, C.J., Jensen, E.L., Hartwell, L.H., Friend, S.H. **2000**. Differential toxicities of anticancer agents among DNA repair and checkpoint mutants of *Saccharomyces cerevisiae. Cancer Res.* 60(2): 328–333.

42. Bender, A., Pringle, J.R. **1991**. Use of a screen for synthetic lethal and multicopy suppressee mutants to identify two new genes involved in morphogenesis in *Saccharomyces cerevisiae*. *Mol Cell Biol.* 11(3): 1295–1305.
43. Alon, U., Surette, M.G., Barkai, N., and Leibler, S. **1999**. Robustness in bacterial chemotaxis. *Nature* 397: 168–71.
44. Kacser, H. and Burns, J. **1981**. The molecular basis of dominance. *Genetics* 97: 639–666.
45. Alon, U., Surett, M.G., Barkai, N. and Leibler, S. **1999**. Robustness in bacterial chemotaxis. *Nature* 397:168–71.
46. von Dassow, G., Meir, E., Munroe, E.M., Odell, G.M. **2000**. The segment polarity network is a robust developmental module. *Nature* 406 (6792): 188–192.
47. Tong, A.H., Evangelista, M., Parsons, A.B., Xu, H., Bader, G.D., Page, N., Robinson, M., Raghibizadeh, S., Hogue, C.W., Bussey, H., Andrews, B., Tyers, M., Boone, C. **2001**. Systematic genetic analysis with ordered arrays of yeast deletion mutants. *Science* 294(5550): 2364–8.
48. Tripodis, N., Hart, Augustinus A.M., Fijneman, Remond J.A., Demant, P. **2001**. Complexity of lung cancer modifiers: Mapping of thirty genes and twenty-five interactions in half of the mouse genome. *JNCI* 93(10): 1484–1491.
49. Hartwell, L.H. **1974**. *Saccharomyces cerevisiae* Cell Cycle. *Bact. Rev.* 38(2): 164–198.

TIM HUNT

I was born in 1943 at Neston in the Wirral, not far from Liverpool where my father, Richard William Hunt was a lecturer in paleography, the study of mediaeval manuscripts. Richard's father was a doctor, and there is still a chemist's shop (i.e. pharmacy) in Winchester that bears the family name. My mother's father, Harry Rowland was a businessman, a timber merchant who imported wood from South America. Actually, as I discovered from letters to my mother, Kit, which I found after they were both dead, Richard was working in London at Bush House, presumably in some kind of intelligence role although he never spoke of it, and I was always afraid to ask. He was the kind of man who, having signed the Official Secrets Act, would have felt bound by it to the grave. Immediately the war was over, Richard accepted a position as Keeper of the Western Manuscripts in the Bodleian Library in Oxford, a post he held for the next 32 years. My earliest memories are of the winter of 1947–8, pushing a pram down to the coal depot about a mile away near the railway station as the delivery trucks having been stuck in snow drifts. Food rationing also made a deep impression on me, and the habit of breaking eggs separately into a cup lest one should be bad and the week's omelette be spoiled persisted for many years. Kind American mediaevalists would send food parcels, which prejudiced me strongly in favour of the USA at an early age.

My education started with latin taught at home by a governess, I can't imagine why, and for some reason I attended the Infants Department of the Oxford High School for Girls before moving to the Dragon School at the dangerous age of 8 or so. All I can remember of the girls is their bossy mothers, who liked to organise parties that I detested because of the competitive spirit that I detected in games like musical chairs. There was also knitting squares for the refugees' blankets, at which I was hopeless – the girls produced neat squares, myself ragged trapezoids. And there was the matter of the garden. My small patch was so barren that I was told to describe an imaginary garden when it came time to write Nature notes. The Dragon was much better, much less regimented, at the same time much more playful and more serious. Woe betide you for grammatical errors. Although mornings were largely devoted to latin and greek, at which I got worse and worse as time went by, there was a science lesson every week conducted by a young German called Gerd Sommerhoff. It was he who showed me that biology was an easy subject, and from then on I really never had to make any more career decisions. I also liked English, was bad at maths and hopeless at history, and fanatical about cricket, though not terribly good. My hero was Denis Compton.

At the age of 14, I moved across town to Magdalen College School, Oxford, where science played a much larger role in the curriculum. I loved Chemistry in particular, largely because the teacher, Colonel Simmons was much more concerned with principles than facts, although a thoroughly practical man himself. We were allowed considerable freedom, and on more than one occasion started fires from distilling volatile flammable solvents. One became adept at avoiding injury. Later on, biology again came to the fore when a young teacher called Terence Doherty took just three of us for Zoology. We dissected my brother's pet rabbit when it died, which was a treat after all the formalin-fixed dogfish. Terence wanted to be a painter, really, and later worked as an assistant with the sculptor, Michael Ayrton. I was also introduced by Terence to the Craftsman Potters' Shop in London, where I still love to browse. Very important during these years were Extramural Lectures given by the University of Oxford and the Christmas lectures in the Oxford Museum. For me, the University appeared to be largely devoted to the classics and history, and our house had a steady flow of mediaevalists of various stripes, but here in corners were revealed real treasures: the shrunken heads of the Pitt Rivers Collection, the dinosaur skeletons, lectures on the Theory of Evolution (on the occasion of the centenary of the publication of the Origin of Species) and on the Krebs Cycle. We also used to visit local factories and research institutes; at Alcan Aluminum they were trying to develop a Coca-Cola resistant lining for the cans, and somewhere we watched the zone refinement of an enormous silicon crystal. Balances that could weigh a human hair were pretty impressive, and even the telephone exchange was fascinating behind the scenes. George Beadle spent a year as Eastman Professor and came to give a

I am sitting on the bench in the middle, flanked by granny (maternal) on the left and my mother on the right who holds my brother Sandy on her knee. My father stands behind me.

lecture to the Scientific Society (his son Redmond attended the school), and I remember asking him to explain the work that led him to win the Nobel Prize.

In the fall of 1961 I went up to Clare College, Cambridge, to read Natural Sciences, with the intention of becoming a biochemist in the end. The courses were mostly excellent, and I did much better in the practical exams than in the theory. We were worked extremely hard, with certain practical classes falling on Saturday afternoons. Sydney Brenner organised seminars in his college rooms, and gave lectures that were officially off-limits to biochemists. It was all tremendously exciting, and we were very conscious, I think, of being surrounded by Very Great Men as well as very much in awe of the weight of the history of scientific discovery, especially in physics.

I started my scientific career in 1964 in the Department of Biochemistry at Cambridge, under the supervision of Asher Korner, who encouraged a great deal of freedom among his students to work on any aspect they chose of DNA, RNA or protein synthesis! Very early on, I decided that I wanted to work on the control of translation of mRNA, and thanks to Louis Reichardt (who spent a year in Cambridge in the same bay as me before going to Stanford to do his Ph.D. with Dale Kaiser) I learned about the use of rabbit reticulocytes for studies of haemoglobin synthesis, and began to appreciate the advantages of simple model systems. Together with Tony Hunter, a fellow student, I became interested in the question of when the haem was inserted into globin to make haemoglobin, and the question of whether ribosomes had to queue up under conditions where iron (and hence haem) was limiting. This was inspired by a talk we hear in 1965 by Vernon Ingram at a meeting in Cambridge on the subject of haemoglobin, at which I also listened to Henry Borsook's talk comparing protein synthesis in sea urchin eggs with protein synthesis in red cells. I thought very little about this at the time, but it planted an important seed. Tony and I found that the ribosomes were evenly spaced along the mRNA, and never formed a queue unless we forced them by other means. In 1966, I went to another meeting about haemoglobin, this time in Thessaloniki in northern Greece, where I met Irving London. I persuaded him that it would be fun to come and work in his lab during the summer, and by the time I was back in Cambridge (the train through Yugoslavia went very slowly), tickets to New York awaited me. I spent July through October 1966 living in a very hot dormitory out in the far east Bronx, and thoroughly enjoyed myself. The lab was the coolest place to be.

When I finished my Ph.D. in 1968, I went to New York to work with Irving, who had long been interested in the haem question. These were turbulent times, but I was fortunate in having a number of collaborators at the Albert Einstein College of Medicine. In particular, with Nechama Kosower and her husband Ed, we discovered that tiny amounts of oxidized glutathione were extremely inhibitory to protein synthesis in reticulocytes, and with Ellie Ehrenfeld that even tinier amounts of double-stranded RNA killed protein synthesis. Most striking was the finding that depletion of haem, and addition of GSSG or dsRNA all seemed to have similar effects on cell-free protein synthesis in the reticulocyte lysate, as though these disparate agents all acted in a similar manner. I also discovered that simple gel-filtration of the lysate caused a drastic fall in its activity. All these effects needed several more years to work out. One other thing I learned about during the studies on the inhibition of protein synthesis by dsRNA was that micrococcal nuclease, which requires Ca^{2+} for activity, did not inhibit protein synthesis in the reticulocyte lysate provided no Ca^{2+} was present.

On my return to Cambridge, I found myself reunited with two friends from the Korner laboratory, Richard Jackson and Tony Hunter. They had recently made the important discovery that Met-tRNA$_f$ was used to initiate haemoglobin synthesis, and this enabled us to probe the process of initiation in much more detail, leading to a set of very surprising results: that the initiator tRNA bound to ribosomes before the mRNA, and that all the inhibitory effects I had been studying in New York blocked the very first step, the binding of Met-tRNA$_f$ to 40S ribosomal subunits. It took three or four years to realise that this inhibition was due to the action of at least two inhibitors of the process, one activated by the absence of haem, the other by the presence of dsRNA, and that these inhibitors were protein kinases, which phosphorylated eIF-2, the key initiation factor that catalysed the binding of Met-tRNA$_f$ to ribosomes. Actually, it turned out to be rather complicated, because phosphorylation did not block the reaction the first time round, but rather prevented nucleotide recycling on the initiation factor, so that its re-use was inhibited.

By 1977, the outlines of all this were pretty clear, and we had a very good understanding of the reticulocyte lysate and its properties. The cause of the gel-filtration problem proved to be loss of polyamines, and the GSSG inhibition was a problem with thioredoxin and thioredoxin reductase which were needed to keep something in the reduced state. The micrococcal nuclease calcium dependence was put to good use by Hugh Pelham, a graduate student in the lab, who developed the use of nuclease treatment of the lysate to assay heterologous mRNAs. An important factor in this development was a ready source of messenger RNA, for which we used tobacco mosaic virus. Working with David Zimmern, a student of Aaron Klug's, and John Knowland, who was in John Gurdon's laboratory, it became clear that the RNA in the virions did not direct the synthesis of coat protein. Tony Hunter and I therefore ground up som TMV-infected tobacco leaves in liquid nitrogen and found to our great joy that a small mRNA was present that efficiently directed the synthesis of authentic TMV coat protein.

I therefore turned to new systems for studying translational control, and took the opportunity of teaching summer courses at the Marine Biological Laboratory, Woods Hole, to look at changes in protein synthesis in sea urchin and clam eggs after fertilization. This opened new horizons, not only in learning to deal with new systems, but in the breadth of approaches and interests of the scientists who passed through Woods Hole during the summers. I learned a tremendous amount of developmental and cell biology over the years. In 1979, Joan Ruderman and her student Eric Rosenthal were teaching the embryology course, and I helped them with experiments on the translational control of maternal mRNA; the major mRNAs concerned later turned out to be the A and B-type cyclins and the small subunit of ribonucleotide reductase. We carried out quite a lot of work together on the translational control, and Nancy Standart worked as a postdoc first with Joan and then with me in Cambridge. I believe that these clam mRNAs were the first really well-authenticated case of specific translational control. Together with Nancy, we later identified the key regulatory regions in the 3' untranslated region of ribonucleotide reductase and cyclin A mRNAs that are necessary for translational masking.

By 1982, work on the control of protein synthesis in sea urchin eggs had almost ground to a halt; every idea that my students and I tested proved to be false, and the very basis of the system was essentially flawed in that we were studying a system in which the ribosomes changed from having almost no activity to one in which about 2% of them were active, leaving the 98% majority of them still inert. The one question we did not ask up to that point, because the answer was supposed to be known already, was as to why an increase in protein synthesis occurred at fertilization, although it had been shown many years earlier that inhibiting protein synthesis inhibited cell division. One day in July 1982, with teaching over, I did a very simple experiment that was actually designed to answer the question of whether the proteins made after fertilisation were the same as the proteins made after parthenogenetic activation of the eggs. The eggs were divided into two batches, one of which was fertilized, the other treated with calcium ionophore, A23187. Radiolabelled methionine was added to both, and samples taken until after the fertilized eggs had divided into two-cell embryos. When the autoradiograph of the 1-dimensional SDS slab gel was developed, I noticed a very strange thing in the fertilized samples, which was a labelled band that went away, unlike all the others which, as expected, got stronger with time. Scarcely was the film dry when I ran into John Gerhart at the traditional Friday night lecture wine and cheese party, who told me about experiments he and Marc Kirschner had been doing on the oscillations in frog MPF activity during meiotic maturation of *Xenopus* oocytes. They found that MPF activity went away transiently between meiosis I and meiosis II, but if protein synthesis was inhibited, the MPF never returned. The behaviour of the band I had just seen fitted in beautifully with their physiological findings, and I was sure that this band, which we called cyclin, must have something to do with MPF. The rest of that summer was spent looking for cyclins in other species, and describing the basic behaviour of the strange disappearance, which turned out to occur about 10 minutes before each cell division. The oscillations went on for several hours after fertilization. Clam eggs did the same thing, except that two bands showed this behaviour. It is very surprising that nobody had spotted this before; it could have been done any time after the invention of the slab SDS polyacrylamide gel in about 1970.

It was impossible to work on cyclins back in Cambridge, because there were no clams or sea urchins, and by the time the next summer arrived I thought perhaps the suddenly-disappearing proteins had been a complete fantasy.

Fortunately, on return to Woods Hole next summer the results proved quite reproducible, but it was clearly going to be an uphill struggle to find out what was going on, and whether it had any general application. Progress was slow for various accidental reasons, like the fact that Tom Evans who spent the summer of 1982 as my undergraduate assistant had already signed up with David Secher at the MRC Molecular Biology Laboratory as his Ph.D. supervisor and it did not seem right to alter the arrangements. Jon Pines joined me a year later, but his first year was a terrible one for sea urchin eggs, delaying library construction by a whole year! Not until Christmas 1986 did we succeed in cloning and sequencing sea urchin cyclin B. After that, the pace began to accelerate, and it did not take long to isolate clones for *Xenopus* cyclins A and B, which was very exciting, because it made it clear that these proteins were not restricted to marine invertebrates. We developed the use of oligodeoxynucleotides in conjunction with RNase H as a way of specifically cutting particular mRNAs, and were able to show thereby that B-type cyclin synthesis was necessary for *Xenopus* extracts to enter mitosis, and that *Xenopus* MPF contained B-type cyclins as well as $p34^{cdc2}$.

Today, cyclins and cyclin-dependent protein kinases are recognised as key elements in the regulation of cell cycle transitions, and the family of proteins has grown considerably, thanks to work in many laboratories. I well recall the scepticism in the early days, however, when most people did not believe things could be so simple as making an enzyme to catalyse mitosis, and then destroying the enzyme to leave mitosis. Moreover, in retrospect, we were very slow to realise that cyclins were regulatory and activating subunits of Cdc2 and its relatives. I do not know why this penny took so long to drop.

Moving to ICRF allowed better access to *Xenopus* than had been possible in Cambridge, and important achievements have been the identification of c-mos as the activator of MAP kinase with Angel Nebreda, which also led to the discovery of a new member of the MAP kinase family called p38; the identification of John Shuttleworth's MO15 kinase as the CDK activating kinase with Randy Poon, and the purification of soluble and active fragments of cyclin A that allowed Jane Endicott, Martin Noble and Nick Brown to solve its crystal structure. We have defined a very large family of cyclins in *Xenopus*, and tested the importance of different members of the cyclin B family by making "knockout" mice. Right now, major interests of the laboratory lie in understanding how cyclins are targeted for proteolysis by their "destruction boxes", and in seeking substrates for cyclin dependent kinases. This is no easy matter, and our successes in the past, finding the salient substrates for c-*mos* and MO15 owed more to luck, and keeping our eyes open while doing other things, than any rational approach.

POSTSCRIPT

Of the many letters of congratulation that arrived after the announcement of the Nobel Prize, the one that gave me the purest pleasure was from a Swedish woman called Kerstin Westin in Uppsala, enquiring if I was, by any chance, the Tim Hunt she had known in Oxford in 1950 when she worked as an au pair girl with my family – and I was. Unlike my brother Sandy, who was only 5 at the time, I could clearly recall the young girl who looked after us all those years ago, and we had a happy reunion 51 years later in Stockholm. Apparently, nothing about the 7-year old boy gave any hint of his future. I would say that I was lucky to grow up in a loving family with wonderful teachers. Later, as a young scientist, I had the further good fortune to witness many of the founders of molecular biology at close hand. They set the standards of what was and could be expected. For my generation, Francis Crick was probably the most obviously influential presence. He was often at lunch in the canteen of the Laboratory of Molecular Biology where he liked to explain what he was thinking about, and he was always careful to make sure that everyone round the table really understood. He was a frequent presence at talks in and around Cambridge, where he liked to ask questions. Sometimes, I remember thinking, they seemed slightly ignorant questions to which a man of his extraordinary range and ability ought to have known the answers. Only slowly did it dawn on me that he only and always asked questions when he was unclear or unsure, a great lesson.

For many years, I used to supervise students in Cambridge, preferring either to teach the introductory course in Cell Biology or third year students in their biochemistry laboratory projects. Here again, I had the good fortune to meet a succession of clever people, many much cleverer than I. Supervisions worked best when the group (typically of two or three students and myself) was engaged on a mutual course of discovery, for at the time, not so much was known about how cells worked and one could profitably muse on how things *might* work. Very different, as a friend pointed out, from teaching, say, crystallography where the principles had been worked out years and years before and your job was to clarify them for the newcomer. As more and more became known, I found teaching became less fun, whereas learning became more pleasurable. In particular, I learned a tremendous amount at the Marine Biological Laboratory during the summers I taught and researched there, and when that phase of my life was over, I started working with John Wilson on the problems book that accompanies *Molecular Biology of the Cell* by Alberts *et al.* The ferreting around trying to make sense of unfamiliar territory and the discussions with the other authors provided an even more intense pleasure of learning new things.

But none of these pleasures, great and satisfying though they are, match the joy of discovery.

PROTEIN SYNTHESIS, PROTEOLYSIS, AND CELL CYCLE TRANSITIONS

Nobel Lecture, December 9, 2001

by

TIM HUNT

Cancer Research UK, Clare Hall Laboratories, South Mimms, Herts EN6 3LD, United Kingdom.

Unlike my two distinguished friends, I did not set out on my scientific career with the intention of studying the cell cycle, and had no idea that the winding road of discovery would lead in that direction. On the contrary, I was interested in the control of protein synthesis, thanks to my Ph.D. advisor, Asher Korner, who studied the effects of insulin on protein synthesis in the rat. When I joined his laboratory in 1964, Asher used to perform one experiment every week, comparing the activity of ribosomes from the livers of control and hypophysectomised rats. On my first day, Asher suggested that I spend some time in the library to look for an interesting project, which I did, finding papers about nephrotic rats whose livers greatly increased their synthesis of serum albumin to compensate for its loss through leaky glomerular membranes. How, I wondered, did the ribosomes "know" that the albumin levels were low, and how was that "knowledge" translated into enhanced albumin synthesis? As far as I know, this would still make a good research project, so it is probably lucky that my first attempts to make some rats nephrotic by injecting them with a kidney homogenate were completely unsuccessful.

IN THE BEGINNING: RABBIT RETICULOCYTES

I was also very fortunate that Louis Reichardt, recently graduated from Harvard, had decided to spend the year in Cambridge, and we shared a bench together. Louis had learned how to work with rabbit reticulocytes, and was trying to make a cell-free system for globin synthesis that would respond to added haem by increasing the rate of globin synthesis. Ironically, while Louis was not successful in his project, either, Hildegard Lamfrom and Paul Knopf working up the road at the MRC Laboratory for Molecular Biology reported that they were able to make a cell-free extract of reticulocytes that briefly maintained a high rate of protein synthesis; I don't believe we were aware of their efforts at the time. And in the spring of 1965, several of us attended our first scientific meeting, a few hundred yards away in the Chemistry Laboratory. The Symposium was devoted to haemoglobin synthesis and was introduced, very appropriately, by Henry Borsook from Caltech. Borsook had shown in the middle 1950s that reticulocytes needed to be supplied with iron salts in the medium to maintain a high rate of protein synthesis, but his talk on the 8th April 1965 was entitled "Early development of the echinoid egg compared with erythropoiesis" (1). He contrasted the restriction of developmental options found in red blood precursors with the unfolding pattern of development seen in fertilized sea urchin eggs. What caught my attention was his excellent account of the changes in protein synthesis that follow soon after sea urchin eggs are fertilized. But there was no hope of studying sea urchins in Cambridge, and it was a talk by Vernon Ingram that got me started in research. Ingram spoke about the relationship between haem and globin synthesis, and presented some preliminary data that he interpreted to indicate that ribosomes formed a queue behind the sequence in globin that would form the haem pocket. He used the methodology that had been introduced by Howard Dintzis in his brilliant experiments which showed that proteins were made from N- to C-terminus (2). When we got back to the laboratory, we realised that Ingram had interpreted his data exactly backwards, and that, if his data were correct, they meant that ribosomes raced down the mRNA until the haem was added and then slowed right down. So Tony Hunter and I decided to repeat the experiments, making some improvements to the experimental approach, building our own electrophoresis tank and teaching ourselves the rudiments of protein chemistry, encouraged by Alan Munro. The results were slow in coming, but eventually proved beyond doubt that the ribosomes paid no attention whatsoever to when the haem was added and were normally distributed completely at random along the globin messenger RNA (3). Certain tricks allowed us to show that our methods would have enabled us to see queues, had they existed.

One day, I let a centrifuge run too long and we accidentally discovered that α-globin chains were made by smaller polysomes than β-chains (4), and we set out to discover the cause by comparing the rates of elongation of the two chains. By omitting an extremely elementary control, we wrongly concluded that β-chains took longer to translate than α-chains, a misinterpretation that was later corrected by Harvey Lodish (5). This was a sharp lesson about control experiments. We missed another trick, too, which would have been to discover that both globin chains were initiated with a methionine residue. In this case, our mistake was not to believe our own results and to ascribe them wrongly to contamination. Fortunately, Richard Jackson and Tony Hunter later corrected this oversight before anyone else (6).

In 1968, I left Cambridge and went to work in New York with Irving M. London, who was then the chairman of the Department of Medicine at the Albert Einstein College of Medicine. Irving had worked with David Shemin, the man who had elucidated the biosynthetic pathway of haem, and was passionately interested in the question of how haem regulated globin synthesis. About this time, Howard Schulman discovered that addition of haemin to Lamfrom and Knopf's lysate sustained prolonged high rates of protein synthesis (7) and it was a simple matter to confirm this result and start a programme designed to find out how the haem was working in molecular terms. Many people contributed to this endeavour, in particular Marco Rabinowitz at the NIH who discovered a dominant inhibitor of protein synthesis that formed when the reticulocyte lysates were incubated for long periods at 37° in the absence of added haemin (8). The molecular nature of this inhibitor and the basis of its mode of action were to remain obscure for another five or six years, and as so often, in my experience of research, the clues and answers came from indirect attacks in unexpected quarters, rather than a full frontal assault. For Marco's inhibitor proved very difficult to purify.

The first clue came from a collaboration with a neighbour in the Department of Medicine, Nechama Kosower. Nechama's husband Ed was a chemist who had suggested the use of a mild oxidizing agent called "diamide" to cure sickle cell anaemia (I cannot reconstruct the train of thinking). The diamide rapidly converted glutathione to its oxidized form and (reversibly) inhibited protein synthesis in intact reticulocytes. We assumed that the effect on protein synthesis was somehow connected with the loss of GSH, and started to use the cell-free lysate system to investigate. One day it occurred to me that the inhibition could be due to the formation of GSSG rather than loss of GSH, and I tested the effects of adding GSSG to the lysate in the presence of optimal levels of haem. To my surprise and joy (because it had not occurred to the others), minute amounts of GSSG caused an inhibition of protein synthesis that was very reminiscent of that seen when haem was omitted (9). Protein synthesis at first occurred at a high rate, and then initiation shut off after a few minutes. Just as with omitting haem, protein synthesis could be restored after its shut-off, in this case by addition of a reducing agent like dithiothreitol. At some point, it was found that addition of N-ethyl maleimide was a very good way of turning on the Rabinowitz inhibitor, so a finger of suspicion pointed at critical –SH groups; but this apparent clue never really led anywhere. Richard Jackson and I (Figure 1) returned to study the GSSG effect some years later, and discovered that the inhibition of protein synthesis was due to loss of reduced NADP and glucose-6-phosphate, which took us off into studies of thioredoxin and thioredoxin reductase. Affinity chromatography using 2',5' ADP Sepharose (which binds many NADP-requiring enzymes) proved extremely useful in this analysis and the whole episode was a very good and enjoyable series of lessons in practical biochemistry of a rather old-fashioned kind (10, 11).

At about the time of the collaboration with the Kosowers, research on the mechanism of action of haem on protein synthesis was not going well, and as a diversion I decided that it would be fun to see if the reticulocyte lysate could translate added poliovirus RNA, and if so, to work out the gene order using essentially the same methods that Tony Hunter and I had employed in our studies of globin synthesis. I discovered that there was a group in Harry Eagle's Department of Cell Biology who worked on poliovirus, and went and begged some RNA from Don Summers. Analysis of the reaction on the newly-invented SDS-polyacrylamide gel showed no sign of any newly-synthesized high molecular weight product(s), however, and I suspected (correctly, as it turned out many years later) that factors from the nucleated cells that poliovirus normally infected were required for the initiation of polio protein syn-

Figure 1. Richard Jackson and Tim Hunt.

thesis. To test this idea, I collaborated with Ellie Ehrenfeld, and made the exciting discovery that poliovirus-infected HeLa cell cytoplasm contained a potent inhibitor of reticulocyte lysate protein synthesis (12). The inhibition looked almost exactly like that seen when haem was omitted or GSSG added. What was different, however, was the identity of the inhibitor, which had the remarkable properties of heat stability, sedimentation at 20S in a fairly sharp peak, and resistance to digestion with DNase, RNase and proteases. We were baffled for several weeks by this mysterious substance, whose assay was very tedious (and non-linear), because of course RNase and proteases were deadly inhibitors of protein synthesis and every assay involved re-isolating the inhibitor on sucrose gradients to separate it from the various agents used to probe its identity. In the end, after a depressing period when we deduced that we must be dealing with a giant complex carbohydrate, whose analysis we knew ourselves unable to conclude, we realised that double-stranded RNA had all the properties of our unknown. (13). We consulted with Jerry Hurwitz on the 9th floor, who suggested we used micrococcal nuclease, and not to forget to add the calcium, which sowed the seed of the nuclease-treated reticulocyte lysate assay system for eukaryotic mRNA that Hugh Pelham eventually developed (14).

The micrococcal nuclease made the inhibitor disappear completely, and the next step was to see if the effect was specific for polioviral dsRNA. I bought some poly I:C from Sigma and tested it from 1 mg/ml down in 3-fold steps and was puzzled to find that only the very lowest concentration seemed to cause significant inhibition. It was reasonable to suppose that I had simply placed the samples back-to-front in the scintillation counter, so I repeated the experiment with a much wider range of concentrations of the synthetic dsRNA and was extra-careful with labelling of the samples. The results were startling. Not only did high levels of dsRNA *not* inhibit protein synthesis, but the levels required for inhibition were staggeringly low: we later calculated that it would only take one molecule of polivirus replicative form dsRNA to completely inhibit protein synthesis by the millions of ribosomes present in a HeLa cell. The dsRNA therefore had to be acting catalytically and not as we at first imagined by mimicking the 5' end of mRNA and titrating out some initiation factor. We could explain the lack of inhibitory effect of high levels of dsRNA by supposing that two molecules of the putative pro-inhibitor had to bind side-by-side on the RNA, and later back in Cambridge, Hugh Robertson and Tony Hunter found that a stretch of about 50 nucleotides of perfectly double-stranded RNA was required to activate the inhibitor (15), a pre-echo of much more modern times that explain why it is important for 'RNAi' to be short: otherwise, protein synthesis is generally inhibited.

The similarity of the kind of inhibition caused by lack of haem or addition of GSSG or dsRNA strongly suggested that they must have an underlying mechanism in common, and the similarities only increased when I left the Bronx and returned to Cambridge at the end of 1970 to be reunited with my old friends Tony Hunter and Richard Jackson. By then, they had discovered that protein synthesis was initiated by a special methionine tRNA, Met-tRNAf, which provided an important new tool for investigating the pathway of initiation of protein synthesis and its control by the strange collection of disparate inhibitors. Our next great insight came when Steve Legon and Chris Darnbrough found that a previously ignored complex between Met-tRNAf and 40S ribosomes disappeared before protein synthesis shut off (16). It was then fairly simple to show that these complexes formed without the need for mRNA, and we realised that ribosomes must find the start of mRNA by scanning the message with the anticodon of the initiator tRNA. Since it was established dogma that tRNA only bound to ribosomes at the instruction of mRNA, our paper met with a hostile and sceptical reception on first submission, and we needed help from Mike Mathews, another ex-Kornerite, who was by then working at the MRC Laboratory of Molecular Biology and had developed a method for assaying mRNA.

INTRODUCTION TO PROTEIN KINASES

But here again we got stuck for a while, largely because we were using the reticulocyte lysate as an assay system, and either total protein synthesis or the formation of the 40S-Met-tRNA$_f$ complexes as the read-out. We did not seriously attempt to dissect partial reactions, largely because none of us had proper biochemical training, and we had got a long way simply analysing this wonderfully active cell-free system. At this stage, I don't think we seriously tried even to purify any of the inhibitors, and did not seriously speculate on what the underlying molecular explanations might be. It would have been too far. Instead, Steve Legon followed up a clue that cyclic AMP might have some connection with the various inhibitory conditions, and began to test all manner of purines, many of which alleviated the effects of leaving out the haem or adding dsRNA (17). These phenomenological effects were further pursued by Ken Balkow, who had joined the laboratory shortly before a tremendous fire consumed the entire laboratory, and forced our departure to a teaching laboratory in Herman Lehmann's clinical biochemistry Department at the Addenbrooke's Hospital. The destruction of all previous results and the new close contact with molecular biologists, in particular with John Gurdon's group, were highly beneficial, and once the laboratory was set up and running again, progress was quite rapid. Richard Jackson began to get evidence that ATP might be necessary for the inhibitor of initiation to form, and realised that it should be possible to test whether ATP was also necessary for the inhibitor to act by using a partial reaction, the binding of Met-tRNA$_f$ to 40S ribosomal subunits, which required GTP but not ATP. This proved successful, and inhibition clearly required ATP, something that would never have emerged from studies of bulk protein synthesis. From there it was a very short step to adding labelled ATP to the inhibition reactions and seeing a new phosphorylated band form. This was the work of a talented new graduate student, Paul Farrell. We also discovered that both the haem-regulated inhibitor and the dsRNA-activated inhibitor tended to undergo self-phosphorylation. Moreover, the identity of the target for phosphorylation emerged almost at once, because the molecular weight of the major target band corresponded suspiciously closely to one of the subunits of eIF-2 which had recently been isolated by Theo Staehelin and his colleagues Hans Trachsel and Bernhard Erni at the Basel Institute for Immunology. This protein formed a complex with Met-tRNA$_f$ and GTP, and delivered the tRNA to the 40 S ribosome. It was a very satisfying explanation for the previous 5 years of accumulated phenomenology, and it simply had to be right (18). I went to Basel for a week or two to learn how to make and handle the factor, in the process for the first time seeing at first hand how proteins were purified. It was not long before we had convinced ourselves that phosphorylation of eIF-2 was responsible for the inhibition of protein synthesis, but there was one remaining puzzle. Phosphorylation of eIF-2 did not seem to impair its ability to form complexes with Met-tRNA$_f$ and GTP, nor its ability to deliver initiator tRNA to the 40S ribosome. Gradually it emerged that what went wrong was the ability of the eIF-2 to recycle, specifically to exchange its GDP for new GTP. Once again, our lack of proper biochemical skills prevented us from making any significant contribution to the identification of the GDP/GTP exchange factor known as eIF-2B. Similarly, despite considerable efforts, I failed miserably to purify the haem-regulated protein kinase or to understand its mode of activation, and we fared no better with the dsRNA-activated protein kinase now known as PKR.

Getting to the bottom of this kind of regulation of protein synthesis was very satisfying. We had been faced with a long-standing mystery and worked it out. The answers made sense and looked as though they might have rather wider applicability. During my stay in New York, I had spent a couple of weeks one summer in Gordon Tompkins's laboratory at UCSF looking at "the pleiotypic program' (19) close up, and had seen how extremely easy it was to inhibit polysome formation by all kinds of mild abuse of nucleated cells. Could it be that the global control of protein synthesis seen in cultured cells starved of amino acids or glucose, or subjected to sudden changes in temperature, were also simply due to phosphorylation of eIF-2? I grabbed an opportunity presented by Tom Humphreys to help teach the Embryology Course at the Marine Biological Laboratory, Woods Hole, in the summer of 1977 in order to get my hands on some sea urchin eggs, and back home in Cambridge we set new graduate students working on making cell-free systems from nucleated cells to see if what was true of reticulocytes was universal. Progress was very disappointing, in every case rather seeming to suggest that the answer was no. About the only interesting thing to emerge from these studies was another protein kinase, which Tony Walker and Carl Anderson discovered by adding DNA to cell-free extracts of HeLa cells (20). We failed to appreciate that this needed free ends of DNA, as Steve Jackson later found (21), such was our inexperience with DNA. It was in any case quite clear that this had nothing to do with the control of protein synthesis.

The sea urchin eggs were more difficult to work with than I had imagined, too, and this was not helped by the equipment and layout of the MBL's embryological classroom. The idea was to study fertilization in as many different phyla and organisms as possible, using the simplest possible equipment and a microscope. Biochemical approaches were not much in vogue, and running gels impossible at first. There was no liquid nitrogen or –80° freezers, and Eppendorf tubes and Gilson pipettors were nowhere to be seen. Moreover, the season for sea urchin eggs was short, and I had no experience. For example, I had always thought that blood and sea water had similar ionic compositions, but this turned out to be quite false, sea water being much more concentrated. So what should one use as a homogenisation medium to have any hope of obtaining an active cell-free system in which to study protein synthesis? By then, Richard Jackson and I had almost completely defined the salient ions and low molecular weight compounds that were present in the reticulocytes (22), and we had a keen appreciation of how important they were for the maintenance of high-rate protein synthesis, to say nothing of its physiological control.

SEA URCHIN AND CLAM EGGS

So the next summer was spent in Cambridge, and I missed both the sea and the exhilaration of learning new things, for though Woods Hole had been a disappointing research environment for what I wanted to do, it was the perfect place to learn cell and molecular and developmental biology because of the stream of expert lectures on a wide variety of topics, together with the opportunities for discussion in labs, bars and beaches. So I had no problem in agreeing to return there in 1979 (Figure 2), except that I now insisted on a

Figure 2. The Embryology Course, MBL, 1979. Included in the picture are the author, Tom Humphreys, Joan Ruderman, Eric Rosenthal, Andrew Murray, Ed Southern and Gerry Rubin.

–80° freezer being provided and went with a suitcase full of all the little things I might need – tubes and tips and gel-plates and even a peristaltic pump – so that I could set up a lab away from home. During that summer, Dennis Ballinger and I followed up a hint from two years before from Tom Aune that changes in phosphorylation could be seen in sea urchin eggs after fertilization (23). This proved to be very useful experience, in fact, because running SDS polyacrylamide gels of fertilized sea urchin eggs was not completely trivial, owing to the powerful proteases present in the head (acrosome) of the sperm. But even more rewarding was helping Eric Rosenthal and Joan Ruderman with their analysis of the changing patterns of protein synthesis found in clam oocytes. Bruce Brandhorst had recently made detailed studies of sea urchin eggs, and had concluded that there was little if any evidence of a change on the kinds of proteins made before and after fertilization (24). But clams were quite another matter, probably because whereas sea urchin eggs are shed in a kind of G0-like state, having completed meiosis in the ovary, clam eggs are in fact G2-arrested oocytes, and fertilization causes them to complete meiosis, rather as progesterone acts on frog oocytes except with a different time-scale. Joan and Eric had beautiful gels showing a spectacular change in the synthesis of at least three prominent bands on a 1-dimensional gel after fertilization, and it was easy to show that the mRNA encoding these proteins was present before fertilization, by cell-free assay in the reticulocyte lysate. It was also a simple matter to demonstrate that the mRNAs were in the untranslated "mRNP" compartment in the oocytes, and shifted onto polysomes by the time the oocytes had completed meiosis, about 40 minutes after fertilization (25). This was quite an important advance, being one of the first really striking demonstrations of authentic translational control, a field that had been somewhat lagging behind the general control of protein synthesis in reticulocytes for want of clear examples to study. But we also began to wonder why these cells needed to upregulate protein synthesis or to change the kinds of proteins they were making after fertilization. Considering what a long time the eggs had had to fill up with ribosomes, polymerases and so on, and considering that they were not going to increase in mass until they began to eat, many hours if not days later, it was puzzling to know what these new proteins might be doing.

My introduction to cell cycle control was provided by a clear, scholarly and beautiful seminar given by John Gerhart one afternoon in the summer of 1979. He told the embryology class about the properties of MPF and of his struggles to purify it (26). I was captivated by the idea that a biochemical approach might provide insights into a cell cycle transition, that there might actually be an enzyme which literally catalysed mitosis. However, outrageous such an idea seemed in many ways, it seemed from all that was known that this was rather likely to be true. But although attractive, there did not seem any obvious way to get from what I was studying to MPF, and the appearance of the new bands in the clams occurred too late to correspond to what was known about MPF in frogs, the focus being on meiosis rather than the cleavage divisions. Like Borsook's lecture twelve years before, this talk made a deep and lasting impression without having the least influence on the work I was doing at the time.

I spent the next two summers teaching and researching in Woods Hole as a member of the faculty of the Physiology Course, first with Ken van Holde and then Joel Rosenbaum as director. Eric Rosenthal and Andrew Murray helped with the teaching, and Dennis Ballinger spent time in the lab working on the ribosomal phosphorylation, although we gradually came to appreciate that this was not the secret of the increase in protein synthesis: for one thing, this was not a universal concomitant of fertilization in sea urchin eggs, and for another, the phosphorylated ribosomes did not enjoy privileged access to the mRNA. So, although the teaching was fun – showing the students how to handle reticulocyte lysates, make and assay mRNA from strange sea creatures, and run beautiful SDS-polyacrylamide gels – no frontiers were being pushed back. I was envious of Andrew Murray, who together with his advisor Jack Szostak was defining the minimal elements of chromosomes in yeast (27).

THE DISCOVERY OF CYCLIN

Thus it was in late July, 1982, with the teaching over and the sea urchin season drawing to a close (Figure 3), that I planned and executed the experiment that changed my life and got me straight into studying the cell cycle. It was, as far as I can recall (and even three weeks later its original purpose was already unclear, as judged by a rather detailed letter home to Richard Jackson) de-

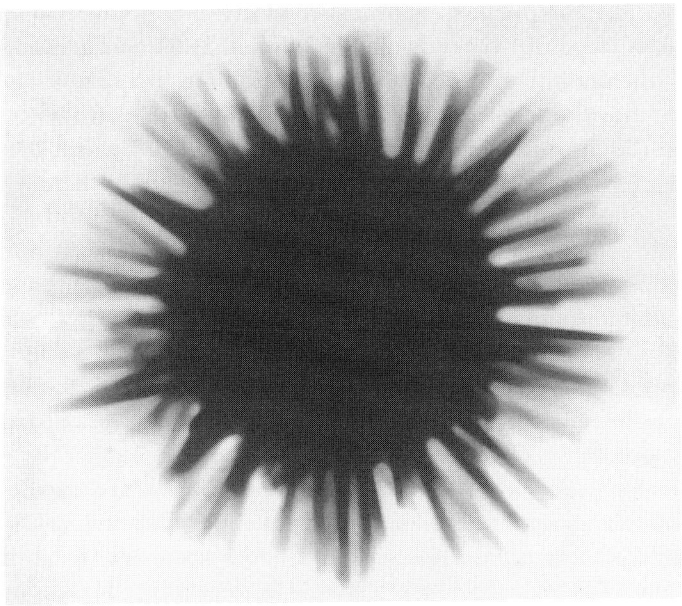

Figure 3. Arbacia punctulata.

signed to see if there were any differences in the pattern of protein synthesis in properly fertilized sea urchin eggs compared to eggs activated by the calcium ionophore, A23187. By this time, I knew exactly how to do the experiment and to prepare the samples for the gel analysis. I added [^{35}S]methionine to suspensions of the eggs in Millipore-filtered sea water in a 50-ml beaker. Samples were removed at intervals into trichloracetic acid, and the protein precipitates washed with acetone before dissolving in sample buffer. I did not want to miss any transients, and also wanted to build some redundancy into the experiment. In retrospect, it is amazing that this simple experiment hadn't been done before – after all, we had done exactly the same thing using $^{32}PO_4$ as the label – why we had never used [^{35}S]methionine, if only as a kind of class demonstration, is a mystery. I will return to the question of why nobody else did it, either, at the end of this account, in a footnote about gels.

The autoradiogram showed something very odd and unexpected, for although most of the bands got stronger and stronger as time went by, one band did not show this, the expected behaviour. It started as one of the strongest, but at a certain point it faded away. It was difficult to think of any other explanation than that it underwent specific (in the sense that no other bands were affected) proteolysis at some point in the early development of the fertilized egg. The parthenogenetically activated samples did not show such clear disappearance.

That very evening, I chanced to meet John Gerhart again at the wine and cheese party that follows after the Friday evening Lecture. He told me about the experiments on MPF that he, Mike Wu and Marc Kirschner had been doing (28), and the priceless information that, although the first appearance of MPF in frog eggs provoked by the injection of MPF did not require protein synthesis, the second wave of MPF corresponding to the second meiotic division *did* require new proteins to be made. This was electrifying, because it suggested that a protein was consumed at the end of meiosis I so that new protein synthesis was required to replenish it. This kind of behaviour was, of course, exactly what I had seen that morning in the behaviour of cyclin. It is worth noting that between the two meiotic divisions there is no replication of DNA, so that the explanation for a protein synthesis requirement during the cell cycle that was much in the air at the time, that of making new histones to package up the DNA, did not apply. It seemed far more likely that if MPF was turned on by some kind of post-translational modification, it was turned off by the most drastic of all such modifications, proteolysis. Yet we stopped short of concluding that MPF must contain cyclin as one of its subunits: that would be only one of many possible explanations, and it seemed extremely improbable that we should have hit so lucky by chance. Nevertheless, in the course of this, easily the most exciting conversation of my scientific life, the outline of the now conventional view of the cell cycle as depending in part on programmed proteolysis was born, more or less fully fledged. It was on July 22nd, 1982. I must say that such "eureka" moments are very, very rare in my experience. It normally takes several weeks of experiments to tease out the truth even when you have a really pretty good idea of what is going on.

It was of course clear that the experiment must be repeated, and that we should check cell cycle progression as best we could at the same time, using a nice healthy batch of eggs that underwent synchronous division. We needed to check the comings and goings of cyclin in relation to the cell cycle. Tom Evans, a Cambridge project student, had come with me that summer to act as an assistant, and he and I set to work to do as much as we possibly could before the *Arbacia* egg season ended. We soon confirmed the behaviour of cyclin, and showed that it was continuously synthesized and periodically almost completely degraded, about 10 minutes before the fertilized eggs divided. We tried but failed to assess the metaphase to anaphase transition, and we added every known inhibitor of cell cycle progression (as we naively understood them at the time) to see if they affected the oscillations of cyclin. Dan Distel, another student, was doing a project that wasn't going well, so I suggested that he take a look in clam eggs to see if any proteins came and went. Much to our surprise and delight, his lovely gel of August 17th showed that the two larger translationally regulated proteins A and B were both cyclins, whereas C behaved conventionally, and accumulated steadily. By labelling later embryos and then "chasing" with the protein synthesis inhibitor, emetine, it was possible to see that cyclin was still disappearing well into cleavage. We announced our findings at the Annual General Meeting of the MBL (29), but later kept quiet about this short note, lest it prevent publication in a "proper" journal.

The apparent size of sea urchin cyclin judged by its gel mobility was very close to that of tubulin, and fortunately for us there were a number of real experts in the course that summer. John Kilmartin, who was a friend of Joel

Rosenbaum's, had recently obtained an excellent anti-tubulin monoclonal antibody which he had attached to Sepharose, so it was easy to see if cyclin did or did not bind, and my graduate student Sarah Bray found first, that cyclin was by no means entirely soluble, second, that it did not bind to the column, and third, that another of the labelled proteins, corresponding to band C of the clam as it later turned out, was quantitatively retained by the column, along with the tubulin. We assumed at first that it was a newly-synthesized microtubule-associated protein, but in fact, its C-terminus fortuitously matched the epitope that the antibody recognised, an early lesson in the use of antibodies in research. Later, Nancy Standart made good use of this antibody and we found it could be used for affinity chromatography of what turned out to be the small subunit of ribonucleotide reductase (30). Ironically, both this enzyme subunit and the cyclins shared the property that they were just one subunit of a two-subunit protein of which the other subunit was laid down in eggs as part of the maternal endowment, but we were not to know this until much later.

Although there seemed to be a suspiciously close relationship between cyclin and MPF, and the behaviour of cyclin easily explained how MPF was turned off at the end of mitosis, we were miles from proving any such thing. Indeed, the universality of MPF was by no means established at that time. The only way forward was to clone, sequence and express cyclin to find out what it was, as well as to investigate the consequences of its non-appearance, which we imagined might be possible using antibodies, for example. But there were no people or grants or indeed sea urchins in Cambridge and any progress had to wait until the next summer. I wrote up the initial observations and sent them to Cell, who replied that they would publish the paper, but in "nothing like its present form". One of the reviewers said that it represented "wild speculation based on dubious logic", and I still blush when I look at even the revised version that appeared in the late spring of 1983 (31), although the Discussion ended on a deliberately guarded note: "The parallels between the behavior of MPF and cyclin are striking, but whether there is a direct correspondence between the physiological entity and the chemical one remains to be determined". We did not over-interpret the data.

By the next Spring, the whole episode felt like a dream that seemed increasingly too good to be true. I tried to explain how exciting this was to everyone I met, but the general reaction was sceptical. It had never (as far as I know, or have been able to find in fairly extensive reading of the old literature) been suggested before that cell cycle control might involve proteolysis and it was inconceivable that such specific intracellular proteolysis was even possible, for the recently-described ubiquitin system had been defined in terms of a mechanism for degrading unfolded or denatured proteins. Not until Michael Glotzer left an autoradiograph to expose over a long ski weekend in 1990 did the first signs appear that programmed proteolysis could occur by the ubiquitin pathway (32).

Jonathan Pines did his "Part II Project" with me in the spring of 1983, and agreed to join me as a graduate student to work on cyclin in the autumn.

Meanwhile, another project student, Richard Cornall, came with me to Woods Hole as my assistant that summer, Jonathan having decided to bicycle across the U.S.A., ending up briefly in Woods Hole at the summer's end. When Richard and I got to the MBL, the first thing we tried was simply to repeat the famous experiment. Much to my relief, cyclin was still there and still showed exactly the same behaviour, and Richard set about trying to find out exactly when it started to disappear: before or after the metaphase to anaphase transition? We were helped by Yoshio Masui, the master of MPF, who was on the faculty of the Embryology Course downstairs and very kindly showed us how to do orcein staining. Elayne Bornslaeger looked at the effects of inhibiting DNA synthesis on the behaviour of cyclin, and confirmed that this prevented the cells entering mitosis, completely stabilizing cyclin. It became clear that cyclin disappeared very close to the time of the metaphase-anaphase transition, but frustratingly we had no way to specifically block its disappearance and examine the consequences.

By this time, of course, I was trying to find out everything that was known about cell cycles and their control. I read reviews and books and talked to all sorts of people, but none of them shed much light on what I was trying to understand. Nobody had ever suggested that something might have to go away in order to keep the cycle turning, as far as I could see. In discussions with people, everyone agreed that this wasn't such a ridiculous idea. Morevoer, it fitted in very well with what was known about the role of protein synthesis in early development. As Wagenaar and Mazia and Wagenaar reported (33), following on Hultin's original observations (34), protein synthesis was needed for entry into mitosis during early cleavage, and was required in the first half of every cell cycle, quite consistent with the idea that protein or proteins needed to be replaced by new synthesis after they had been degraded, although of course the proteins in question could have been used up in other ways – assembly into chromatin, for example.

During that summer, Andrew Murray looked hard for cyclins in budding yeast, but was unable to detect any by the kind of pulse-chase methods that worked so well in sea urchins. By this time, too, Eric Rosenthal had succeeded in identifying clones for clam cyclin A in cDNA libraries (35), whereas we hadn't even started to make libraries from sea urchins, and material was in short supply. Back in Cambridge, nobody in the Biochemistry Department had any experience of recombinant DNA work, and indeed, the rules and regulations were still rather tight: special rooms and dedicated equipment were required for this kind of work, and in any case, surprisingly few people yet appreciated the revolution that was taking place. We were regarded as hopelessly trendy, and when I ventured to ask Marion Purvis, the Departmental Administrator if there was some way to keep frogs, she replied helpfully "Over my dead body". It did not look very promising, and by the end of Jon's first year, we were no nearer to cloning cyclin. Then Jeremy Minshull (Figure 4) started as a graduate student in the laboratory, and we began to investigate the "hybrid arrest of translation" that we were going to need to identify the cyclin clones, and also perhaps to knock out cyclin synthesis in sea urchin

Figure 4. Jeremy Minshull and Jonathan Pines flank Paul Nurse at a Conference Jacque Monod in Roscoff.

eggs. We got M13 shotgun clones for TMV that Philip Goelet had used in his determination of the sequence of this classic virus (36), and began to test if antisense clones could block synthesis of TMV protein programmed by viral RNA. We very quickly discovered two things. First, the circular M13 DNA needed to be cut into pieces in order to work (we used HaeIII, which can cut single-stranded DNA). Second, we discovered that although a clone that corresponded to the extreme 5' end of the viral RNA was pretty efficient at preventing protein synthesis, ones that started further along the mRNA hardly had any effect at all – unless one added RNase H to cut the RNA strand of the DNA-RNA hybrids that had formed. It turned out that the levels of RNase H in reticulocyte lysates were extremely low, and that the hybrid arrest that had previously been reported used wheat germ extracts, which were rich in this enzyme (37). We got side-tracked into doing rather detailed studies on the inhibition of globin synthesis by short oligonucleotides, defining times, concentrations, temperatures and specificities (for example, what would a single mismatch do?); these were indeed to prove useful later.

By this time, Jonathan had made a small cDNA library in M13 virus from *Arbacia* mRNA and using Jeremy's protocols found a clone that would specifically ablate cyclin synthesis among the first 50 or so that he tried. This was not really surprising, considering that cyclin looked as though it comprised about 5% of the total synthesis after fertilization. From there, of course, it was a fairly simple matter to obtain full-length *Arbacia* cyclin clones by more conventional means, and the complete sequence of cyclin B emerged at Christmas 1986 (38), by which time it was possible to compare sequences with Spisula (clam) cyclin A, for clone 1T55 had been sequenced by Kevin Farrell and Joan Ruderman in 1983-4. The two cyclins showed strong homology over their last 200 or so residues, but looked like nothing in any of the sequence databases at the time. They were clearly not protein kinases, although immunoprecipitates of clam cyclin A seemed to possess protein kinase activity. This was difficult to interpret without much fuller characterization of the early antisera, and I do not think the idea that cyclin might be associated with another protein that was a protein kinase seriously – or even playfully – entered our minds at this stage. I knew all about these developments, because I had stopped teaching in Woods Hole in 1983 and worked with Joan during the summers of 1984 and 1985, characterising the comings and goings of (labelled) cyclins during meiosis in clam oocytes. We learned how to fix cells and observe chromosomes and to define the points of no return for protein synthesis in relation to the various meiotic and mitotic landmarks. I was entranced by the chromosomes, and understood meiosis properly for the first time in my life. We often used parthenogenetic activation to look at meiosis, partly because it was more reliable than adding sperm and partly because it was easier to watch how the female pronuclei behaved without the confusion of the sperm, which tended to stick to the surface of the eggs and make for less aesthetically pleasing views. We discovered, however, that it was important to wash out the calcium ionophore quite early, or strange things happened that we never fully understood. And the A23187-activated oocytes got stuck at the first mitotic division. Without a second spindle pole, a monaster formed, and the cells remained in M-phase for very long periods, interestingly with high levels of cyclin B yet low levels of cyclin A. This was very reminiscent of what happened when colchicine was added. What was very gratifying was that inhibiting protein synthesis eventually led to a reduction in the level of cyclin B, and exit from M-phase, the only time that inhibiting protein synthesis ever speeded up cell cycle progression. I thought this was pretty spectacular, but I don't think many people understood. Moreover, I was pleased by the thought that colchicine's stabilisation of mitotic chromosomes was really due to stabilisation of cyclin, and only indirectly due to inhibition of tubulin polymerisation. Not many people knew that! Less pleasing, however, was the discovery that there was not a perfect correlation between the time of disappearance of the cyclin in these experiments and the decondensation of the chromosomes or the re-formation of the nucleus. There were clearly other things going on that we were not seeing, and interpreting all the curious phenomenology surrounding the meiotic divisions in relation to the requirements for protein synthesis was quite impossible. In many ways the nicest experiments we did in those two summers was measuring how long the proteolysis window stayed open in mitosis in these lovely objects (39). As we had previously found in the sea urchin egg, inhibiting DNA synthesis completely stabilised cyclins A and B alike. Once cells entered mitosis, however, cyclin A went away a minute or two before cyclin B, which in turn went down about 30 seconds or so before the chromosomes visibly parted at anaphase. Proteolysis then remained on for about 5 minutes, after which the cyclins could accumulate again. We couldn't see any other labelled proteins showing this kind of behaviour, although we began to suspect that if Nature had invented such an exquisitely specific proteolysis machine, it was unlikely that it was used exclusively for cyclins, and we wondered, quite rightly as it has much more recently turned out, if the "glue" that held sister chromosomes together might not suffer the same fate as the cyclins. But this really was wild speculation. I should add, too, that we never thought to check if this 5-minute proteolysis window applied to anything else than clam and sea urchin eggs, for the simple reason that we had no idea if yeast or human cells contained cyclins. There was a real possibility that these proteins had evolved to control the rapid early cleavage cell cycles of marine invertebrate eggs, and were found nowhere else.

THE LINK WITH MPF

So the next great advance came from a simple experiment by Katherine Swenson and Joan Ruderman in the Spring of 1986 (40). They asked what would happen if synthetic mRNA made from clone 1T55, encoding clam cyclin A, was injected into a frog oocyte. The answer was extremely gratifying, not to say electrifying, for the oocytes behaved exactly as if they had been injected with MPF. This made people sit up and take note of cyclin. At the same time, the result was puzzling in terms of MPF for two reasons. First, it was well-known at the time that stage VI frog oocytes contained "pre-MPF" whose activation by a starter-dose of active MPF did not require protein synthesis. Indeed, the pathway of activation was quite mysterious, although it was suspected to involve changes in the phosphorylation of something to do with MPF. Thus, although Katherine's spectacular result implied that cyclin was indeed intimately involved in the control of entry into M-phase, it hinted that cyclin was rather peripheral to MPF itself. This tentative conclusion was only reinforced by data in the paper showing that clam oocytes, which also contained pre-MPF, did not contain any detectable cyclin A until well after the first appearance of MPF. Probably it was thinking along these lines that made us favour the idea that cyclin was "anti-anti-MPF", and a jotting from a notebook of 1985 shows an attempt to make sense of things in these terms (Figure 5).

The summer of 1986 found me in Berkeley, California with the authors of *Molecular Biology of the Cell*, beginning to work seriously on the companion *Problems Book* with John Wilson. In those days, one had to go to libraries to read the literature, and I was lucky to get John Gerhart's blessing to use the excellent little library up the hill in Stanley Hall. Even better was Mike Wu's offer one day to show me a real live MPF assay, which I eagerly accepted, and the thought that I could get some maternal mRNA from Eric Rosenthal, who was by then a postdoc with Fred Wilt in the main Zoology Department. We quickly confirmed Katherine and Joan's result using poly A^+ RNA from *Urechis*

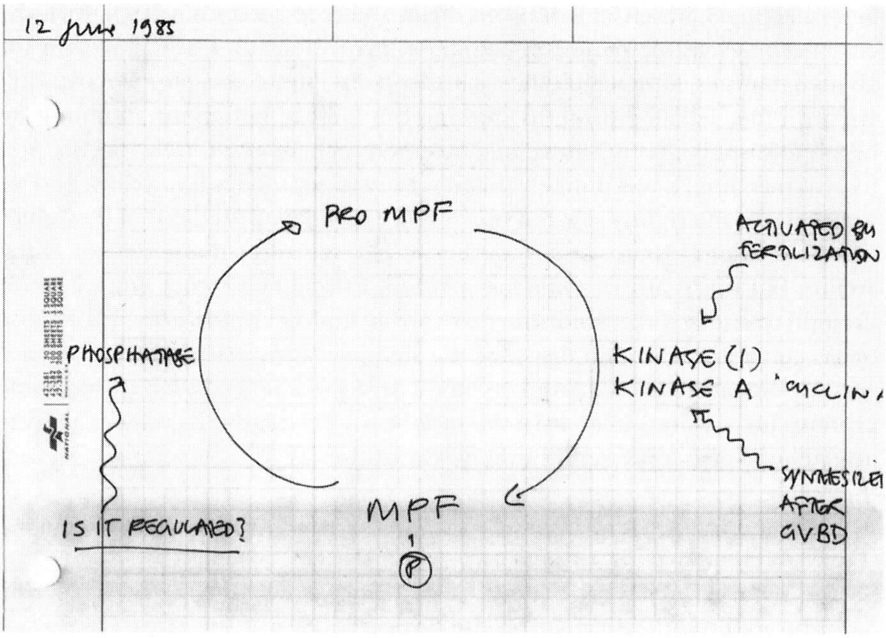

Figure 5. The relationship between cyclin and MPF in 1985.

caupo and Mike was surprised, because in all his time doing mRNA assays for the Bay Area Community, he had never once observed oocyte maturation. We followed this up with an experiment to see if mRNA from *Xenopus* eggs made *Xenopus* oocytes mature, which it did. Creeping back to the incubator at 10 in the evening to see if it worked was extremely naughty and thrilling. Here was suspicious evidence that vertebrates probably had cyclins, too; a great encouragement to Jeremy Minshull after I returned to Cambridge in September. It was a little bit surprising that nobody had thought of looking for the mRNA for MPF earlier, and rather frustrating that there was no easy way to take matters further without very much more work along the cloning and sequencing path. During that summer, both Andrew Murray and Marc Kirschner came to the rented house where we were staying, and we discussed how to unravel the relationship between cyclins and MPF. Andrew's idea was to make a cell-free system based on the one first described by Lohka and Masui (41) that would test the ability of cyclin synthesis to "drive" the cell cycle. It was impossible to imagine doing this in Cambridge, and we agreed that he could have a cyclin clone to test as soon as we had one. Jeremy and I would concentrate on the antisense approach, at which we felt we were getting pretty competent.

Careful comparison between the sea urchin cyclin B and clam cyclin A sequences showed a short stretch of very high homology even at the DNA level, and we designed a minimally redundant antisense oligonucleotide based on this sequence. A test of this oligonucleotide on a variety of starfish, sea urchin and frog mRNA preparations showed very promising results. Bands of the right size went away. After Christmas, Jeremy came back and within weeks we had our first (short) frog cyclin B clones. At this point, I sent him and Jon back to Berkeley to try the effects of cyclin B and of the antisense oligonucleotides on frog oocyte maturation, under the expert tuition of John Gerhart and Mike Wu. Jon's mRNA worked beautifully to induce oocyte maturation, but Jeremy's antisense experiments were not as convincing and we began to suspect that frogs might possess a larger repertoire of cyclins than we had imagined. Indeed, back in England, Jeremy started collaborating with Alan Colman in Birmingham, who was very interested in the MPF story (and in whose laboratory John Shuttleworth cloned the Cdc2 homologue "MO15", which we later found to be the missing CDK activating kinase (42)). They soon found that although the antisense oligonucleotides were working perfectly, the oocytes' ability to mature in response to progesterone was unimpaired. This was a setback, but not a terminal setback. After all, if cyclin were MPF, we knew that the oocytes already contained the protein, so ablating the mRNA would not make any difference. And we had only obtained cyclin B clones: doubtless there would be cyclin A as well, whose synthesis might be vital for the activation of MPF. It was to take almost 10 years to sort out this point (43). In fact, Jeremy kept on isolating and sequencing cyclin clones from *Xenopus* and then testing their role in oocyte maturation by antisense ablation. We stopped after B1, B2 and A1, thereby missing B3, B4, B5 and A2 to say nothing of cyclins D, E and F. We knew then that there were more B-type cyclins to be found. Jon Pines finished his thesis in March 1987 and left Cambridge for Tony Hunter's laboratory at the Salk Institute with the mission of identifying the human cyclins. He was soon writing home with news of success (44). Meanwhile, Nancy Standart went to Roscoff to see if starfish, the other great source of MPF, had cyclins. Sure enough, there were two rounds of synthesis and destruction corresponding to the two meiotic divisions, although protein synthesis was not necessary for starfish oocyte maturation. It is obvious from the discussions of the papers during that period that we were completely obsessed by the question of the relationship between cyclins and MPF (45). Equally obvious is our bafflement.

THE CONNECTION WITH YEASTS

As soon as the signature sequences of cyclin were known, I began scanning protein sequences in papers about the cell cycle, and also asking Paul Nurse if any of their genes looked like ours. The answers were always negative, and DNA sequencing was not usually very high up the priority lists of geneticists at the time. In fact, it was Mark Solomon in Marc Kirschner's laboratory who spotted that Bob Booher's sequence of fission yeast Cdc13 corresponded to a B-type cyclin. This caused great excitement and a certain amount of skulduggery, as it was forbidden to tell Paul Nurse of the news, although as luck would have it I visited Oxford shortly thereafter. As usual, I enquired if any of their genes corresponded to cyclins, and was told that they did not. I went off on holiday and came back to find Booher's paper in the August number of the

EMBO Journal (46), where the sequence was plain for all to see, although there was no mention of cyclins. Shortly after, a couple of notes drawing attention to the homology appeared in Cell (47, 48), and the whole story began to make sense.

In 1987, Alan Colman was teaching in Woods Hole and organised a small symposium about MPF, where Jim Maller showed the famous picture of his and Fred Lohka's purification of *Xenopus* MPF (49). Their purest fractions contained two bands and displayed histone kinase activity. The lower band had a molecular weight around 32,000 and the larger band was a shade small for cyclin B, but I promised Jim that as soon as we had antibodies, we would see if it was cyclin, a promise we later kept with the help of Jean Gautier (50). It turned out that Jim's lab and mine used different recipes for SDS-polyacrylamide gels that largely accounted for the difference in mobility.

The key developments as far as I was concerned took place in parallel in Cambridge and in San Francisco. In Cambridge, Jeremy and I enlisted the help of Julian Blow to help with frog extracts, and succeeded in blocking cyclin synthesis with carefully chosen oligonucleotides that had minimal effects on the synthesis of other proteins in the lysates. We found that these reagents did not affect DNA replication, but did block entry into mitosis (51). Andrew Murray's approach was much bolder and the results more impressive. Jon Pines's sea urchin cyclin clone could "drive" several trains of cell cycles in an RNase-treated *Xenopus* egg extract, and moreover, if the N-terminus was removed by genetic engineering, the cyclin (famously known as Δ90) could drive the extract into mitosis, but such a cyclin was stable and the extracts were stuck in M-phase (although there was confusion – which remains to this day, in fact – about whether the extracts were in metaphase or anaphase) (52, 53).

Marcel Dorée and I have recently discussed elsewhere how it emerged that cyclins and Cdc2 were partners (54), and indeed, Marcel's purification of starfish MPF, which he projected as a dried-down gel at a meeting in St. Andrew's University in Scotland in April 1989 (55), set the seal on the matter after a period of confusion during which there had been claims that Cdc2 alone was required for MPF activity. Full resolution of the matter came later from biochemical studies of the activation of CDK2 involving in parallel my laboratory, in particular Randy Poon, Jörg Adamczewski and John Shuttleworth, and that of Marcel Dorée. We found that GST-CDK2, provided by Li-Huei Tsai, could be slightly activated by cyclin A, and that the activation was tremendously enhanced by immunoprecipitates of MO15 (56, 42). It was not too long before the structures of the components and of the whole complex were determined, which explained very clearly why CDKs required cyclins for activation (Figure 6).

I have said very little about what we knew from the yeast genetic approaches, even though we were well aware of them. I had met Lee Hartwell when I was still at Einstein, and knew Paul Nurse quite well from the time he was in Sussex. But as I have made clear, it was the aquatic creatures, the clams, sea urchins, frogs and starfish that really inspired me, and the allure of MPF

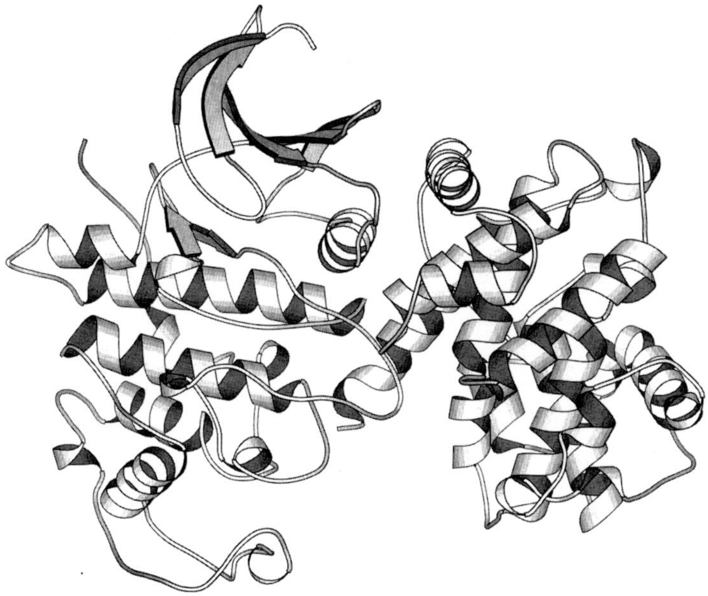

Figure 6. The structure of CDK2 bound to a C-terminal fragment of cyclin A. Coordinates from 1JST by A. A. Russo, P. D. Jeffrey & N. P. Pavletich.

which attracted me to studying the cell cycle. It is interesting to note that none of the classical *cdc* mutations in budding yeast corresponded to cyclins (although Steve Reed, Fred Cross and Bruce Futcher quickly identified them by various ingenious approaches). Nor did the crucial CDK activating kinase reveal itself through genetics. In fact, the cell cycle field in general provided an extremely useful meeting place for the two cultures.

One final comment. The decade starting in about 1986 was a fantastic experience for anyone working on the cell cycle. Discoveries emerged from all sides and unexpected quarters at a headily bewildering rate. The culture was generous and open, and the field attracted extremely talented scientists who were very much fun to work with and talk to. I would like to thank them. This Nobel prize honours them all.

A FOOTNOTE ABOUT GELS

Why did nobody run a timed series of ^{35}S-methionine labelled fertilized sea urchin or clam eggs on a 1-dimensional SDS-polyacrylamide gel between 1970 and 1982? One answer, of course, is that very few people worked on these cells; but more important, perhaps, are simple technical reasons. Thus, in the late 1960s, Paul Gross and his colleagues were interested in patterns of protein synthesis in cleaving sea urchin eggs and in mammalian cell cycles (57-59), but because the SDS gel had not yet been developed, and the slab gel not quite invented, the resolution of individual proteins and timed comparisons were not possible. The next serious investigator of the question of whether fertilized sea urchin eggs synthesized different proteins from unfertilized eggs was Bruce Brandhorst in the mid-1970s (60), by which time the high-resolution 2-dimensional gel system had been invented by Pat O'Farrell (61). Two serious obstacles stood in the way of detecting cyclins on 2-dimensional gels. First, as we later discovered, cyclins are difficult, if not impossible to focus in the isoelectric dimension of the system, and they appear as ugly smears if they appear at all. Second, it is not easy to compare multiple 2-dimensional autoradiograms, and the idea of taking many closely-spaced time points would be both difficult to achieve and rather pointless in order to address the questions being asked at the time. I was very lucky.

REFERENCES

(1) Borsook, H. **(1966)**.. Early development of the echinoid egg compared with erythropoiesis. *Biol. Rev. 41*: 259–274.

(2) Dintzis, H.M. **(1961)**. Assembly of the peptide chains of haemoglobin. *Proc. Natl. Acad. Sci USA 48*: 247–261.

(3) Hunt, T., Hunter, T. & Munro, A. **(1968)**. Control of haemoglobin synthesis: distribution of ribosomes on the messenger RNA for alpha and beta chains. *J. Mol. Biol. 36*: 31–45.

(4) Hunt, R.T., Hunter, A.R. & Munro, A.J. **(1968)**. Control of haemoglobin synthesis: a difference in the size of the polysomes making alpha and beta chains. *Nature. 220*: 481–483.

(5) Lodish, H.F. & Jacobsen, M. **(1972)**. Regulation of hemoglobin synthesis. Equal rates of translation and termination of α- and β-globin chains. *J. Biol. Chem. 247*: 3622–3629.

(6) Jackson, R. & Hunter, T. **(1970)**. Role of methionine in the initiation of haemoglobin synthesis. *Nature 227*: 672–676.

(7) Zucker, W.V. & Schulman H.M. **(1968)**. Stimulation of globin-chain initiation by hemin in the reticulocyte cell-free system. *Proc. Natl .Acad. Sci .USA 59*: 582–589.

(8) Maxwell, C.R. & Rabinovitz, M. **(1969)**. Evidence for an inhibitor in the control of globin synthesis by hemin in a reticulocyte lysate. *Biochem. Biophys. Res Commun. 35*: 79–85.

(9) Kosower, N.S., Vanderhoff, G.A., Benerofe, B., Hunt, T. & Kosower, E.M. **(1975)**. Inhibition of protein synthesis by glutathione disulfide in the presence of glutathione. *Biochem Biophys Res Commun.* 45: 816–812.

(10) Jackson, R.J., Herbert, P., Campbell, E.A. & Hunt, T. **(1983)**. The roles of sugar phosphates and thiol-reducing systems in the control of reticulocyte protein synthesis. *Eur. J. Biochem. 131*: 313–324.

(11) Hunt, T., Herbert, P., Campbell, E.A., Delidakis, C. & Jackson, R.J. **(1983)**. The use of affinity chromatography on 2'-5'ADP- Sepharose reveals a requirement for NADPH, thioredoxin and thioredoxin reductase for the maintenance of high protein synthesis activity in rabbit reticulocyte lysates. *Eur. J. Biochem.* 131, 303–311.

(12) Hunt, T. & Ehrenfeld, E. **(1971)**. Cytoplasm from poliovirus-infected HeLa cells inhibits cell-free haemoglobin synthesis. *Nature New Biol.* 230, 91–94.

(13) Ehrenfeld, E. & Hunt, T. **(1971)**. Double-stranded poliovirus RNA inhibits initiation of protein synthesis by reticulocyte lysates. *Proc. Nat. Acad. Sci. USA* 68, 1075–1078.

(14) Pelham, H.R. & Jackson, R.J. **(1976)**. An efficient mRNA-dependent translation system from reticulocyte lysates. *Eur. J. Biochem. 67*: 247–256.

(15) Hunter, T., Hunt, T., Jackson, R.J. & Robertson, H.D. **(1975)**. The characteristics of inhibition of protein synthesis by double-stranded RNA in reticulocyte lysates. *J. Biol. Chem. 250*: 409–417.

(16) Darnbrough, C.H., Legon, S., Hunt, T & Jackson, R.J. **(1973).** Initiation of protein synthesis: evidence for messenger RNA- independent binding of methionyl transfer RNA to the 40S ribosomal subunit. *J. Mol. Biol. 76*: 379–403.

(17) Legon, S., Brayley, A., Hunt, T. & Jackson, R.J. **(1974)**. The effect of cAMP and related compounds on the control of protein synthesis in reticulocyte lysates. *Biochem. Biophys. Res. Comm. 56*: 745–752.

(18) Farrell, P.J., Balkow, K., Hunt, T., Jackson, R.J. & Trachsel, H. **(1977)**. Phosphorylation of initiation factor eIF-2 and the control of reticulocyte protein synthesis. *Cell 11*: 187–200.

(19) Hershko, A., Mamont, P., Shields, R. & Tomkins, G.M. **(1971).** Pleiotypic response. *Nat. New Biol.* 232: 206–211.

(20) Walker, A.I., Hunt, T., Jackson, R.J. & Anderson, C.W. **(1985)**. Double-stranded DNA induces the phosphorylation of several proteins including the 90 000 mol. wt. heat-shock protein in animal cell extracts. *EMBO J. 4*, 139–145.

(21) Jackson, S.P., MacDonald, J.J., Lees-Miller, S. & Tjian, R. **(1990)**. GC box binding induces phosphorylation of Sp1 by a DNA-dependent protein kinase. *Cell 63*: 155–165.

(22) Jackson, R.J., Cambell, E.A., Herbert, P. & Hunt, T. **(1983)**. The preparation and properties of gel-filtered rabbit reticulocyte lysate protein synthesis systems. *Eur. J. Biochem.* 131, 289–301.

(23) Ballinger, D. & Hunt, T. **(1981)**. Fertilization of sea urchin eggs is accompanied by 40S ribosomal subunit phosphorylation. *Dev. Biol.* 87, 277–285.

(24) Brandhorst, B.P. **(1976).** Two-dimensional gel patterns of protein synthesis before and after fertilization of sea urchin eggs. *Dev. Biol. 52*: 310–317.

(25) Rosenthal, E.T., Hunt, T. & Ruderman, J.V. **(1980)**. Selective translation of mRNA controls the pattern of protein synthesis during early development of the surf clam, *Spisula solidissima. Cell 20*: 487–494.

(26) Wu, M. & Gerhart, J.C. **(1980).** Partial purification and characterization of the maturation-promoting factor from eggs of *Xenopus laevis. Dev. Biol.* 1980 *79*: 465–477.

(27) Murray, A.W. & Szostak, J.W. **(1983).** Construction of artificial chromosomes in yeast. *Nature 305*: 189–193.

(28) Gerhart, J., Wu, M. & Kirschner, M. **(1984).** Cell cycle dynamics of an M-phase-specific cytoplasmic factor in *Xenopus laevis* oocytes and eggs. *J. Cell Biol. 98*: 1247–1255.

(29) Evans, T., Hunt, T. & Youngblom, J. **(1982).** On the role of maternal mRNA in sea urchins: studies of a protein which appears to be destroyed at a particular point in each cell division cycle. *Biol. Bull. 163*: 372.

(30) Standart, N.M., Bray, S.J., George, E.L., Hunt, T. & Ruderman, J.V. **(1985).** The small subunit of ribonucleotide reductase is encoded by one of the most abundant translationally regulated maternal RNAs in clam and sea urchin eggs. *J. Cell Biol. 100*: 1968–1976.

(31) Evans, T., Rosenthal, E.T., Youngbloom, J., Distel, D. & Hunt, T. **(1983)**. Cyclin: a protein specified by maternal mRNA in sea urchin eggs that is destroyed at each cleavage division. *Cell 33*: 389–396.

(32) Glotzer, M, Murray, A.W. & Kirschner, M.W. **(1991).** Cyclin is degraded by the ubiquitin pathway. *Nature* 349: 132–138.

(33) Wagenaar EB. **(1983).** The timing of synthesis of proteins required for mitosis in the cell cycle of the sea urchin embryo. *Exp. Cell. Res. 144*: 393–403.

(34) Hultin, T. **(1961)**. The effect of puromycin on protein metabolism and cell division in fertilized sea urchin eggs. *Experientia* 17, 410–411.

(35) Rosenthal, E.T., Tansey, T.R., & Ruderman, J.V. **(1983).** Sequence-specific adenylations and deadenylations accompany changes in the translation of maternal messenger RNA after fertilization of *Spisula* oocytes. *J. Mol. Biol.* 166: 309–327.

(36) Goelet, P., Lomonossoff, G.P., Butler, P.J., Akam, M.E., Gait, M.J., & Karn, J.**(1982).** Nucleotide sequence of tobacco mosaic virus RNA. *Proc. Natl .Acad. Sci. USA.* 79: 5818–5822.

(37) Minshull, J. & Hunt, T. **(1986)**. The use of single-stranded DNA and RNase H to promote quantitative "hybrid arrest of translation" of mRNA/DNA hybrids in reticulocyte lysate cell-free translations. *Nucleic Acids Res. 14*, 6433–6451.
(38) Pines, J. & Hunt, T. **(1987)**. Molecular cloning and characterization of the mRNA for cyclin from sea urchin eggs. *EMBO J., 6*: 2987–2995.
(39) Hunt, T., Luca, F.C. & Ruderman, J.V. **(1992)**. The requirements for protein synthesis and degradation, and the control of destruction of cyclins A and B in the meiotic and mitotic cell cycles of the clam embryo. *J. Cell Biol. 116*: 707–724.
(40) Swenson, K.I., Farrell, K.M. & Ruderman, J.V. **(1986)**. The clam embryo protein cyclin A induces entry into M phase and the resumption of meiosis in *Xenopus* oocytes. *Cell 47*: 861–870.
(41) Lohka, M.J. & Masui, Y. **(1983)**. Formation *in vitro* of sperm pronuclei and mitotic chromosomes induced by amphibian ooplasmic components. *Science* 220: 719–721.
(42) Poon, R. Y., Yamashita, K., Adamczewski, J. P., Hunt, T. & Shuttleworth, J. **(1993)**. The cdc2-related protein $p40^{MO15}$ is the catalytic subunit of a protein kinase that can activate $p33^{cdk2}$ and $p34^{cdc2}$. *EMBO J.* 12, 3123–3132.
(43) Hochegger, H. Klotzbücher, A., Kirk, J., Howell, M., le Guellec, K., Fletcher, K., Duncan, T., Sohail, M. & Hunt, T. **(2001)**. New B-type cyclin synthesis is required between meiosis I and II during *Xenopus* oocyte maturation. *Development,* 128, 3795–3807.
(44) Pines, J. & Hunter, T. **(1989)**. Isolation of a human cyclin cDNA: evidence for cyclin mRNA and protein regulation in the cell cycle and for interaction with $p34^{cdc2}$. *Cell.* 58: 833–846.
(45) Standart, N., Minshull, J., Pines, J. & Hunt, T. **(1987)**. Cyclin synthesis, modification and destruction during meiotic maturation of the starfish oocyte. *Dev. Biol. 124*: 248–258.
(46) Booher, R. & Beach, D. **(1988)**. Involvement of *cdc13*⁺ in mitotic control in *Schizosaccharomyces pombe*: possible interaction of the gene product with microtubules. *EMBO J.* 7: 2321–2327.
(47) Solomon, M., Booher, R., Kirschner, M. & Beach, D **(1988)**. Cyclin in fission yeast. *Cell 54*: 738–740.
(48) Goebl, M & Byers, B. **(1988)**. Cyclin in fission yeast. *Cell 54*: 738–740.
(49) Lohka, M.J., Hayes, M.K. & Maller, J.L. **(1988)**. Purification of maturation-promoting factor, an intracellular regulator of early mitotic events. *Proc. Natl. Acad .Sci .USA. 85*: 3009–3013.
(50) Gautier, J., Minshull, J., Lohka, M., Glotzer, M., Hunt, T. & Maller, J.L. **(1990)**. Cyclin is a component of maturation-promoting factor from *Xenopus. Cell 60*: 487–494.
(51) Minshull, J., Blow, J. & Hunt, T. **(1989)**. Translation of cyclin mRNA is necessary for extracts of activated *Xenopus* eggs to enter mitosis. *Cell 56*: 947–956.
(52) Murray, A.W. & Kirschner, M.W. **(1989)**. Cyclin synthesis drives the early embryonic cell cycle. *Nature 339*: 275–278.
(53) Murray, A.W., Solomon, M.J. & Kirschner, M.W. **(1989)**. The role of cyclin synthesis and degradation in the control of maturation promoting factor activity. *Nature 339*: 280–286.
(54) Dorée, M. & Hunt, T. **(2002)**. From Cdc2 to Cdk1: when did the cell cycle kinase join its cyclin partner? *J. Cell Sci. 115*: 2461–2464.
(55) Labbé, J. C., Capony, J. P., Caput, D., Cavadore, J. C., Derancourt, J., Kaghad, M., Lelias, J. M., Picard, A. & Dorée, M. **(1989)**. MPF from starfish oocytes at first meiotic metaphase is a heterodimer containing one molecule of cdc2 and one molecule of cyclin B. *EMBO J. 8*: 3053–3058.
(56) Fesquet D, Labbé, J.C., Derancourt, J., Capony, J.P., Galas, S., Girard, F., Lorca, T., Shuttleworth , J., Dorée, M. & Cavadore JC. **(1993)**. The MO15 gene encodes the catalytic subunit of a protein kinase that activates cdc2 and other cyclin-dependent kinases (CDKs). through phosphorylation of Thr161 and its homologues. *EMBO J.* 12: 3111–3121.

(57) Fry, B.J. & Gross, P.R. **(1970).** Patterns and rates of protein synthesis in sea urchin embryos. II. The calculation of absolute rates. *Dev Biol. 21,* 125–146.

(58) Fry, B.J. & Gross, P.R. **(1970).** Patterns and rates of protein synthesis in sea urchin embryos. I. Uptake and incorporation of amino acids during the first cleavage cycle. *Dev. Biol. 21,* 105–124.

(59) Kolodny, G.M. & Gross, PR. **(1969).** Changes in patterns of protein synthesis during the mammalian cell cycle. *Exp. Cell Res. 56,* 117–121.

(60) Brandhorst, B.P. **(1976).** Two-dimensional patterns of protein synthesis before and after fertilization of sea urchin eggs. *Dev. Biol. 52,* 310–317.

(61) O'Farrell P.H. **(1975).** High resolution two-dimensional electrophoresis of proteins. *J. Biol. Chem. 250,* 4007–40021.

Paul Nurse

PAUL M. NURSE

My parents were born in Norfolk and spent their early years working in the big houses of that rural English county, my mother as a cook and my father as a handyman and chauffeur. After the 1930s recession they moved to Wembley, North-West London, where my father worked as a mechanic in the local H.J.Heinz food processing factory, and my mother brought up their four children and was a part-time cleaner. I was by far the youngest of the family, and at times it was like being an only child. My parents were neither wealthy nor academic, but we lived comfortably and they were always extremely supportive of my academic efforts and aspirations, both at school and university.

My primary school was a considerable distance from where we lived and so I had long walks, often alone, to and from school. This walk took me through a park and some rough land where I could not fail to notice the animals, insects and plants there and how they changed during the seasons. During the winter my attention was attracted to the changes in the stars and planets in the sky. I think it was this curiosity about the natural world which awoke my early interest in science. Two incidents from this time that I remember, were, wondering why leaves were larger on plants growing in the shade compared with the same plants growing in sunlight, and watching Sputnik 2, the second ever artificial satellite and the first with a living cargo (a dog called Laika), as it sped across the skies of London. My life-long interest in astronomy started then and I still regularly use a telescope for astronomical observations, although very much as an amateur.

I enjoyed my time at primary school because my teachers made the world seem such an interesting place and encouraged my innate curiosity. At age 11 in 1960, I moved to an academic state secondary school, Harrow County Grammar School for Boys. This was a mixed experience for me. It was a good, well-resourced school, but was very exam oriented and most of the other boys came from wealthier and more academic families which sometimes made me feel like a fish out of water. I was never very good at exams, having a poor memory and finding the examination process rather artificial, and there never seemed to be enough time to follow up things that really interested me. But there were good things about the school. I had an excellent Biology teacher Keith Neal, who encouraged his pupils to study natural history and to do real experiments. I had a great time investigating the pigments of different mutant fruit flies by following experimental protocols published in Scientific American, and I also remember making my own beetle collection when it was still acceptable to make such collections. There were also good extra-curricu-

My parents and myself with my motorbike in the late 1960s.

lar activities, particularly hill and mountain walking and more surprisingly, learning to fly. I am still a keen mountain walker and an enthusiastic glider pilot. I also made some very good friends who remained important to me into adulthood. It was during my time at secondary school that I abandoned religion. My mother was a Baptist, and as a young teenager I was also a committed believer. But I had real difficulties reconciling a literal belief in Genesis with evolution, and my attempts to accommodate the biblical account of creation by viewing it as a poetic metaphor suitable for an unsophisticated nomadic people was completely rejected by my church. I gradually slipped away from religion over several years and became an atheist or to be more philosophically correct, a sceptical agnostic.

By the end of my time at school I had achieved examination grades which allowed me to go to University, but did not have a basic foreign language qualification which was compulsory for all University entrants. This meant that when I left school I had to work as a technician in a microbiological laboratory associated with the local Guinness brewery. This was a great experience for me because I had a very sympathetic lab head Vic Knivett, who rapidly realised that I could complete the routine requirements of my job in a couple of days each week, and encouraged me to carry out research experiments for the rest of my time. Unfortunately I continued to fail my French Exam and it was only the intervention of Professor Jinks at Birmingham that

My biology teacher, Keith Neal.

got me into a University. He had noticed my application for entry and asked me to visit his Genetics Department. After an extensive interview he arranged for my weaknesses in foreign languages to be over-looked and so I started a Biology degree at Birmingham in 1967.

My time as an undergraduate at Birmingham was extremely stimulating both as a biologist and also for my more general intellectual development. It was the heady times of the sixties when everything could be challenged and everything seemed possible. I met my wife Anne who was a sociology student, and her influence together with activities associated with the student movement of the time opened up my interests amongst other things into the theatre, art, music, politics and philosophy. For the first time I fully recognised the excitement of intellectual endeavour and realised that this was what I wanted to do with my life.

The lecturers on my courses were generally enthusiastic about their subjects and encouraged my interest in the biological world. At first my inclinations turned towards the subject of ecology with its then new theoretical models of ecological systems, but a field course collecting marine specimens in freezing cold weather taught me that I was too soft for practical ecology, and I realised that I had a preference for the more controlled and warmer environment of the laboratory. Gradually my interests settled on developmental biology with an emphasis on plants and a molecular approach to the problem. I had an eccentric zoology tutor Jack Cohen who was hugely stimulating and entertaining, and although frequently wrong was always wrong in an interesting way. He taught me the value of the alternative view and also was the first to introduce me to the cell cycle with a project on the respiration rate of dividing fish eggs, a project which ended in complete disaster. This time at Birmingham turned me into a general biologist, and ever since then I have always tried to take a biological approach to any research project that I have undertaken.

Murdoch Mitchison washing up after a Burns Night party in Edinburgh during the 1970s.

Urs Leupold with his alpine horn in Bern.

A key issue in developmental biology at that time was the problem of how cells underwent differentiation, with most workers concentrating on explanations in terms of changes in enzyme and gene regulation. To me the cell cycle seemed to be a good and simple model for such problems, because the cell underwent molecular changes *as* it proceeded through its cell cycle. So when my thoughts turned towards a PhD I looked for a laboratory where I could study molecular changes during the cell cycle. Tony Simms at the University of East Anglia (UEA) in Norfolk was just beginning such studies by looking at the enzymes of amino acid metabolism during the cell cycle of the fungus *Candida utilis.* I went to his lab as a graduate student because I was joining a project at its inception and thought that I would be able to make a more important contribution to the work. Tony was a great experimentalist and I rapidly learnt the need for good experiments to make any progress at all in a research project. Like many students, I found the drudgery of real experiments and the slowness of progress a complete shock, and at my low points I contemplated other alternative careers including study of the philosophy or sociology of science. However, the atmosphere at UEA was very supportive and my colleagues made me feel I was making a useful contribution so I survived the difficulties of carrying out real research. In fact I am very much an experimentalist and an empiricist, so it would have been a major mistake for me to have abandoned this type of work. But this experience did

Anne, Sarah and Emily in Brighton during the early 1980s.

Anne, Sarah and Emily in 2000.

teach me the need for sympathetic support of scientific colleagues, because at the forefront of research there are so many difficulties that depression and low motivation are a constant danger.

The next step was what to do as a post-doctoral worker and this question exercised me greatly in my final year as a graduate student. I felt strongly that since the pursuit of good science was so difficult it was essential that the problem being studied was an important one to justify the effort expanded. Rather grandly I argued to myself that the process of reproduction was a central property of life, and that this was seen in its simplest form with the reproduction of cells. Therefore, I reasoned that study of the cell cycle responsible for the reproduction of cells was important and might even be illuminating about the nature of life. In particular the control of these processes would be crucial, just like control of flux through a pathway was important for amino acid metabolism. But how could such processes be investigated given so little was known about them? The answer came to me in 1972 when I read two papers from Lee Hartwell who showed how genetics could be used to study the budding yeast cell cycle. I thought this was a beautiful approach to the problem, one I wanted to use as well. The difficulty was that Anne was by now a teacher and neither of us wanted to move to the USA, and so I needed to find a UK laboratory for this work. Murdoch Mitchison in Edinburgh was the UK authority on the cell cycle and worked on fission yeast but he was not a geneticist. I went to Edinburgh on a wonderful blue day with the city under snow to discuss my aspirations, and was immediately attracted both to Murdoch's laboratory and to this beautiful city. He suggested that I spent a couple of months in Bern Switzerland with Urs Leupold, the father of fission yeast genetics, so that I could gain experience that would allow me to begin a genetic analysis of the fission yeast cell cycle. Urs was a careful but inspirational teacher and within six months I was able to introduce genetics into Murdoch's cell cycle laboratory.

My 6 years with Murdoch were pivotal for my entire research career. He gave me both complete support and total freedom, spending hours each week talking with me but never instructing me what to do. An astonishingly generous supervisor, he never once was a co-author on any of the papers I produced during my time in Edinburgh. He considered, quite wrongly of course, that he had not made a sufficient contribution to justify inclusion! His laboratory was an exciting environment where I had many interesting colleagues. During my first year, two new recruits came, Peter Fantes as a postdoc and Kim Nasmyth as a graduate student. Peter was of a mathematical and theoretical bent and it was through discussions and joint work with him that the importance of cell mass in regulating progression through the fission yeast cell cycle first became clear to me. Murdoch asked me to look after Kim who was extra-ordinarily bright, and he extended the cell cycle mutant collections during his thesis work. My main efforts focussed on trying to identify the rate controlling steps during the cell cycle. Crucial for this analysis were wee mutants that were advanced prematurely through the cell cycle and so divided at a reduced cell size. I would like to claim that I reasoned abstractly that such mutants would be useful and then tried to find them, but in reality I noticed them only by accident whilst searching for completely different mutants. As is often the case it was nature that provided the best lead to be followed. A third new recruit was a friend of mine from Bern, Pierre Thuriaux who worked with me on showing that *cdc2* was a rate limiting factor controlling the onset of mitosis. A second key advance was showing that *cdc2* was also required for the onset of S-phase. This emerged from work with Yvonne Bissett while we were trying to identify a G1 control start similar to the one defined by Lee Hartwell in budding yeast. As a negative control for this experiment we used *cdc2* mutants which we thought would block cell cycle progression in G2. In fact the results we obtained were ambiguous with the negative controls always giving a small but significant positive response. For some time we thought the experiment flawed and then in desperation hypothesised that *cdc2* was required twice in the cell cycle explaining the mixed results. This turned out to be correct establishing that *cdc2* was a controlling factor for the onset of both S-phase and mitosis.

It was now 1980 and Anne and myself had two little children Sarah and Emily, and we were wondering whether to stay permanently in Edinburgh. This possibility bothered me as I thought it was not advisable to remain in one academic environment, and the long dark winters in Edinburgh could be rather dismal. I also thought that the next stage in cell cycle analysis required molecular genetics, and fission yeast was not developed for these types of experiments, and so I looked for an environment which would make this possible. In the end I decided that the University of Sussex in Brighton was a good place for this work because it had a strong tradition in bacterial molecular genetics and an excellent reputation in biology. It was also almost as far south as it was possible to go in the UK and so had a reasonable chance of having decent summers! I set up my first laboratory there with a technician and myself and began working on fission yeast molecular genetic manipulation. In this project I had great assistance from an excellent post-doc from the next door lab, David Beach, who was already skilled in DNA cloning procedures. There followed a very fruitful collaboration that led to the cloning of the more important cell cycle genes. This allowed me to recruit more workers, in particular my first graduate student Jacky Hayles, who went on to contribute immeasurably to many subsequent projects in my laboratory and still is a stimulating and imaginative colleague with me today. Another very important recruit was Paul Russell who cloned the more important regulators of *cdc2*.

This progress in the molecular analysis of the cell cycle led to more interest being taken in my work and as a consequence to greater competition. My job was not secure in Sussex as I only had a time limited position and the financial support for my lab was limited. I had failed to secure a permanent appointment at Sussex or the major MRC research institutes in Cambridge and London, and was on the verge of moving to the EMBL in Heidelberg where the Director Lennart Philipson had offered me a post. However, we wanted to stay in the UK and just in time Walter Bodmer, Scientific Director of the Imperial Cancer Research Fund (ICRF) offered me a permanent lab head position at his main laboratories in Lincoln's Inn Fields and I moved there in 1984. For the first time I found myself in a very well funded laboratory and a hot-house scientific environment. There were pros and cons to this; the financial support meant we could do any experiment we wished but there was not the same collegiate support that I was used to in the Universities. Perhaps, not unreasonably, some wondered what a yeast researcher was doing in a cancer research institute. Despite this I was now able to set up an effective laboratory which could take on many of the problems that I could not have addressed before. Lincoln's Inn was a wonderful scientific institute with nearly every molecular biology procedure being carried out by someone within its walls. I quickly built up a laboratory of excellent post-docs who brought in expertise that I did not have, expertise which was essential for a proper molecular analysis of the cell cycle control genes. These included Viesturs Simanis, Sergio Moreno and Kathy Gould, who between them worked out that *cdc2* encoded a protein kinase that was regulated during the cell cycle and was controlled by tyrosine phosphorylation. Another post-doc Melanie Lee cloned the human *CDC2* gene by rescue of a fission yeast *cdc2* mutant, and so established that the cell cycle in humans was likely to be regulated in the same way as yeasts. This was a major step forward, all the more so because she persevered with a project that many argued was highly unlikely to succeed.

At the end of the 1980s as a complete surprise my old Edinburgh friend, Ed Southern offered me the Chair of Microbiology at the University of Oxford. It was a difficult decision to move to Oxford as things were going so well at Lincoln's Inn, but when my daughters realised that they would be able to ride ponies almost daily the debate was over and we moved to Oxford in 1988. I had a lot more space and my lab grew in size probably beyond a level that I could properly supervise. Many of my lab colleagues moved with me to Oxford and others including Chris Norbury, Iain Hagan and Tamar Enoch joined me there. During this time the links of *cdc2* with cyclins and maturation promoting factor (MPF) were established, as well as its involvement with checkpoint controls. My administrative work load increased dramatically as the head of a University Department and also as President of the UK Genetical Society. This together with being the supervisor of a large laboratory meant that I failed to take sufficient advantage of the broader academic environment that a University has to offer. I had hoped to recreate my times at Sussex and Edinburgh but this was not to be. However, I did learn how to manage things and how to deal with finances which led to me being approached about returning to London as Scientific Director to ICRF. The temptation of the properly founded laboratory was too great to resist and I returned to Lincoln's Inn Fields in 1993.

This return to ICRF made me a colleague of Tim Hunt with whom I had had close contacts for the previous decade although we had never worked together. He was and is a most delightful companion and our very different backgrounds meant that our conversations, for me at least, were always stimulating and productive. We had the closeness and mutual respect which meant we could utterly disagree without the slightest risk of the other becoming upset. As a biochemist he was more grounded in reality, whilst as a geneticist I was more abstract in my thought. This meant I tended to lose contact with what was really possible within the constraints of the Laws of Thermodynamics as Tim was always ready to point out to me. We were also both natural performers on the lecture stage and not infrequently we entertained as a double act at conferences. There were many other scientists on the world stage that had great influence on the cell cycle field during the crucial times of the 1980s to the early 1990s. Important for me were those working on frog and marine invertebrate egg cell cycles, work started by Yoshio Masui. Marc Kirschner provided immense clarity to this sometimes murky field, whilst Jim Maller successfully achieved the very important task of purifying maturation promoting factor or MPF. My lab collaborated with the French Scientist Marcel Dorée who not only brought well needed biochemistry to my work but great gaiety as well. Closer to my own experimental system were the yeast geneticists, in particular Mitsuhiro Yanagida, a towering powerhouse of rigorous experimental activity, and Andrew Murray, a worker of great imaginative ability. My interactions with these and many others greatly shaped and corrected my thinking about the cell cycle and its control.

The1990s saw my children grow up and in their early 20s they left home. Sarah is an assistant producer for a major TV soccer programme in the UK, whilst Emily has just started a PhD in theoretical particle physics. My heavy responsibilities as Director General of ICRF have taken their toll on my experimental work. It is a constant worry to me that I have not supervised my graduate students and post-docs properly during this time. It has only been their excellent quality which has kept the work going reasonably well. Their main focus has been elucidating the controls regulating the onset of S-phase and ensuring that there is only one S-phase each cell cycle, a new set of projects dealing with the controls during the meiotic cell cycle, and genomic analysis of fission yeast. These problems all interest me and hopefully in the not too distant future I will be able to devote more of my time to their study. Two other projects are also being pursued. The first is a Science in Society Initiative of the Royal Society which I chair and has as its objective an improvement in dialogue between scientists and the public. Its agenda bears an uncanny similarity to that discussed in the late 1960s during student occupations at the University of Birmingham. The second is the merger of the ICRF with the CRC to generate a single organisation Cancer Research UK, which will form the largest cancer research organisation in the world outside of the USA.

The award of this prize is a water-shed in my life, forcing me to look back over my past and to consider what I should do for the next 15–20 years. I have an idealistic view of science as a liberalising and progressive force for humanity. Better understanding of the natural world not only enhances all of us as human beings, but can also be harnessed for the better good, leading to improved health and quality of life. It is also a truly international activity which breaks down barriers between the peoples of the world, an objective that always has been necessary and never more so than now. Scientific understanding is often beautiful, a profoundly aesthetic experience which gives pleasure not unlike the reading of a great poem. It has been a privilege to pursue knowledge for its own sake and to see how it might help mankind in more practical ways. I hope that the future will allow me to continue that pursuit for as long as I am able.

CYCLIN DEPENDENT KINASES AND CELL CYCLE CONTROL

Nobel Lecture, December 9, 2001

by

PAUL M. NURSE

Cancer Research UK, PO Box 123, 61 Lincoln's Inn Fields, London WC2A 3PK, United Kingdom.

The cell is the basic structural and functional unit of all living organisms, the smallest entity that exhibits all the characteristics of life. Cells reproduce by means of the cell cycle, the series of events which lead to the division of a cell into two daughters. This process underlies growth and development in all living organisms, and is central to their heredity and evolution. Understanding how the cell cycle operates and is controlled is therefore an important problem in biology. It also has implications for medicine, particularly cancer where the controls of cell growth and division are defective. In this account I describe the contributions my laboratory has made to understanding cell cycle control, focussing on the major regulators of the eukaryotic cell cycle, the cyclin dependent kinases.

EVENTS AND CONTROL OF THE CELL CYCLE

The most important events of the cell cycle are those concerned with replication of the genome and segregation of the replicated genomes into the daughter cells formed at division (Mitchison 1971). In eukaryotic cells these events are separated in time; chromosome replication occurs during S-phase early in the cell cycle, and segregation of the replicated chromosomes occurs during M-phase or mitosis at the end of the cell cycle. The phase before S-phase is called G1 and the phase before mitosis is called G2. The replication of the chromosomes is based on the double helix structure of DNA which unwinds during S-phase to generate two templates used for the synthesis of two new complementary DNA strands. During mitosis a bipolar spindle is formed and the two double helix DNA molecules making up each replicated chromosome become condensed and oriented towards opposite poles of the cell. The DNA molecules attach to microtubules emanating from the spindle poles and move away from each other toward opposite poles to be segregated at cell division. Thus the formation of two genomes during the cell cycle occurs at the molecular level during S-phase and at the cellular level during mitosis.

To ensure that each newly formed daughter cell receives a complete genome the onset and progression of S-phase and mitosis are controlled so that they occur in the correct sequence once each cell cycle, are corrected for errors in their execution, and are co-ordinated with cellular growth. My laboratory has worked on how these cell cycle controls operate in the single celled eukaryote fission yeast or *Schizosaccharomyces pombe,* and has also extended these studies to metazoan cells.

FISSION YEAST AND CELL CYCLE CONTROL

The fission yeast was first developed as an experimental model for studying the cell cycle by Murdoch Mitchison in the 1950s (Mitchison 1971). It is a cylindrically shaped cell, 12–15 μm length and 3–4 μm diameter, typically eukaryotic and yet with a genome of less than 5000 genes (Wood, Gwilliam *et al.* 2002). Murdoch used fission yeast to study how cells grow during the cell cycle, devising procedures for physiological analysis and to synchronise cells so they proceeded together through the cycle. Another approach to studying the cell cycle in yeasts was taken by Lee Hartwell in the early 1970s, using the budding yeast *Saccharomyces cerevisiae* (Hartwell 1974). He isolated temperature sensitive cell division cycle (cdc) mutants which were unable to complete the cell cycle when incubated at their restrictive temperature. A similar approach was also possible in principle with the fission yeast, because Urs Leupold working in Bern, Switzerland had established the techniques needed for classical genetic analyses of this organism. Thus it was straightforward for me to follow Lee's approach by isolating cdc mutants in fission yeast when I joined Murdoch's Edinburgh laboratory in 1973, having had a brief period of post-doctoral training to learn genetics in Bern with Urs.

The first mutants collected were mainly defective in the events of mitosis and cell division and subsequent screens carried out together with Kim Nasmyth identified more mutants defective in S-phase (Nurse, Thuriaux *et al.* 1976). These cdc mutants identified genes required for the events of S-phase, mitosis and cell division, but it was not possible to determine which, if any, of these genes were involved in controlling these events. However, the chance observation that mutants could be isolated which divided at a reduced cell size provided an approach to identify such cell cycle controlling genes. The reason such wee mutants (wee is the Scottish word for small) were useful is because progression through the fission yeast cell cycle is co-ordinated with cell growth so that in constant growth conditions division occurs at a fixed cell size. Mutants altered in gene functions which are rate limiting for cell cycle progression result in more rapid progression through the cell cycle, and as a consequence cells undergo division before the normal amount of growth has taken place and divide at a small size. All the initial wee mutants isolated were found to map to a single gene *wee1* (Nurse 1975). One of these was temperature sensitive, being of almost normal cell size at a low temperature and wee at a high temperature. Shift experiments from low to high temperature demonstrated that the *wee1* gene acted in G2 and controlled the cell cycle timing of mitosis (Fig 1). Experiments with Peter Fantes analysing this and other mutants and wild type cells in different growth conditions (Fantes and

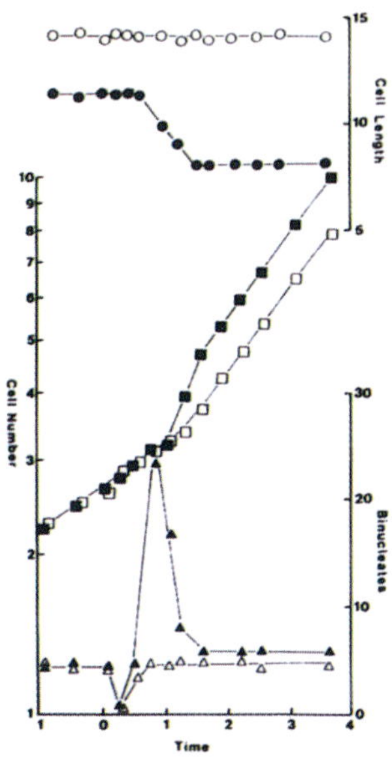

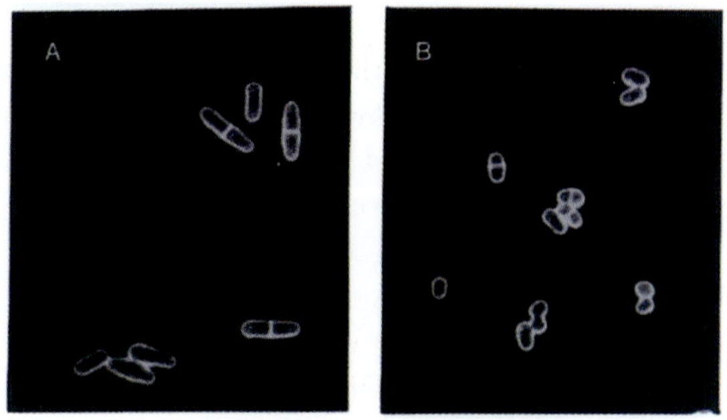

Peter Fantes

Figure 1. Wee Mutants in Fission Yeast

The photomicrograph shows fission yeast cells dividing at wild type size (A) and at a small size as in a wee mutant (B). The graph shows a temperature sensitive wee1 mutant and wild type cells shifted to high temperature at time 0. The filled triangles follow the percentage of wee1 mutant cells undergoing division and the filled circles their cell size at division. Peter Fantes was an Edinburgh colleague who helped work out the relationship between cell size and cell cycle progression in fission yeast.

Nurse 1977), (Fantes and Nurse 1978), revealed that the onset of both S-phase and mitosis were co-ordinated with attainment of a critical cell size.

Pierre Thuriaux using classical genetic procedures showed that the *wee1* gene product acted as an inhibitor of mitotic onset (Nurse and Thuriaux 1980). The genetic procedures included suppression of the wee phenotype by nonsense suppressors and the analysis of dominant and recessive mutants, and led to the conclusion that the wee mutant phenotype was associated with loss of the *wee1* gene function. As well as the large numbers of *wee1* mutants there was one dominant mutant which mapped to a second gene called *wee2* which was shown by fine structure mapping to be identical with the gene *cdc2*. The *cdc2* gene function had previously been shown to be required for mitosis (Nurse, Thuriaux *et al.* 1976), and so these new experiments established that *cdc2* could be mutated in one of two ways: (1) to a loss of function blocking mitosis, and (2) to a gain of function advancing mitosis at a small cell size. We concluded that the *cdc2* gene product functioned as an activator of mitotic onset and proposed that *wee1* and *cdc2* acted together in a regulatory network controlling the onset of mitosis.

Next it was shown that *cdc2* had a role controlling the onset of S-phase. This came about as a consequence of a survey of cdc mutants to look for those which blocked in G1 phase prior to commitment to the cell cycle. The approach followed was that of Lee Hartwell, who had reasoned that budding yeast mutants blocked early in the cell cycle prior to commitment would still be able to conjugate if challenged to do so. Cdc2^{ts} mutants were used as negative controls for the fission yeast experiments because they blocked in G2 and therefore it was assumed that they would be committed to the cell cycle. A low but significant percentage of these cdc2^{ts} mutant cells did conjugate, a result initially thought to be due to some mutant cells leaking past the block point. However, a significant percentage of conjugation continued to be observed with the cdc2^{ts} mutant and so the alternative but unlikely explanation that some cells were blocking in G1 prior to S-phase was tested. Surprisingly these tests established that *cdc2* was unusual in being required twice during the cell cycle, first in G1 for onset of S-phase and then again in G2 for onset of mitosis (Nurse and Bissett 1981).

These experiments showed that *cdc2* had a central role controlling the fission yeast cell cycle. In G1 it was required to commit the cell to onset of S-phase, and in G2 it acted as a major rate limiting step determining the onset of mitosis. This was unexpected because the biochemical processes of S-phase and mitosis are very different and yet appeared to be controlled by the same gene function. From this time *cdc2* became my major topic of study.

MOLECULAR CHARACTERISATION OF CELL CYCLE CONTROL

The above genetic experiments were abstract in their approach and had revealed nothing about the molecular role of *cdc2* in cell cycle control. This could only be established by cloning the *cdc2* gene but before these experiments could be carried out a DNA transformation procedure needed to be developed for fission yeast. These procedures would enable gene libraries to be introduced into cdc2^{ts} cells allowing the *cdc2* gene to be cloned by rescue or complementation of the temperature sensitive mutant phenotype. Developing a transformation procedure and other methods for molecular genetics became my first priorities on setting up my own laboratory at the University of Sussex near Brighton, where I worked in collaboration with David Beach. The DNA transformation procedure developed was based on methods already available for budding yeast using both ARS (origins of replication) elements and selectable markers from that organism (Beach and Nurse 1981). Gene replacement procedures were developed, although the efficiency of homologous recombination in fission yeast is less than in budding yeast. The *cdc2* gene was cloned by complementation and the cloned gene was found to fully rescue the cdc2^{ts} mutant defects at both the G1/S and G2/M boundaries (Beach, Durkacz *et al.* 1982).

To check if a gene related to *cdc2* was also present in budding yeast we transformed a cdc2^{ts} mutant with a budding yeast library and found a segment of DNA that could rescue the cdc2^{ts} mutant. Steve Reed had cloned four cdc genes from budding yeast and provided these to us prior to their publication so we were able to check if the cdc2^{ts} complementing segment of

DNA was one of these genes. Hybridisation was found with the budding yeast *CDC28* gene, indicating that the fission yeast *cdc2* gene and the budding yeast *CDC28* gene were functionally equivalent (Beach, Durkacz *et al.* 1982). This was another unexpected result because *CDC28* was thought only to act at the G1/S boundary in budding yeast, although a cdc28^{ts} mutant had been described which became blocked at mitosis (Piggott, Rai *et al.* 1982). We proposed that *cdc2/CDC28* acted at both the G1/S and G2/M transitions in both yeasts, but in budding yeast it had been difficult to detect the G2/M block point because it occurred very early in the cell cycle just after the G1/S block point due to the budding mode of cell division (Nurse 1985). The similarity of the controls between the rather distantly related yeasts also encouraged us to speculate that there might be similar controls in mammalian cells (Beach, Durkacz *et al.* 1982).

The movement of my laboratory from Sussex to the Imperial Cancer Research Fund's Lincoln's Inn Fields laboratories in central London in 1984 provided the environment and resources for a proper molecular characterisation of the *cdc2* gene function. The gene was shown to encode a protein kinase by two post-doctoral workers Viesturs Simanis and Sergio Moreno. Viesturs developed antibodies against the Cdc2p protein and showed that immunoprecipitates had protein kinase activity and that this activity was temperature sensitive *in vitro* in extracts made from cdc2^{ts} mutants (Simanis and Nurse 1986). This result confirmed earlier work from Steve Reed's laboratory showing that the budding yeast *CDC28* gene encoded a protein kinase (Reed, Hadwiger *et al.* 1985). Sergio optimised the protein kinase assay and demonstrated that activity varied considerably during the cell cycle (Fig. 2), peaking just at the onset of mitosis (Moreno, Hayles *et al.* 1989).

The molecular basis of the periodic regulation of the Cdc2p protein kinase was worked on by two further post-doctoral workers, Paul Russell and Kathy Gould. Paul cloned both the *wee1* and *cdc25* genes by complementation, and showed that they acted upstream of *cdc2*. Wee1p had sequence similarity with protein kinases, suggesting that it might phosphorylate Cdc2p directly to inhibit Cdc2p protein kinase activity (Russell and Nurse 1987). The *cdc25* gene was shown to act in a positive manner antagonistically to the Wee1p inhibitor (Fig. 3), but the failure to find any sequence similarities with previously identified genes meant that it could only be speculated that Cdc25p was necessary for a protein phosphatase activity that countered the Wee1p protein kinase (Russell and Nurse 1986).

After my laboratory moved to the Biochemistry Department at Oxford University Kathy Gould carried out a biochemical analysis of Cdc2p phosphorylation. A major phosphorylation site identified was tyrosine 15, the first time tyrosine phosphorylation had been detected in a microbial eukaryote (Gould and Nurse 1989). Phosphorylation of this residue was associated with the G2 phase of the cell cycle when Cdc2p protein kinase activity was at a low level. Tyrosine 15 is located near the ATP binding site of the protein kinase, suggesting that phosphorylation of this residue might influence catalytic activity. The physiological relevance of this phosphorylation was confirmed by

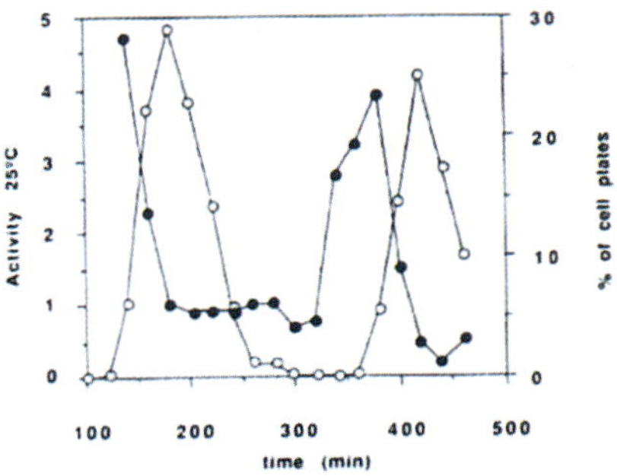

Sergio Moreno

Figure 2. The Cdc2p CDK Activity Through the Cell Cycle

The graph shows CDK activity in a synchronous culture of wild-type cells. The open circles are the percentage of dividing cells and the closed circles the Cdc2p protein kinase activity peaking at mitosis. Sergio Moreno worked in laboratory for over six years contributing much to Cdc2p and its regulation.

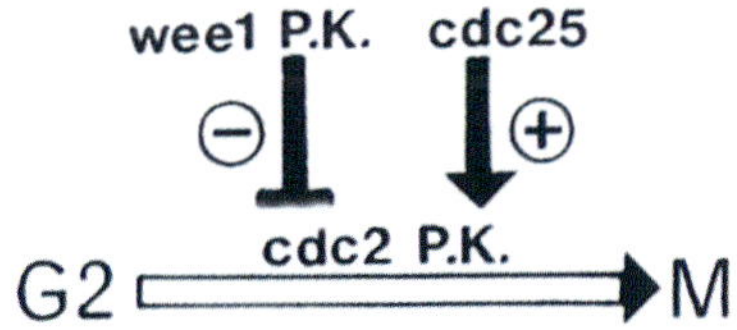

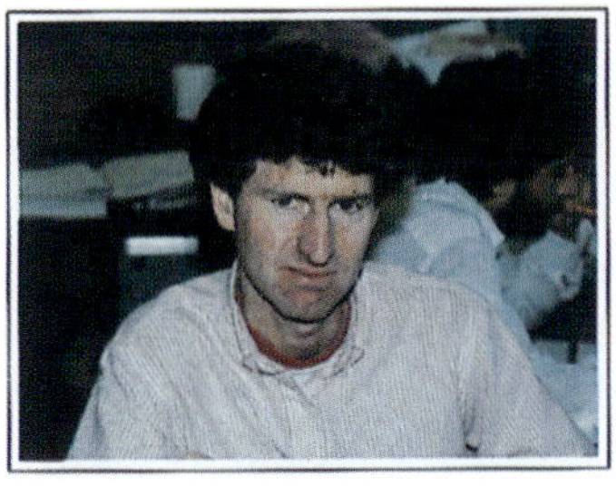

Paul Russell

Figure 3. The G2/M Regulatory Network

Wee1p acts as a negative regulator and Cdc25p as a positive regulator of the Cdc2p protein kinase at the G2/M transition. The *wee1* and *cdc25* genes were cloned and their regulatory relationships were determined by Paul Russell when he was a post-doctoral worker with me in Brighton and London.

constructing an unphosphorylatable phenylalanine 15 mutant which advanced cells prematurely into mitosis. Cdc25p was also shown to be required to remove the phosphate from the tyrosine 15 residue in Cdc2p. This work suggested that the Cdc2p protein kinase activity was regulated by tyrosine 15 phosphorylation, and that the level of phosphorylation was regulated by the balance of activities between the Wee1p protein kinase mitotic inhibitor and Cdc25p mitotic activator.

One further gene important for *cdc2* regulation is *cdc13*. This was cloned by Booher and Beach (1988) and by my laboratory (Hagan, Hayles *et al.* 1988), and the putative gene product was shown to have significant similarity with cyclins. Tim Hunt and Jon Pines had characterised sea urchin cyclin (Pines and Hunt 1987), and Tim had proposed cyclin as a cell cycle regulator during early embryonic cleavage. Cdc13p cyclin varied in level during the fission yeast cell cycle, and was required for Cdc2p protein kinase activation (Moreno, Hayles *et al.* 1989) establishing that the Cdc13p cyclin is necessary for Cdc2p to bring about the G2/M transition.

UNIVERSAL ROLE FOR CDC2P IN CELL CYCLE CONTROL

In parallel with these studies on the molecular characterisation of Cdc2p, my laboratory was also attempting to establish if there was a Cdc2p in metazoan cells. Two major approaches were used, the first being the cloning of the human *cdc2* gene achieved by Melanie Lee (Lee and Nurse 1987). Initially Melanie had tried to clone a human homologue of *cdc2* on the basis of structural similarity. These approaches identified protein kinases, but as there are at least 500 protein kinases in the human genome it was difficult to know whether the cloned genes were *cdc2* candidates or not. Because of this difficulty Melanie tried a different approach of cloning the gene by complementation of a $cdc2^{ts}$ fission yeast mutant using a human cDNA library from Hirota Okayama. Complementing clones were isolated (Fig 4), and a tense period of a month or so followed whilst alternative less interesting explanations of this result were eliminated. The discovery of a human homologue of *cdc2* had important implications and so we were all worried that our hopes were being raised only to be dashed at the final hurdle! Melanie carefully completed the necessary controls, sequenced the human cDNA, and one exciting morning we were all huddled round the computer when the sequence comparisons between the human and yeast proteins indicated that there was a 63% identity between them. The human *CDC2* gene could fully substitute for the defective fission yeast *cdc2* gene, despite the evolutionary divergences of these organisms of between 1000–1500 M years. This result strongly supported the idea that cell cycle control was conserved in yeast and humans, and therefore probably in all other eukaryotes. We speculated that human *CDC2* might act at two points in the cell cycle, at the G1 restriction point known to operate in mammalian cells, and at the G2/M transition where it acted as maturation promoting factor (MPF) known to control M-phase in metazoan eggs and oocytes.

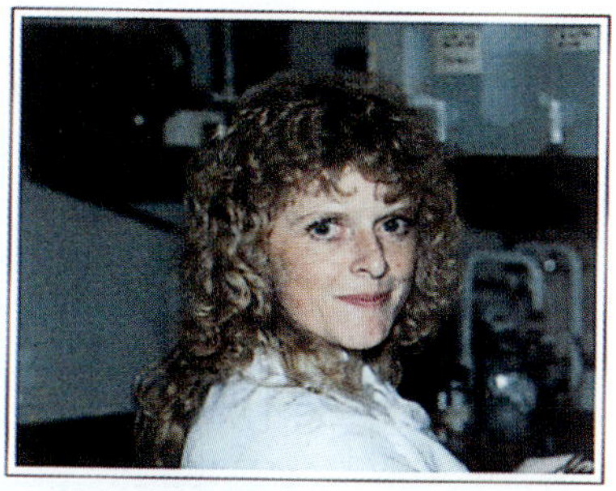

Melanie Lee

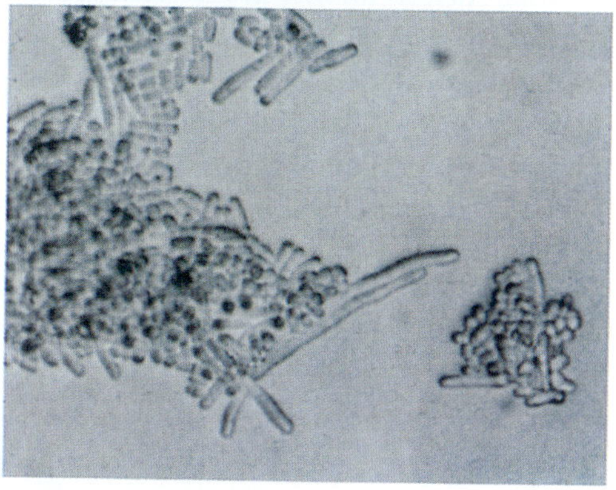

Figure 4. The Cloning of Human CDC2

The photomicrograph shows cdc2^{ts} mutant cells being complemented by the human *CDC2* gene. The human gene is on a plasmid and when this is lost from cells they are unable to divide and become highly elongated. This bold experiment was carried out by Melanie Lee at the ICRF (now Cancer Research UK) Lincoln's Inn laboratories.

The second approach directly involved MPF itself. Yoshio Masui working in Toronto had identified MPF as a factor which induced frog egg maturation which involved meiotic M-phase (Masui and Markert 1971). Yoshio also developed cell free assays to monitor MPF (Lohka and Masui 1983) which were further developed by Jim Maller and Fred Lokha in Denver who purified MPF from the Xenopus frog (Lohka, Hayes *et al.* 1988). The purified preparation contained two proteins one of which was 32kD, a molecular mass rather similar to Cdc2p. Western blot and immunoprecipitation experiments using antibodies against a conserved Cdc2p peptide demonstrated that the 32kD component of MPF was indeed a homologue of Cdc2p (Gautier, Norbury *et al.* 1988). Subsequent collaborative work with Marcel Dorée's group in Montpellier established that a periodic mitotic kinase in starfish embryos, also contained a Cdc2p homologue (Labbe, Lee *et al.* 1988) as did starfish MPF (Labbe, Capony *et al.* 1989). Marcel's work also greatly helped our own biochemical investigations of Cdc2p in fission yeast.

The link with MPF was important because it established that the biochemical mechanisms underlying mitotic onset were the same in yeasts, starfish and frogs. MPF had been shown to 'biochemically' advance starfish and frog eggs into meiotic M-phase whilst Cdc2p had been shown to 'genetically' advance yeast cells into mitotic M-phase. Evidence from this and other workers was sufficiently strong to propose that there was a universal control mechanism regulating the onset of M-phase in all eukaryotes (Nurse 1990). This operates through a G2/M CDK with a catalytic CDK sub-unit complexed with a cyclin sub-unit, regulated by a Wee1p protein kinase phosphorylating a tyrosine residue (and sometimes the adjacent threonine) near the catalytic site, and a Cdc25p protein phosphatase which removed the phosphates to activate the CDK. CDKs were also found to regulate the G1 to S-phase transition in multicellular eukaryotes, but a different CDK to the one acting at G2/M is used in these organisms. In contrast fission yeast CDK can bring about both the G1/S and the G2/M transitions.

The universality of cell cycle controls in eukaryotes should have been anticipated given the high conservation already noted for biochemical pathways and for many processes of molecular biology. Possibly the rather different appearance of cells and cell division in microbial eukaryotes, plants and Metazoa made universality seem less likely than these other processes. However, Schwann, one of the early proponents of the cell theory had already recognised this possibility in 1839 when he stated "We have seen that... cells are formed and grow in accordance with essentially the same laws; hence, that these processes must everywhere result from the operation of the same forces" (Wilson 1925).

FURTHER ROLES FOR CDKs

In more recent years two further roles for CDKs have emerged. The events of the cell cycle usually occur in a fixed sequence, and if an early event such as S-phase is incomplete then a later event such as mitosis becomes blocked. In principle there are two general types of mechanism that can account for these dependencies. There could be a hard wiring of the two events such that the later event is unable to occur physically or chemically without the earlier event. This is analogous to a metabolic pathway, with the product from the first enzyme acting as the substrate for the second. Alternatively the two events could be linked by a regulatory loop that inhibits the second until the first is complete. Lee Hartwell developed the second of these two alternatives into the idea of checkpoints whereby the cell monitors or "checks" cell cycle progression at certain "points" in the cell cycle, and if events prior to that point are incomplete then further progression is delayed (Hartwell and Weinert 1989). The checkpoint idea is also extremely useful for thinking about cell cycle delays in response to DNA damage, and has helped understanding of how genome stability is maintained during cell reproduction.

Tamar Enoch investigated the dependency of mitosis upon completion of S-phase in fission yeast. She showed that this dependency was lost in cells with specific *cdc2* mutations, or in mutants with high levels of the CDK activator Cdc25p (Enoch and Nurse 1990). These mutants could undergo mitosis when DNA synthesis was inhibited with hydroxyurea, and so we concluded that the checkpoint control monitoring the completion of S-phase led to inhibition of the G2/M CDK, preventing mitosis until S-phase was complete. This established that CDKs were a part of the checkpoint control ensuring that mitosis only takes place when the genome is fully replicated.

Another role for CDKs is to ensure that there is only one S-phase in each cell cycle. When a cell completes S-phase and enters G2, another S-phase does not take place until the mitosis of that cell cycle is complete. To investigate this control, fission yeast mutants were sought which underwent more than one round of S-phase each cell cycle generating cells of higher ploidy. These were found to be altered in the G2/M CDK Cdc2p/Cdc13p (Broek, Bartlett *et al.* 1991), suggesting that the state of this CDK is important for restraining S-phase during G2. Consistent with this, over-expression of the CDK inhibitor Rumlp was found to inhibit the G2/M CDK, and also to bring about repeated rounds of S-phase (Moreno and Nurse 1994). Finally Jacky Hayles showed that deleting the Cdc13p G2/M cyclin from cells resulted in repeated rounds of S-phase establishing that the presence of the G2/M CDK in G2 cells inhibited S-phase (Fig 5). Only after mitosis when this CDK was destroyed

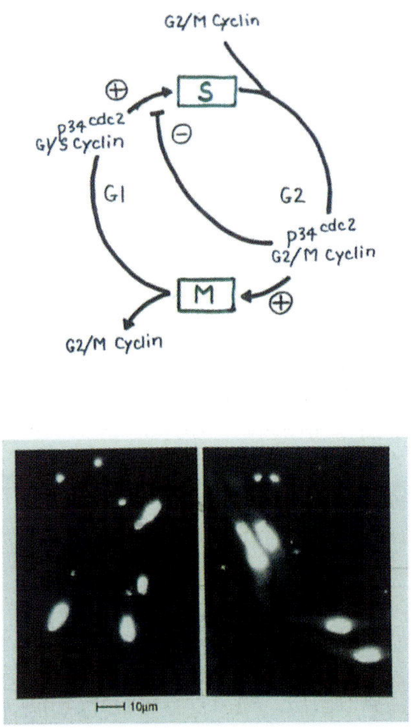

Jacky Hayles

Figure 5. Repeated S-phase in Cells Lacking G2/M CDK Activity

The photomicrograph shows cells lacking the G2/M cyclin Cdc13p. Their nuclei are stained with DAP1 and are the large nuclei with high DNA content. The nuclei are smaller in wild type cells. The presence of the G2/M CDK prevents a further round of S-phase during G2. Jacky Hayles has worked in my laboratory for 20 years, and has contributed much to many of the different projects important for understanding how CDKs regulate the cell cycle in fission yeast.

could another S-phase take place (Hayles, Fisher *et al.* 1994), implicating CDKs in the control mechanism maintaining one S-phase per cell cycle.

These two roles further emphasise the importance of CDKs in regulating the orderly progression through S-phase and mitosis during the cell cycle. The onset of S-phase is thought to require two sequential steps: the first of these only take places if no CDK activity is present whilst the second requires the presence of CDK activity (Wuarin and Nurse 1996). In early G1 there is no CDK activity allowing progression through step one. Later in the cell cycle at the G1/S boundary CDK activity appears which allows progression through step two and brings about the initiation of S-phase. During G2 the continued presence of CDK activity prevents step one from occurring again and this blocks onset of a further S-phase. At the G2/M boundary there is a further increase in CDK activity which brings about mitosis. Exit from mitosis and the ending of the cell cycle requires destruction of CDK activity, and because the subsequent G1 cells lack CDK activity they are able to carry out step one for S-phase and the whole series of events can be repeated (Stern and Nurse 1996).

What lies in the future for CDKs and cell cycle control (Nurse 2000)? It remains an embarrassment that so few CDK substrates have yet been identified. Until this situation improves understanding of the molecular mechanisms underlying the onset of both S-phase and mitosis will remain incomplete. Solution of this problem will need the development of new procedures to identify *in vivo* substrates for protein kinases. CDK regulation has been relatively well characterised but needs to be further refined given the importance of tightly regulated kinase activity at different stages of the cell cycle. Theoretical modelling will be required to understand how the temporal changes of CDK activity and spatial location are regulated through the cell cycle. Regulation of CDKs during development and the role this may play in generating tissue and organ form is another interesting problem. The meiotic cell cycle is modified from the mitotic cell cycle so S-phase is suppressed between M-phase I and M-phase II, an altered regulation likely to be due to differences in CDK behaviour between the two types of cell cycles. Such differences may also be relevant for the switch to reductional chromosomal segregation during meiosis.

Working out how CDKs act as major regulators of the cell cycle has been an exciting endeavour and I feel fortunate to have been at the right place and time to have contributed to this enterprise. As is clear from my account this has been a collaborative venture involving many friends and colleagues, some working in my laboratory and others working in other cell cycle laboratories around the world. Without their efforts the work described here would not have been possible. Finally it is a real pleasure to acknowledge my two co-awardees Lee Hartwell and Tim Hunt and my long-term colleague Jacky Hayles, and to thank them for their collegiality and inspiration for more than two decades.

REFERENCES

Beach, D., B. Durkacz, *et al.* **(1982).** "Functionally homologous cell cycle control genes in budding and fission yeast." Nature 300(5894): 706–9.

Beach, D. and P. Nurse **(1981).** "High frequency transformation of the fission yeast *Schizosaccharomyces pombe.*" 290 (*Nature*): 140 142.

Booher, R. and D. Beach **(1988).** "Involvement of cdc13+ in mitotic control in *Schizosaccharomyces pombe:* possible interaction of the gene product with microtubules." *EMBO J* 7(8): 2321–2327.

Broek, D., R. Bartlett, *et al.* **(1991).** "Involvement of $p34^{cdc2}$ in establishing the dependency of S phase on mitosis." *Nature* 349(6308): 388–393.

Enoch, T. and P. Nurse **(1990).** "Mutation of fission yeast cell cycle control genes abolishes dependence of mitosis on DNA replication." *Cell* 60(4): 665–673.

Fantes, P. and P. Nurse **(1977).** "Control of cell size at division in fission yeast by a growth-modulated size control over nuclear division." *Exp Cell Res* 107(2): 377–86.

Fantes, P. A. and P. Nurse **(1978).** "Control of the timing of cell division in fission yeast. Cell size mutants reveal a second control pathway." *Exp Cell Res* 115(2): 317–29.

Gautier, J., C. Norbury, *et al.* **(1988).** "Purified maturation-promoting factor contains the product of a Xenopus homolog of the fission yeast cell cycle control gene cdc2+." *Cell* 54(3): 433–9.

Gould, K. L. and P. Nurse **(1989).** "Tyrosine phosphorylation of the fission yeast cdc2+ protein kinase regulates entry into mitosis." *Nature* 342(6245): 39–45.

Hagan, I., J. Hayles, *et al.* **(1988).** "Cloning and sequencing of the cyclin-related cdc13+ gene and a cytological study of its role in fission yeast mitosis." *J. Cell Sci.* 91: 587–595.

Hartwell, L. (1974). "*Saccharomyces cerevisiae* cell cycle." *Bacteriol Rev* **38**: 164–198.

Hartwell, L. and T. Weinert **(1989).** "Checkpoints: controls that ensure the order of cell cycle events." *Science* 246: 629–634.

Hayles, J., D. Fisher, *et al.* **(1994).** "Temporal order of S-phase and mitosis in fission yeast is determined by the state of the $p34^{cdc2}$/mitotic B cyclin complex." *Cell* 78: 813–822.

Labbe, J., J. Capony, *et al.* **(1989).** "MPF from starfish oocytes at first meiotic metaphase is an heterodimer containing one molecule of cdc2 and one molecule of cyclin B." *EMBO J* **8**: 3053–3058.

Labbe, J. C., M. G. Lee, *et al.* **(1988).** "Activation at M-phase of a protein kinase encoded by a starfish homologue of the cell cycle control gene cdc2+." *Nature* **335:** 251–254.

Lee, M. G. and P. Nurse **(1987).** "Complementation used to clone a human homologue of the fission yeast cell cycle control gene cdc2." *Nature* 327: 31–35.

Lohka, M. and Y. Masui **(1983).** "Formation *in vitro* of sperm pronuclei and mitotic chromosomes induced by amphibian ooplasmic contents." *Science* 220: 719–721.

Lohka, M. J., M. K. Hayes, *et al.* **(1988).** "Purification of maturation-promoting factor, an intracellular regulator of early mitotic events." *Proc. Natl. Acad. Sci. USA* 85: 3009–3013.

Masui, Y. and C. Markert **(1971).** "Cytoplasmic control of nuclear behaviour during meiotic maturation of frog oocytes." *J. Exp. Zool.* 177: 129–146.

Mitchison **(1971).** *The biology of the cell cycle.* London & New York, Cambridge University Press.

Moreno, S., J. Hayles, *et al.* **(1989).** "Regulation of $p34^{cdc2}$ protein kinase during mitosis." *Cell* 58: 361–372.

Moreno, S. and P. Nurse **(1994).** "Regulation of progression through the G1 phase of the cell cycle by the *rum1*$^+$ gene." *Nature* 367: 236–242.

Nurse, P. **(1975).** "Genetic control of cell size at cell division in yeast." *Nature* 256: 547–551.

Nurse, P. **(1985).** "Cell cycle control genes in yeast." *Trends in Genetics* 1: 51–55.

Nurse, P. **(1990).** "Universal control mechanism regulating onset of M-phase." *Nature* 344(6266): 503–8.

Nurse, P. **(2000).** "A long twentieth century of the cell cycle and beyond." *Cell* 100(1): 71–8.

Nurse, P. and Y. Bissett **(1981).** "Gene required in Gl for commitment to cell cycle and in G2 for control of mitosis in fission yeast." *Nature* 292: 558–560.

Nurse, P. and P. Thuriaux **(1980)**. "Regulatory genes controlling mitosis in the fission yeast *Schizosaccharomyces pombe.*" *Genetics* 96: 627–637.

Nurse, P., P. Thuriaux, et al. **(1976)**. "Genetic control of the cell division cycle in the fission yeast *Schizosaccharomyces pombe.*" *Molec. Gen. Genet.* 146: 167–178.

Piggott, J. A., R. Rai, *et al.* **(1982)**. "A bifunctional gene product involved in two phases of the cell cycle." *Nature* 298: 391–394.

Pines, J. and T. Hunt **(1987)**. "Molecular cloning and characterization of the mRNA for cyclin from sea urchin eggs." *EMBO J.* 6: 2987–2995.

Reed, S. I., J. A. Hadwiger, *et al.* **(1985)**. "Protein kinase activity associated with the product of the yeast cell division cycle gene CDC28." *Proc. Natl. Acad. Sci. USA* 82: 4055–4059.

Russell, P. and P. Nurse **(1986)**. "cdc25+ functions as an inducer in the mitotic control of fission yeast." *Cell* 45: 145–153.

Russell, P. and P. Nurse **(1986)**. "*Schizosaccharomyces pombe* and *Saccharomyces cerevisiae:* a look at yeasts divided." *Cell* 45(6): 781–2.

Russell, P. and P. Nurse **(1987)**. "Negative regulation of mitosis by wee1+ a gene encoding a protein kinase homolog." *Cell* 49: 559–567.

Simanis, V. and P. Nurse **(1986)**. "The cell cycle control gene $cdc2^+$ of fission yeast encodes a protein kinase potentially regulated by phosphorylation." *Cell* 45(2): 261–268.

Stern, B. and P. Nurse **(1996)**. "A quantitative model for the cdc2 control of S phase and mitosis in fission yeast." *Trends Genet* 12(9): 345–50.

Wilson, E. **(1925)**. *The cell in development and heredity.* New York, Macmillan.

Wood, V., R. Gwilliam, *et al.* **(2002)**. "The genome sequence of *Schizosaccharomyces pombe.*" *Nature* 415: 871–880.

Wuarin, J. and P. Nurse **(1996)**. "Regulating S phase: CDKs, Licensing and proteolysis." *Cell* 85: 785–787.

Physiology or Medicine 2002

SYDNEY BRENNER, H. ROBERT HORVITZ and JOHN E. SULSTON

"for their discoveries concerning genetic regulation of organ development and programmed cell death"

THE NOBEL PRIZE IN PHYSIOLOGY OR MEDICINE

Speech by Professor Urban Lendahl of the Nobel Committee at Karolinska Institutet.
Translation of the Swedish text.

Your Majesties, Your Royal Highnesses, Ladies and Gentlemen,

We have all begun our lives in a seemingly modest way – as the fertilised egg cell, a tenth of a millimetre in size. From this small cell, the adult human being develops, with its hundred thousand billion cells, through cell division, cell differentiation and by formation of the various organs. To only make new cells is however not sufficient, certain cells must also die at specific time points as a natural part of the growth process. Think for example about how we for a short period during foetal life have web between our fingers and toes, and how this is removed by cell death.

The importance of cell differentiation and organ development was understood by many, but progress was slow. This was largely an effect of our complexity, with the large number of cells and many cell types – the forest could not be seen because of all the trees. Could the task to find the genetic principles be made simpler? Were there a species simpler than humans, but still sufficiently complex to allow for general principles to be deduced?

Sydney Brenner in Cambridge, UK, took on the challenge, and his choice was the nematode *Caenorhabditis elegans*. This may at first seem odd, a spool-shaped approximately one millimetre long worm with 959 cells that eats bacteria, but Brenner realised in the early 1960s that it was, what we today would call, "loaded with features". It was genetically amenable and it was transparent, so that every cell division and differentiation could be directly followed in the worm under the microscope. Brenner demonstrated in 1974 that mutations could be introduced into many genes and visualised as distinct changes in organ formation. Through his visionary work, Brenner created an important research tool. The nematode had made into the inner circle of research.

John Sulston came to Brenner´s laboratory in 1969. He took advantage of that cell divisions could be followed under the microscope and assembled the cell lineage in the worm, showing which cells are siblings, first and second cousins. He found that cell divisions occurred with a very high degree of precision, the cell lineage was identical between different individuals. He also realised that certain cells in the lineage always died at a certain time point. This meant that programmed cell death was not a stochastic process, but rather occurred with a very high degree of precision. During the course of this work Sulston identified the first gene important for the cell death process: *nuc-1*.

Robert Horvitz came to work with Brenner and Sulston in 1974. Horvitz started a systematic search for genes controlling programmed cell death. He identified the key genes for the cell death process proper. The discovery of these central death genes, *ced-3*, *ced-4* and *ced-9*, changed the view on programmed cell death from something rather obscure to a process with a strict genetic programme. Horvitz also showed that there are human homologues to the death genes in the worm and that those have corresponding functions – the cell death machinery had deep evolutionary roots.

This year's Nobel Prize celebrates the Joy of Worms. Brenner´s almost prophetic visions from the early 1960s of the advantages of this model organism have been fulfilled. It has given us new insights into the development of organs and tissues and why specific cells are destined to die. This knowledge has proven valuable, for instance, in understanding how certain viruses and bacteria attack our cells, and how cells die in heart attack and stroke.

Sydney Brenner, Robert Horvitz and John Sulston,

Your discoveries concerning the genetic regulation of organ development and programmed cell death have truly opened new avenues for biological and medical research. On behalf of the Nobel Assembly at Karolinska Institutet I wish to convey to you our warmest congratulations and I ask you to step forward to receive the Nobel Prize from the hands of His Majesty the King.

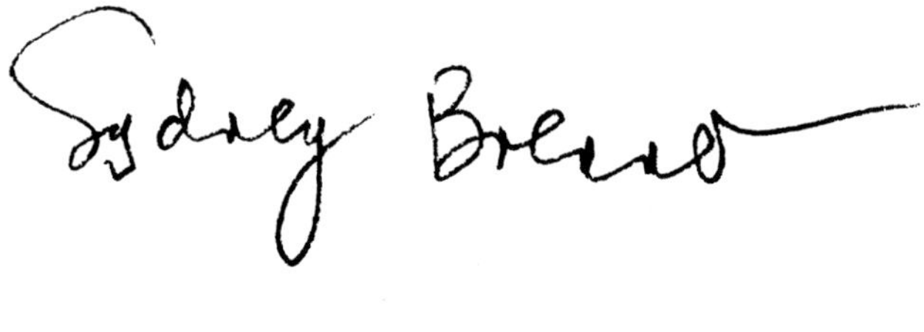

SYDNEY BRENNER

An autobiography published a few years ago records many of the salient events in my life which led me to become a scientist and my Nobel Lecture covers the intellectual background and consequences of the research work for which the Prize has been awarded.

In this somewhat compressed version, I start with my birth on the 13th January 1927 in a small town, Germiston, in South Africa. My parents were Jewish immigrants from Eastern Europe; my father came to South Africa from Lithuania in 1910, my mother, from Latvia, in 1922. My father was a shoe repairer and our first home was in some rooms at the back of his shop. He never learnt to read or write but, in addition to English, Yiddish and Russian, he learnt to speak Afrikaans and Zulu. I learnt to read at an early age, and a customer of my father, Miss Walkinshaw, persuaded my father to allow me to go at the age of five without charge to her kindergarten. I completed the first three years of primary school in one year and was admitted to the local school the age of six directly into the fourth year, some two years younger than all my contemporaries. After 4 years in primary school, I went to Germiston High School where I matriculated in December 1941, just before turning 15.

During this time I discovered the Public Library in Germiston, one of the many libraries set up all over the world with funding from Andrew Carnegie's endowment. It was here that I found a source of knowledge and the means to acquire it by reading, a habit of learning which I still follow to this day. I also became interested in chemistry and gradually accumulated enough test tubes and other glassware to do chemical experiments, using small quantities of chemicals purchased from a pharmacy supply house. I soon graduated to biochemistry and tried to discover what gave flowers their distinctive colours. I made the (to me) astounding discovery that the pigments I extracted changed their colours when I changed the pH of the solution.

I was fortunate in that the Town Council of Germiston gave me a bursary of 60 pounds per year that allowed me to go to the University of the Witwatersrand in Johannesburg to study medicine and, in 1942, at the age of fifteen, I began the course studying Physics, Chemistry, Botany and Zoology. I lived at home and I cycled every morning to the railway station to travel by train to Johannesburg followed by a walk to the University, carrying sandwiches for my lunch and returning in the evening the same way. My Uncle Harry had given me a microscope as a present which allowed me to continue my personal explorations of the living world. This was the beginning of my contacts with the real science. In my second year, after moving to the Medical

School, I began the courses of Anatomy and Physiology. I had begun to see that I was interested in cells and their functions.

It was noted then that I would be too young to qualify for the practice of medicine at the conclusion of my six year medical course and I was allowed to deviate and spend one year in a Medical B.Sc. course in Anatomy and Physiology. This was heaven. The small group of about a dozen students inhabited a small room in the Anatomy Department where we each had a small laboratory bench. We learnt how to do research by working in small groups with more advanced researchers in the Department. I learnt physical chemistry with Joel Mandelstam (later Professor at Oxford University), microscopy with Alfred Oettle, and neurology with Harold Daitz who became a close friend and who died at a very early age in Oxford. Raymond Dart and Robert Broom taught me anthropology and paleontology, and the man who inspired all this activity was Joseph Gillman, a histologist who had created a centre of research in that isolated place. He invited me (I needed no persuasion) to continue with research and I stayed on for two more years doing an Honours degree and then an M.Sc., supporting myself by working part-time as a laboratory technician. I read many books and taught myself many subjects during this period, learnt how to build equipment and how to do experiments, and had many arguments and discussions with Joe Gillman. I also began to publish papers. My scientific bibliography begins in 1945 with a paper published with Joe Gillman and his brother, Teddy, but my first paper as sole author appeared in 1946. This paper dealt with a histochemical reaction, and it was the first of several which reflected my growing interest in a subject which I later called cell physiology. My M.Sc. thesis was in the field of cytogenetics, another self-taught subject, and this was the beginning of my research in genetics. This background was to serve me well in later years when I became a molecular biologist.

In 1946, W. Le Gros Clark visited South Africa and invited me to come to his Department of Anatomy in Oxford, but I was advised by everybody to finish my medical course, because it was believed then that a medical qualification would be essential for a research position later. I did go back to my medical studies but I continued working in the Department of Anatomy and moved to the Department of Physiology when Joe Gillman became Professor there. I was not a good medical student and had an erratic career, brilliant in some subjects, absolutely dismal in others. In my final year I failed Medicine, scraped through Surgery but got a First Class in the third subject, Obstetrics and Gynecology. I had to go back and repeat Medicine and Surgery and six months later, in July 1951, I finally received the degrees of MB BCh. I had already decided that I would do research and that I needed to go abroad. CH Waddington, who had earlier visited South Africa had advised me to go to Cambridge. I applied to the Department of Biochemistry and never even received a reply. I had decided that the subject I was interested in was molecular biology which, of course, did not exist at the time, and when I was awarded a scholarship by the Commission for the Royal Exhibition of 1851, H. Raikes, head of the University of Witwatersrand, who was originally an Oxford trained chemist, advised me to write to C.N. Hinshelwood, the Professor of Physical Chemistry at Oxford University, who had interests in the applications of physical chemistry to biology. That sounded closer to what I wanted to do. Hinshelwood had written a book called the "Chemical Kinetics of the Bacterial Cell" which I read and thought was in the direction I wanted to go. He accepted me and suggested I work on bacteriophage resistance in bacteria. I immediately began to read about bacterial viruses and in October 1952 I arrived in Oxford to do a Ph.D. in the Physical Chemistry Laboratory.

There was still food rationing in England and life was difficult all through my 2 year stay in Oxford. In addition, I and the others were outsiders three times over; we were scientists, we were research students and we were colonials. Many of my friends in Oxford shared these stigmata and the only compensation was the opportunity to join Halifax House and lunch there. This was where I met Jack Dunitz, a crystallographer, and through him Leslie Orgel, a theoretical chemist, both of whom have remained lifelong friends and colleagues.

We had many discussions on DNA, for I had come to Oxford with two half ideas both of which were more than half wrong. One was a way of working out the structure of DNA using dyes and the other was how nucleic acids could participate in the synthesis of proteins. I can remember in November 1952 Jack telling me about two fellows in Cambridge who were going to solve the structure of DNA. When in April 1953, Jack told us that these two fellows in Cambridge, Francis Crick and Jim Watson, had indeed solved the structure of DNA, Jack, Leslie and I drove to Cambridge on a day between the 16 and 18th April to see the model.

This was the watershed in my scientific life. The moment I saw the model and heard about the complementing base pairs I realized that it was the key to understanding all the problems in biology we had found intractable – it was the birth of molecular biology. It was a revelation reinforced by conversing with Jim Watson at greater length during a walk we took together, when I realized that working with bacteriophage had put me on the right road to enter this exciting new field, even though what I was doing was trivial.

My wife and I were married in London in December 1952 and she was also engaged in doing a Ph.D. in Psychology in London. She was allowed to move to Oxford and, until June 1954, we lived in a flat in Woodstock Road working on our theses, had a child in addition to my stepson, Jonathan, dreaming all the time of food and the warm climate of our native South Africa.

An opportunity to visit the United States came about when Dr. M. Demerec, the director of the Carnegie Institution Laboratory at Cold Spring Harbor visited Sir Cyril Hinshelwood. It was Demerec who invited me to Cold Spring Harbor and who helped me to obtain a Carnegie Corporation Travelling Fellowship which enabled me at the end of the summer in Cold Spring Harbor to make a trip across the United States to visit other laboratories. I drove across America with Jim Watson to Cal Tech and I then had a period of research at the Virus Laboratory in Berkeley, working with Gunther Stent. I also had made a short visit to Washington, D.C.

My visit to America was important because through it I met many of the then important workers of the Phage School, Seymour Benzer, who became a lifelong colleague and friend, Max Delbruck, the founder of the phage school, Salvador Luria and many others who were destined to play important roles in the development of of the new science of molecular biology.

I had also met Francis Crick again in Woods Hole and Cold Spring Harbor and visited him in Cambridge in December 1954 on my way back to South Africa. At this meeting we discussed how I might join him in the future and I returned to South Africa at the end of 1954 as I was committed to do so by the terms of the Carnegie Corporation Fellowship.

I set up a laboratory in the Department of Physiology in the Medical School in South Africa and begin to try to find a bacteriophage system which we might use to solve the genetic code. I also continued to work on some theoretical aspects of the genetic code and during this period was able to prove the impossibility of all overlapping triplet codes which was circulated in an RNA Tie Club note and later communicated to the Proceedings of the National Academy of Sciences by George Gamow. Francis worked hard to get me an appointment at the Medical Research Council Unit in Cambridge and in December 1956 we left South Africa for a new career in England. I did all of my work on molecular genetics in the Cavendish Unit and its successor, the MRC Laboratory of Molecular Biology, where the work on *C.elegans* was initiated and developed. I spent 20 years sharing an office with Francis Crick and many new and exciting ideas (both right and wrong) were generated from our conversations. The centre point of our interests had begun to diverge and whereas we were both interested in the nervous system, I was far more interested in finding a simple experimental system which might tell me how brains were constructed, whereas Francis wanted to know about the complex activities of higher nervous systems. He left Cambridge in 1976 to join the Salk Institute where he pursued an entirely new career in neuroscience.

In 1977, I was appointed proleptic Director of the MRC Laboratory to succeed Max Perutz on his retirement in 1979. However, I immediately took over the financial management of the laboratory and spent several years in trying to get the finances on a proper basis. These were times of hyperinflation in Britian and not enough attention had been given to the rising costs of our research. I often referred to myself as the epileptic Director. During this period, I became interested in how the new techniques of cloning and sequencing DNA could influence the study of genetics and I was an early and active proponent of the Human Genome Sequencing Project. By 1985, I found that the administrative load of the Director was becoming tiresome and interfering with what I still wanted to do in research and so when I was asked whether I wanted to continue as Director after 1987 when my term of appointment ended, I jumped at the opportunity and left the laboratory in 1986 when my successor took office. The MRC gave me a small Unit and, with some added resources, I set up a Unit of Molecular Genetics based in the Department of Medicine where space was provided for me by Professor Keith Peters. It was in this Unit that the pufferfish project began. The Unit was closed by the MRC in 1992 when I was 65 but I continued the laboratory with other support for some years thereafter. In the meantime, it became imperative for health reasons that I spend the winter months in a warmer climate and Richard Lerner made this possible by giving me a part time appointment in The Scripps Research Institute in La Jolla, California. Here I found I could pursue new interests in chemistry and especially in the interface between chemistry and biology. I also became involved in a company in the San Francisco Bay Area called Lynx where together with another friend, Sam Eletr, we developed a new massively parallel method for sequencing DNA. In 1995, I founded The Molecular Sciences Institute with a gift from the Philip Morris Company where I hoped that we could create an environment where young people could pursue science in an atmosphere of harmonious purpose and high intellectual challenge. I retired from the Institute in 2000 and in 2001 was appointed a Distinguished Professor in the Salk Institute in La Jolla where I rejoined Francis Crick.

I owe a great debt to the many people who have helped me in my life. My parents would have preferred me to become a surgeon or a physician but were most understanding of the ambitions of their son. My wife and family have borne the burden of a preoccupied husband and father for fifty years. Living most of the time in a world created mostly in one's head, does not make for an easy passage in the real world. Throughout my scientific life and in all my projects I have been joined by many scientists, young and old, whose work was absolutely essential for the success of our scientific endeavours. Many have gone on to do important scientific work but all remember those wonderful times when we and our science were young and our excitement in meeting new challenges knew no bounds. I believe that a scientist should be judged by the quality of the people he has helped to produce and not by prizes or other honours bestowed on him. Let my works speak for themselves.

I am still, at the age of 76, excited by scientific research and the prospect of what can be done in biology. Science is something one is tied to for life and one should never retire from anything until one has secured one's next job. The endless quest for knowledge will continue as long as humans exist.

NATURE'S GIFT TO SCIENCE

Nobel Lecture, December 8, 2002

by

SYDNEY BRENNER

Salk Institute, 100010 N. Torrey Pines Road, La Jolla, California, USA, and King's College, Cambridge, England.

The title of my lecture is "Nature's gift to Science." It is not a lecture about one scientific journal paying respects to another, but about how the great diversity of the living world can both inspire and serve innovation in biological research. Current ideas of the uses of Model Organisms spring from the exemplars of the past and choosing the right organism for one's research is as important as finding the right problems to work on. In all of my research these two decisions have been closely intertwined. Without doubt the fourth winner of the Nobel prize this year is *Caenohabditis elegans*; it deserves all of the honour but, of course, it will not be able to share the monetary award.

I intend to tell you a little about the early work on the nematode to put it into an intellectual perspective. It bridges, both in time and concept, the biology we practice today and the biology that was initiated some fifty years ago with the revolutionary discovery of the double-helical structure of DNA by Watson and Crick. My colleagues who follow will tell you more about the worm and also recount their incisive research on the cell lineage and on the genetic control of all death.

To begin with, I can do no better than to quote from the paper I published in 1974 (1). The paper was unhesitatingly entitled: "The genetics of *Caenorhabditis elegans*" and the opening sentence reads: "How genes might specify the complex structures found in higher organisms is a major unsolved problem of biology." This is still true today. The paper outlined how a genetic approach coupled with detailed studies at the cellular level might be a way of studying this important question. It introduced *C. elegans* as the organism of choice for this work.

This choice had a long history. Twenty years earlier, we had posed a different question. Then, the central problem in biology was how the one-dimensional sequence of nucleotides in DNA specified the one-dimensional sequence of amino acids in proteins. Today, any student would give this question a very simple answer. 'All you have to do is to find a gene and have it sequenced and then make some protein using the gene and get someone to determine its amino acid sequence.' In those early days, the techniques for determining amino acid sequences of proteins were primitive and needed large amounts of proteins which had to be purified first. There were no methods to isolate genes and no techniques for the chemical determination of their sequences. Our analysis of genes was limited to genetics. Indeed, the only way we could assert that there was a gene in an organism was by finding a mutant allele for it. Like Mendel, we could not say that there was a gene for the character of tallness until dwarf mutants were discovered suffering from a heritable lack of tallness. Genetic analysis of linkage used recombination to analyse the structure of chromosomes, to determine the locations of genes and their linear order along a chromosome. But in order to probe the structure of the gene something special was needed. That was provided by bacteriophages, viruses with tiny genomes which grew on bacteria and which showed recombination. In 1954, Seymour Benzer developed a system using the rII gene of bacteriophage T4, in which powerful selection could be exerted for the r+ phenotype. Forward mutants could be easily picked and large numbers of phage crosses could be readily performed. The selection method allowed recombinants to be measured at very high resolution, limited only by the rates of reverse mutations. These experiments revealed that the gene contained hundreds of sites at which mutation could occur and that the scale of separation was of the order of the distances between adjacent base pairs on the DNA structure. The experiments not only exploded the classical idea of the gene as an indivisible unit of function, mutation and recombination, but the fine structure map could now be viewed as a picture of the nucleotide sequence of the DNA. It therefore provided an approach to studying how the sequence of bases in DNA might correspond to the sequence of amino acids in proteins by using genetics for the first and chemistry for the latter. This programme of molecular genetics was never quite completed, but the exploitation of the properties of T4 bacteriophages played important roles in studies of mutagenesis, the general nature of the genetic code and genetic suppression, and in the demonstration of the existence of messenger RNA.

For our purposes today we need to reflect on the properties that made phages that ideal 'model organisms' for this phase of research in molecular biology. They were easy to grow and maintain in a laboratory, large numbers could be readily generated, and many experiments could be conducted in parallel. The final readout was the presence or absence of lysis of bacteria and this assay could be applied to single particles each of which left a plaque – an area of lysis in a lawn of bacteria. For mapping purposes, scoring was digital – yes or no – and single particles could be easily counted to obtain accurate frequencies.

After the basic principles of information transfer from genes to proteins had been established with the identification of messenger RNA, the discovery of the mechanism of protein synthesis and the structure of the genetic code, it was natural for some of us to ask whether the lessons learnt in molecular biology could be applied to the genetics of more complex phenotypes. All questions in genetics involve asking how the phenotype is represented in the genotype or, reflexively, how the genes map onto the phenotype. It was clear that one could not have understood how metabolism or biosynthesis was represented on the genome until one had conceived of the 'one gene-one enzyme relationship', that a gene specified the amino acid sequence of a polypeptide chain, which after folding correctly, was able to perform a specific catalytic function in a metabolic pathway. In the same way, the elaborate structure of the protein coat of a bacteriophage could not be understood without knowing that it was built of many different kinds of subunits and that these were able to self-assembly into the final particle. We have to know how the gene specifies the construction of the entity which carries out the function. The same is true at a higher level of organization. In studying the genetics of behaviour, it is difficult, if not impossible, to go directly from the gene to behaviour, because there is no simple mapping that connects the two. In my paper, I put it in this way: "Behaviour is the result of a complex ill-understood set of computations performed by nervous systems and it seems essential to decompose the question into two: one concerned with the question of the genetic specification of nervous systems and the other with the way nervous systems work to produce behaviour." Thus, just as the structure and function of protein molecules is the necessary connection between the genes and metabolism, the link between genes and behaviour resides in understanding the structure of nervous systems and how they are constructed. These are questions of anatomy and embryological development. This set the requirements for the experimental organism as one which was not only suitable for genetical study in the laboratory but also allowed the structure of the nervous system to be accurately defined. Since a nervous system is essentially a cellular network, we had to be able to observe junctions between cells and their processes and this could only be achieved with the electron microscope which has the necessary resolution. Since the electron microscope provides only a small window because of its high magnification, we needed a small animal which would also need to have a nervous system with a small number of cells. After some searching, my choice finally settled on the small nematode, *Caenorhabditis elegans*. This was a self-fertilizing hermaphrodite with rare spontaneous males. The adults are about 1 mm in length and the life cycle is completed in 3 ½ days. The animals live in a two-dimensional world feeding on *E. coli* on the surface of agar plates. They are easy to grow in bulk, each animal producing about 300 progeny during a cycle. My paper reported the isolation of several hundred mutants together their complementation and mapping to define about one hundred loci. In parallel with the genetic work, I launched a program with Nichol Thompson to determine the complete structure of the worm by serial section electron microscopy (2). This project was completed by John White and Eileen Southgate who joined the group soon after it started and whose work resulted in the determination of the complete structure of the nervous system (and much more besides) in *C.elegans* (3).

The paper I referred to summarized the work on genetics but, in the early seventies, trying to understand the functions of genes specifying the development of the nervous system in molecular terms seemed impossibly remote. However, going to the molecular level was inevitable, and we very early initiated studies of mutants which affected the movement of the worm by disrupting muscle structure function (4,5,6). The defects were easily characterized by electron microscopy and we began to work on two of the genes, *unc*-15 and *unc*-54, which had abnormal thick filaments in the body muscles. We chose muscle because the structural proteins constitute a considerable fraction of the total protein of the worm and could be easily isolated and characterized by biochemical analysis. Nonetheless, this seemed a rather limited and oblique way to tackle a very large problem.

Our work and indeed the whole field was revolutionized by the discovery of DNA cloning and by the invention by Sanger and Gilbert a little while later of methods for sequencing DNA. It seemed possible that we could obtain the whole genome as a collection of DNA fragments and then proceed to characterize the genes directly by chemical means. At the Asilomar Conference, in 1975, I gave a talk on our work on *C.elegans* and on the promise of applying the new methods for obtaining insights into the molecular basis of gene action in complex organisms. We had already characterised the DNA complement of *C.elegans* and, in a paper (7) accompanying the genetics paper, J. Sulston and I showed that the genome consisted largely of unique DNA sequences and contained 20 times the amount of DNA in *Escherchia coli.* This estimated the haploid amount of DNA as 80 megabases, which was an underestimate partly because the value taken for *E.coli* was too low. We began to work in earnest to try to clone the genes we had assigned to components of thick filaments, the heavy chain of myosin and paramyosin. This was accomplished by Jonathan Karn and Alexander MacCleod (8), who used a myosin mutation previously shown to be a small internal deletion of the protein (9) to validate that the correct gene had been cloned. The great power of the new genetic approach was revealed by subsequent work of Jonathan Karn. Very quickly three other loci were identified for myosin heavy chains, and all were sequenced. It became clear that the best way to obtain the amino acid sequence of a protein was to sequence the gene specifying it; indeed, for the heavy chain of myosin the work involved had been reduced about a thousand fold.

The cloning of genes involved in the development of *C.elegans* and the identification of their proteins led quickly to insights into the molecular basis of the defects in the mutants that had been isolated. The irony was that, while *C.elegans* had been proposed as a model organism to understand genetical specification in the more complex mammalian organisms, by the time we were able analyze our mutant genes, a few thousand genes had been cloned from humans, rats and mice, and these were frquently the sequences that provided instant identification of many *C.elegans* genes. These early studies led John Sulston and others to mapping the genome of *C.elegans* into ordered clones, and this was followed by obtaining first a draft sequence and, more recently, the complete sequence of the genome of the worm. At the same time John Sulston, as you will hear later, was able to derive the complete lineage of the cells in the hermaphrodite and the male. Thus there had became available a model organism in which we knew the positions of every cell in the body and how they were connected to each other, the complete cell lineage which told us where every cell came from in development, and the total DNA sequence together with a host of methods to alter the DNA and its expression. An almost a similar path has been followed by those working on *Drosophila*, but the fly is much more complex than the worm and the anatomy of its nervous system has not reached the level of completeness achieved for *C.elegans*. The complete genome sequence of the fly is now also available and both the fly and the worm will continue to be important models for the study of function in complex organisms.

The new techniques of cloning and sequencing, which greatly enlarged the scope of genetical analysis in the nematode, led to the question of whether they could be extended to other organisms in which standard genetical analysis was difficult or even impossible. The new technology had liberated genetics from the tyranny of the reproductive cycles of organisms and, in principle, could be now applied to any organism: Its power stems from the properties of DNA which enable its isolation, amplification and expression in simple microbes, and from the uniqueness of DNA sequencing, a technology that allows us to extract the essential information – the linear sequence of nucleotides – in DNA from any source, viruses, microbes, plants and animals, as well as from molecules that can be copied into DNA.

In 1985, when the first suggestions were made to sequence the human genome, I thought that the sequencing techniques, even with incremental improvements, would not be equal to the task, and would require a factory scale operation to do it. I had also come to the conclusion that most of the human genome was junk, a form of rubbish which, unlike garbage, is not thrown away. My view at the time was that we should treat the human genome like income tax and find every legitimate way of avoiding sequencing it. It could therefore be asked whether a genome existed in Nature which perhaps had very much less junk but nevertheless had the full repertoire of vertebrate genes? It is easy to ask the question if one already has the answer. Towards the end of the 1960's I spent time in Woods Hole and took advantage of the library where I first discovered the papers of Hinegardner (10). At the time, I was puzzled by the enormous variations in the amounts of DNA in different organisms. Indeed, whereas most physicists thought that organisms did not have enough DNA to specify their complexity, it was clear to me that many organisms had too much. I discovered from Hinegardner that one group of fish, the Tetraodontidae, which included the Japanese pufferfish, *Fugu*, had very small genomes, with a haploid content of about 400 megabases as opposed to the 3000 megabases of mammalian genomes. Although teleost fish are distant from humans they are still vertebrates, with the same body plans, development and physiological systems as ourselves. Because of these basic similarities it seemed unlikely that *Fugu*, with a haploid DNA content one-eighth that of mammals would have eight times fewer genes, making it much more probable that what was missing in *Fugu* was junk DNA. If *Fugu* had the same gene repertoire as humans, then its genome would be more compact giving us the human gene inventory for eight times less work and expense. We proved that this was indeed the case and proposed the genome of *Fugu* as the ideal model vertebrate genome (11), with a DNA content only 4 times that of *C.elegans*. I failed to persuade any of the official genome organizations of the virtues of such a model genome and it remained a personal project until quite recently when, with the collaboration of the US Department of Energy and the generous support of the Agency for Science, Technology and Research of Singapore, we were able to produce a first draft sequence which vindicated everything we had been saying for the past decade (12). This project also reunited several members of the group who helped me start this work in Cambridge: S. Aparicio, G. Elgar and B. Venkatesh. Sam Aparicio, originally (13), and B. Venkatesh and I (14,15,16) have used the *Fugu* genome in an interesting application of the new genetics to discover the structure of control sequences in mammalian genomes, asking whether the mouse is able to read the *Fugu* genome in the same way as it reads its own, by seeing whether there are sequences in *Fugu* which control expression with the same cell specificity as those in the mouse. So far apart are these genomes on the evolutionary scale that time has randomised all irrelevant sequences to reveal only those with conserved function unobscured by remanants of common origin. In fact, these experiments are tantamount to crossing a mouse with a fish, exploiting the power of modern DNA technology to penetrate natural reproductive barriers and extending the power of functional genetical analysis. *Fugu* is not a model organism but is rather an organism that possesses a model genome. The mouse is the model organism in this case, but, unlike forty years ago, when I began to work on *C.elegans,* we now do not require a single model on which everything – genetics, physiology and biochemistry – can be done. Today we can take the genome from one source, and cells from another, and we can create unique biological material by moving genes from one organism to another.

What of the future? I want to discuss two ideas briefly about future research which I think will become the challenge for the future. If we take 2020 as the year of good vision, we may note that this is seventeen years in the future and it represents about the same time today from 1985 when the first discussions began about sequencing the human genome. It would not be excessively outrageous to suggest that the projects begun today could be completed by 2020 especially if we continued to work as a community of scientists.

The first of these projects I call **CellMap.** We are all conscious today that we are drowning in a sea of data and starving for knowledge. The biological sciences have exploded, largely through our unprecedented power to accumulate descriptive facts. How to understanding genomes and how to use them is going to be a central task of our research for the future. We need to turn data into knowledge and we need a framework to do it. So genocentric has modern biology become that we have forgotten that the real units of function and structure in an organism are cells and not genes. The genome has gives us the inventory of gene loci and we must now get on to the discovery of the actions of their products and how these are integrated in the physiology of cells. First, we will need to define all of the non-contingent states of gene expression in an organism, which is proposed as the correct way of defining a cell type. How many different cell types do we possess? There may be 200 in the body but there are likely to be many more in the brain. On top of this are all the contingent states; cells respond to outside stimuli and change their patterns of gene expression, but I do not consider a naïve neuron as a different cell type from its near identical neighbor which may have learnt something, although clearly the capacity to learn is a non-contingent property of that cell type. The next task will be to see how the gene loci map onto these cell types, to solve what may be called the instantiation problem. Many gene loci have multiple instantiations distinguished by different promoters which specify the cell type for expression, or by additions or deletions of coding sequences by differential splicing which determine the location in the cell where that instantiation may act. There may be other differentially spliced sequence differences that control the lifetime of the messenger RNA or of the protein product. Not only is the cell the only physical locus for gene action but it is the correct level of abstraction to construct a framework for understanding functions. **CellMap** is seen as a map in many dimensions; it is at once a map of the cells in the organism onto which are projected the map of instantiations, as well as a map of the molecules in the cell. It is also a temporal map connecting cells with their predecessors and successors in development. By studying how such cells are connected with their homologues in different organisms we can see how these maps are layered in evolutionary space and what has been added to or removed from any particular subsystem as we move up and down on the evolutionary scale. The architecture of **CellMap** will be couched in a form that would facilitate computation, so that we can develop it into a predictive system, and, in the future, a system which we could use for the synthesis of new cell types and new organisms.

My second *gedanken* project is called **Humanity's Genes.** It arose in my mind during a discussion of a proposal to take the inbred lines of mice, and extensively intercross them to generate 30,000 different mice representing different mosaics of the initial gene pools. Specially trained mouse phenotypers would then analyse the physiological properties in these mice and correlate them with their individual genomes. Unfortunately, the latter is the difficult task, as today there is no reasonable technology that can achieve this in any depth. However, suppose technology existed which made it easy to characterise 30,000 genomes, perhaps even to the point of resequencing them, would we bother to do this work with mice? We could go directly to humans, where we already have large numbers of diverse genomes, with skilled and expensively trained phenotypers, called doctors, studying them. Thus, since the technology does not exist, it now needs to be invented to provide the means of accurately analyzing large populations of genomes for detailed studies of natural human genetic variation and its correlation with phenotypes of health and disease. I believe that this will be the major challenge in human biology and medicine in the next decade. I am convinced that we will make our significant discoveries in humans and that the mouse will be used to validate the human findings by genetic synthesis, much in the same way as the chemist confirms a structure analysis by chemical synthesis. Chakravarti has shown the way forward by his synthesis of Hirschsprung's disease in mice (17).

CellMap and **Humanity's Genes** are not really separate projects. We need the first to tell us where to look initially in genomes and to interpret what we may discover in the studies of human genetic variation. What will be the significant differences that we may find? Will they be amino acid changes in polypeptides, or changes in control sequences that affect the timing and amount of the products expressed, or in RNA molecules whose functions we are only now beginning to glimpse ? We may also come to understand the enormous changes in structure and function that were brought about by evolution and gave our brains the immense capacities that they possess.

Nature has been generous to Science and has provided us with many model systems. I have mentioned only the few that have been important in my own scientific work. *C.elegans* will continue to yield fruitful discoveries and insights in spite of my argument that we do not need model systems any longer for the study of human biology. However, there are many aspects of humanity that we still need to understand for which there are no useful models. Perhaps we should pretend that morality is known only to the gods and that if we treat humans as model organisms for the gods, then in studying ourselves we may come to understand the gods as well.

ACKNOWLEDGEMENTS

I need to acknowledge the great debt I owe to the large number of scientists who joined me in converting *Caenorhabditis elegans* from what was once regarded as a joke organism into the powerful experimental system it has become today. The recognition given to our work by the award of the Nobel Prize owes much to their labours. I want particularly to record the patient and generous support given to me by the Medical Research Council of Great Britain who allowed me to initiate and develop the research on *C. elegans* in the MRC Laboratory of Molecular Biology in Cambridge. Such longterm research could not be done today, when everybody is intent only on assured short term results and nobody is willing to gamble. Innovation comes only from the assault on the unknown.

My research on *Fugu* was initiated with funding by the Prize awarded to me by the Fondation Louis Jeantet de Medicine of Switzerland, and by generous grants from E.I. du Pont de Nemours & Company, Wilmington USA. It has also enjoyed continued support from the Institute of Molecular & Cell Biology and the Agency for Science, Technology and Research in Singapore. Some of the research was also carried out in The Molecular Sciences Institute, Berkeley.

REFERENCES

1. Brenner, S. The genetics of *Caenorhabditis elegans*. Genetics **77**, 71–94, 1974.
2. Ward, S., Thomson, J.N., White, J.G. and Brenner, S. Electron microscopical reconstruction of the anterior sensory anatomy of the nematode *Caenorhabditis elegans*. J.Comp.Neurology **160**, 313–338, 1975.

3. White, J.G., Southgate, E., Thomson J.N. and Brenner, S. The structure of the nervous system of the nematode *Caenorhabditis elegans.* Phil.Trans.Roy.Soc.Lond.B. **314**, 1–340, 1986.
4. Epstein, H.F., Waterston, R.H. and Brenner, S. A mutant affecting the heavy chain of myosin in *Caenorhabditis elegans.* J.Mol.Biol. **90**, 291–300, 1974.
5. Waterston, R.H., Epstein, H.F. and Brenner, S. Paramyosin of *Caenorhabditis elegans.* J.Mol.Biol. **90**, 285–290, 1974.
6. MacLeod, A.R., Waterston, R.H., Fishpool, R.M. and Brenner, S. Identification of the structural gene for a myosin heavy-chain in *Caenorhabditis elegans.* J.Mol.Biol. **114**, 133–140, 1977.
7. Sulston, J.E. and Brenner, S. The DNA of *Caenorhabditis elegans.* Genetics **77**, 95–104, 1974.
8. MacLeod, A.R., Karn, J. and Brenner, S. Molecular analysis of the *unc*-54 myosin heavy-chain gene of *Caenorhabditis elegans.* Nature **291**, 386–390, 1981.
9. MacLeod, A.R., Waterston, R.H. and Brenner, S. An internal deletion mutant of a myosin heavy chain in *Caenorhabditis elegans.* Proc.Natl.Acad.Sci.USA **74**, 5336–5340, 1977.
10. Hinegardner, R. Evolution of cellular DNA content in teleost fishes. Am.Nat. **102**, 517–523, 1968.
11. Brenner, S. G. Elgar, R. Sandford, A. MacRae, B. Venkatesh and S. Aparicio. Characterization of the pufferfish *(Fugu)* Genome as a compact model vertebrate genome. Nature **366**, 265–267, 1993.
12. Aparicio S., Chapman J., Stupka E., Putnam N., Chia J.M., Dehal P., Christoffels A., Rash S., Hoon S., Smit A., Gelpke M.D., Roach J., Oh T., Ho I.Y., Wong M., Detter C., Verhoef F., Predkip., Tay A., Lucas S., Richardson P., Smith S.F., Clark M.S., Edwards Y.J., Dogget N., Zharkikh A., Tavtigian S.V., Pruss D., Barnstead M., Evans C., Baden H., Powell J., Glusman G., Rowan L., Hood L., Tan Y.H., Elgar G., Hawkins T., Venkatesh B., Rokhsar D., Brenner S. Whole-genome shotgun assembly and analysis of the genome of *Fugu rubripes.* Science **297**, 1301–1310 (2002).
13. Aparicio S., Morrison, A., Gould, A., Gilthorpe, J., Chaudhuri, C., Rigby, P., Krumlauf, R., Brenner S. Detecting conserved regulatory elements with the model genome of the Japanese puffer fish, *Fugu rupripes.* Proc.Natl.Acad.Sci.USA **92**, 1684–1688, 1995.
14. Venkatesh, B., Si-Hoe, S.L., Murphy, D., Brenner, S. Transgenic rats reveal functional conservation of regulatory controls between *Fugu* isotocin and rat oxytocin genes. Proc.Natl.Acad.Sci.USA **94**, 12462–12466, 1997.
15. Rowitch, D.H., Echelard, Y., Danielian, P.S., Gellner, K., Brenner, S., McMahon, A.P. Identification of an evolutionarily conserved 110 base-pair cis-acting regulatory sequence that governs Wnt-1 expression in the murine neural plate. Development **125**, 2735–2746, 1998.
16. Brenner, S. Venkatesh, B., Yap, W.H., Chou, C.F., Tay, A., Ponniah, S., Wang Y., Tan, Y.H. Conserved regulation of the lymphocyte-specific expression of lck in the *Fugu* and mammals. Proc.Natl.Acad.Sci.USA **99**, 2936–2941, 2002.
17. Carrasquillo M.M.., McCallion A.S., Puffenberger E.G., Kashuk C.S., Nouri N., Chakravartim S.A. Genom-wide association study and mouse model identify interaction between RET and EDNRB pathways in Hirschsprung diease.

H. ROBERT HORVITZ

"He can't even throw a baseball!" was the way my cousin Harvey greeted me soon after my birth on May 8, 1947. Looking forward to a fellow athlete, Harvey, age three, found me ill prepared for sports. I suspect the disappointment continued, as Harvey progressed to become one of the star little league players in Chicago. I did not.

FAMILY HISTORY

Chicago is my hometown. My mother Mary Savit Horvitz was born in Chicago, and my father Oscar Horvitz was born in the neighboring town of Joliet. Both of my parents were first-generation Americans, the children of Jews who left Eastern Europe around the turn of the century. My maternal grandfather David Savit (Savitzky) was born in 1879 and came to Chicago in 1904 (via Liverpool, Halifax and Detroit) from Oster, Russia, in present-day Ukraine about 40 miles north of Kiev. He was the second oldest of ten children, and his widowed mother Malke Zolotar Savitzky and his eight younger

My maternal grandparents, Rose and David Savit (1908).

siblings followed him to Chicago. David was a dress manufacturer and a grocer, but I remember him most reading, playing chess (sometimes with me) or standing at his newstand selling newspapers. The family held him in awe as a gentle and caring human being, and as a scholar.

My maternal grandmother Rose Bleiweiss Savit came to Chicago from Galicia, Austria, around 1902. She was from an area that was alternatingly Austrian and Polish, from a town that is now Debica, Poland, about 60 miles east of Cracow. I always thought of her as Polish. She was one of the lucky members of her family, as her father Hersch David Bleiweiss and at least three of her siblings were murdered by the Nazis. Rose married David Savit in 1908, and my mother Mary Savit was born in 1921, the youngest of four girls, after Ann, Sylvia and Esther. In the early 1950s, David and Rose moved to Miami because of Rose's ill health, but after David's death in 1957 Rose returned to Chicago. She lived with us from 1963 until her death in 1968. Like David, Rose played chess, but she enjoyed poker more and sometimes joined poker games with my friends and me. Before I left for college in 1964, I spent one year living with Rose and during this time heard many stories, most of them about Chicago speakeasies during the 1920s.

My father's parents were Samuel and Celia Horvitz. Sam's original name was Solomon ben Mikhail Gourevitch, but, like many other immigrants to the U.S., he found his name altered upon entry to the country. Sam was from a Jewish village called Shchedrin, located near Minsk, in what was then Russia and is now Belarus. Sam was from a wealthy Russian family. His father ran a

My paternal grandparents, Samuel Horvitz (circa 1924) and Celia Horvitz (1904).

lumber business but, being Jewish, was not allowed to own any land himself. Sam spoke Russian, Polish, Yiddish and German. He married his first wife (not my grandmother) in Poland in 1902 and went to Chicago to avoid the 1904–1905 Russo-Japanese War. Once in the U.S., Sam worked in the garment trade as a presser. In 1910, he was blacklisted for participating in the four-month long Hart, Schaffner and Marx labor strike. Hart, Schaffner and Marx was a giant clothing manufacturer, the largest company in Chicago at that time. Employees were subjected to excessive work hours, miniscule wages and subhuman working conditions, and a strike at the company became the basis of a city-wide violent clash between businesses and the unions. Sam kept a hat with a bullet hole in it, a souvenir of a shot taken at him during a union rally. He became an insurance salesman, and it was because he was working for a Polish insurance company that from 1917 to 1920 he lived in Joliet, which had much industry, many steel mills and large numbers of Polish immigrants. Later, Sam operated a Chicago dry cleaning business, which I remember visiting once, just before he died in 1951.

Sam's first wife died in 1908, and he and their two daughters moved in with cousins, where my paternal grandmother Celia Bolotin was a boarder. Celia was from Novgorad-Syovorsk, about 150 miles northeast of Kiev in Russia. Celia was generally called by her Hebrew name of Tzipporeh, which means "little bird." In Russia she had lived in a one-room house, which her family shared with the domestic animals, for mutual warmth. As a young girl, Celia was unusual. First, unlike most girls in her town, she learned to read. Second, she was rather stubborn and independently minded. Because in one of the books she had read a girl was cruelly mistreated by her step-mother, when young Celia's widowed father remarried and was loading a cart to take his family and possessions to another town to live with his new wife, Celia refused to go, and moved in with an aunt. Both a desire for knowledge and a strong will are family characteristics that have endured. In 1905, Celia, not yet 20 years old, was involved in the first Bolshevik uprising and was captured by the Czar's police. Knowing that they were executing people who knew how to read or write, she feigned illiteracy. They released her, but informed her if caught again she would be killed. Celia left Russia, traveling via Rotterdam to Chicago. She and Sam were married around 1909, and she raised Sam's two daughters (Faye and Bess) as well as two daughters (Pearl and Diana) and two sons (Mike and my father, Oscar) from their own marriage. My father, like my mother, was the youngest. My father's family moved a lot – about every three months – because landlords offered three months free rent to attract tenants. I never knew either of my father's parents very well, as Celia died before I was three and Sam when I was four. My father and Sam were not on good terms, both because Sam left Celia for another woman and because Sam failed to provide any support for the large family he deserted during the Depression. I think both Celia and Sam influenced me greatly through my father, who learned about responsibility from his mother and about the costs of irresponsibility from his father.

My father was born on November 3, 1918, just before the signing of the

My parents, Oscar and Mary Horvitz (1987).

armistice ending World War I. His parents were delighted that the war was ending, and named him Oscar Freedom Horvitz. My father never liked and never used his middle name, signing things as Oscar F. or, more often, simply as Oscar. He grew up on the north side of Chicago, but nonetheless became an ardent fan of the south-side Chicago White Sox baseball team. Dad himself wanted to be a professional baseball player, but his eyes went bad when he was young (as did mine) and his family could not afford to buy eyeglasses. Eventually, one of his teachers purchased a pair of glasses for him. As a teenager, my father worked as a milkman's helper, and many weeks the family lived on the butter, eggs, milk and cottage cheese he was allowed to bring home from whatever was left over at the end of the daily run. After my father finished high school, he went to work to support both his mother and himself. He took a job with an advertising agency, clipping ads from daily and weekly newspapers around the country. He found the job tediously dull, and to amuse himself he memorized the names of every town and its newspaper. He remembered these names throughout his life, and I think back fondly how years later people would express amazement that he knew not only the locations of their hometowns but also the names of their hometown newspapers.

My mother Mary Savit was born on June 5, 1921. Not given any middle name, she decided to adopt the middle initial "R," because it gave her the initials "MRS." She lived in a variety of neighborhoods in Chicago's north side. My parents met at a Christmas party in 1938, when my mother was 17 and my father 20. While in high school, my mother worked for 25 cents per hour as a sales clerk in a dimestore. In college, she had a part-time job as a bookkeeper and telephone switchboard operator at a company that made and sold uniforms for military officers. My mother obtained a teaching certificate in elementary education and then a B.A. in English from the Chicago Teacher's College. She began substitute teaching in the Chicago public schools. My father changed jobs, working next at a mail-order house, and attended night school to study chemistry, which he loved. However, he soon realized that he could not afford the continued schooling that would be needed to allow him to make a living as a chemist. He switched to accounting. In 1941, my father passed the civil service exam in accounting and was hired by the General Accounting Office of the U.S. Government for a job in Washington, D.C. To be together, he and my mother decided to get married. Because he could not afford and my mother did not want a diamond ring, my father bought a ring for my mother in a dimestore. He returned to Chicago for the wedding, which was on June 30, 1942. They lived in Washington until 1945, when they moved back to Chicago. While in Washington, shortly after my parent's marriage, my father was hit by a car, soon thereafter leaving him with a steel pin in his shoulder. He later told me that without this accident I might never have existed, as otherwise he would have been drafted into the army and who knows what would have happened to him if he had gone to war. In Chicago my father continued working with the General Accounting Office, while my mother continued the employment she had begun in Washington, D.C., with the Social Security Board of the U.S. Government.

GROWING UP

I was born in 1947 and given the name Howard Robert Horvitz. My first name, which started with an "H," was in memory of my great-grandfather Hersch, who had been shot a few years before by the Nazis. "Robert" my parents simply liked, in fact so much that they decided to call me "Bobby" and

Me, with The Great Pumpkin (1948).

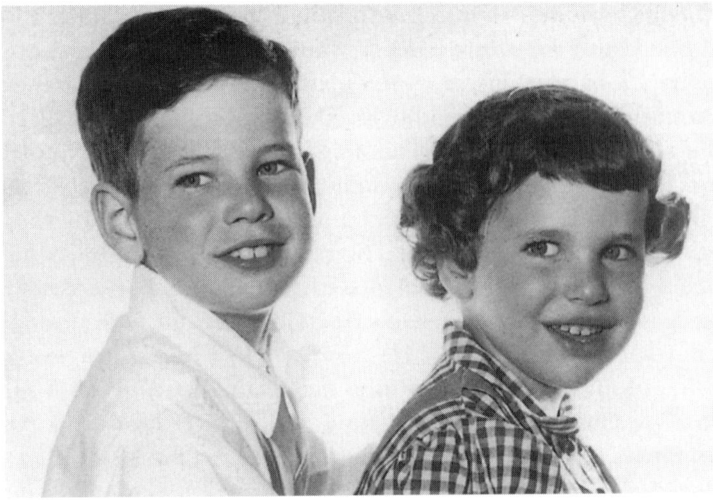

Me, with my sister Carol (1954).

registered me for kindergarten as "Robert H." Later, when I went to university, I reverted to the name on my birth certificate and became known as "H. Robert." As I grew up my parents mostly called me "Bob," reserving "Robert" for times of disapproval and "Rob" or sometimes "Robin," by my father, as terms of love. I begged my parents for a baby sister, and, on October 5, 1950, my sister Carol Cecile was born. She was named for my grandmother Celia, who had died earlier that year. I remember awakening the day my sister was born, surprised to find myself in my parent's bed next to my Aunt Ann, my mother's oldest sister. Ann died of Alzheimer's Disease many years later, one of a number of my personal reasons for becoming interested in neurodegenerative disorders.

We lived on the north side of Chicago, which led to my becoming a fan of the Chicago Cubs baseball team, unlike my father, whose loyalty to the White Sox continued. My parents saved their money and moved repeatedly to find better homes for themselves and their family. When I was born, we lived in a one-room kitchenette near Lake Michigan, and I slept behind a screen while they ate, entertained or slept. We moved some miles west in 1948 to a one bedroom, four-room apartment. After my sister Carol was born, we moved again a few blocks away to a five-room bungalow on Rockwell St. Sometimes at night Carol and I secretly crawled out through her window to play in the back yard with friends who lived next door. One of the great delights of my receiving the Nobel Prize was that I was contacted by many old friends with whom I had lost touch, including one of these next-door neighbors on Rockwell St.

The Rockwell St. house is where I feel that I grew up, living there from ages five through 12. The neighborhood at that time was a mix of 1920s bungalows and apartment buildings, and was mostly Jewish. Author Saul Bellow's father lived a few houses south of us. I was a member of a club we named the Epics, without having any idea what the word meant but simply because we liked its sound. The year after I moved out of Chicago, the Epics became the Commodores, a gang. Rockwell St. was also important for me because it was while living there I became good friends with Ira Zarov, who lived across the street and whose mother was my Den Mother in Cub Scouts. Ira and I remain very close, exchanging e-mails almost daily and spending a week of most summers with our families together at his family home along the south shore of Lake Michigan, near Chicago.

I went to the DeWitt Clinton Elementary School and got into trouble repeatedly for small misdemeanors, particularly in the third grade. One transgression was my insistence upon crossing in the middle of the street to walk to school with Ira. Another involved the fact that Ira and I had acquired vast quantities of gumballs as a consequence of having found a cardboard circle that could substitute for a penny in the gumball machine in the local candy store. A few years later Ira and I were stopped at knife point by a boy trying to rob us. He was no bigger than we were. Ira and I moved apart and pointed out to our assailant that we had him outnumbered. He left.

In 1960, my family moved to the northern Chicago suburb of Skokie, because the schools were better. I attended eighth grade at East Prairie Grammar School. Because I had come from the Chicago school system, I was placed in the lowest of the three academic tracks for all of my classes. Within a few weeks, each of my teachers moved me to the highest track. The consequence was that I got to know, and become friends with, most of the eighth grade students very rapidly. My eighth grade English teacher, Marcia Wachs, is particularly memorable and engaged me in the reading of proper English literature. A year after I left her class, she sent me a book accompanied by a very nice note: "How many times I have thought of you and wondered how you were enjoying high school. ... You seem to have a talent for writing, and I thought you might profit from reading this book. It's a gem ... ." The book was "The Elements of Style" by Strunk and White. I still have both the book and her note, and I use the book to this day (particularly when I try to educate members of my laboratory concerning the usage of "which" and "that").

In 1947, the year I was born, my father became a Certified Public Accountant. His professional life thereafter was that of an accountant. My father loved numbers, a love I learned from him. Eventually, he became the Vice President and Treasurer of a major trucking company based in Chicago. Whenever we went on family trips, Carol and I always watched for Spector trucks, sometimes reminded by my father that the stock he had purchased for us was almost enough to say that we owned one of the tires on the truck. He was President of the Motor Carriers Accountants Society of Chicago. I was always very proud of the fact that he had published a book, which was an income tax guide for farmers. He also authored a chapter in a book about data processing. Published in 1957 and focused on the accounting machines and punched cards of the era, his chapter noted, "Long-range studies of electronic computers are being made at the present time." In 1989, my father died from amyotrophic lateral sclerosis (ALS). His illness moved me to become actively involved in research involving this horrific neurodegenerative disease.

A sense of my father is reflected in a dedication written by my sister and me to an ALS Resource Book, the publication and distribution of which we funded. We wrote: "Dad was a humanitarian, not only in his values, but also in the actions of his everyday life. He cared deeply about people and always had the time to listen and provide insight. He was happiest when he was making someone else happy. In our family, Dad created an environment filled with a love of knowledge and the joy of inquiry and discovery about the world around us. Not a day went by without his learning something new and our learning something from him. So many things interested him: foreign languages, exotic places, history, new words, opera, baseball, people. He was equally excited by an exchange of smiles with little children as by a beautiful piece of music or by some scientific discovery. He also had a way with languages and often surprised people by speaking with them in their native tongue or talking with them about some detail of their country that few would know about. He once dreamed of being a major-league catcher, and then of being a research chemist. Numbers were his friends, and he chose accounting as his profession. However, his love of science continued and was infectious, giving both of us the opportunity and encouragement to study and lead lives dedicated to scientific inquiry. Dad was a wonderful mentor. He always strove to do his best, and he always brought out the best in those around him."

My father's illness and death were an immeasurable loss to me. My father was a major cornerstone of my life. It was my father who came to talk with the elementary school principal with me after my more serious infringements. It was my father who – after at age 15 I had attempted unsuccessfully to drive the family car using a "borrowed" key and knocked down a wall of the garage – convinced me over the telephone not to run away from home and who then came home from work not to punish me but rather to console and comfort me. It was my father who would sit with me and amaze me as he solved diagramless crossword puzzles and double acrostics by simply reading the clues and writing in the answers to each as he read them. It was my father who, throughout my adult life, prepared my income tax returns and who, once ill and dying, taught me how to prepare them myself. I think it was when my father died that I really grew up. I also became more introspective and much closer to both my sister and my mother during his illness. This closeness has continued.

My mother began working as an elementary school teacher in 1948, stopped in 1950 when my sister was born, and then resumed teaching in 1955. She first taught in the third grade and then progressed through the grades with me until she stopped as a seventh and eighth grade teacher of math and science. Her training was in English, and her interest in science and her desire to be a good teacher led her to take part in a program supported by the National Science Foundation in which she spent a summer studying astronomy. Later she returned to school and obtained an M.A. in guidance and counseling, after which she worked as an elementary school guidance counselor. My mother retired in 1988. My mother has been a wonderful model for the professional woman – a loving mother dedicated to both her family and her work. She inspired me, made me proud and developed in me an enormous respect for women in general. My mother chose teaching as a profession in part because it would allow her to be home with her children as they grew up. Nonetheless, to this day, she asks me if she ruined my life by returning to work too soon after my birth. I tell her I do not believe that I turned out so badly.

I could always depend on my mother. She is capable, organized, practical, logical and absolutely reliable. Although she is a worrier, my mother never loses her sense of purpose or her ability to deal with difficult situations. When I was hit in the eye with a baseball, it was my mother who took me for stitches (after leaving my father the succinct note: "Took Bob to the hospital. Don't worry."). When friends and I rode our bicycles to O'Hare Airport (a trip far beyond the confines we were supposed to traverse) and one of the others got a flat tire, it was my mother who came and brought us home. It was also my mother, with her interest and knowledge as a science teacher, who encouraged my first experiences with experimental science. My sixth grade science project, which I think was her idea, was entitled "Electricity Produces Light through Heat" and won a third prize. My ninth grade project, in which I used the fruit fly *Drosophila melanogaster* to replicate Gregor Mendel's famed 3:1 and 9:3:3:1 inheritance ratios, required my mother to relinquish her bathroom for my breeding experiments. She helped me prepare the fly food, which smelled awful, and tolerated the fact that to anesthetize the flies I used ether, which smelled worse. This project also won a prize, and earned me a trip to the Illinois State Science Fair, in Champaign-Urbana. My mother saved the (written) project intact, and when the announcement was made that I had received the 2002 Nobel Prize in Physiology or Medicine, my mother showed a local reporter my 1961 science fair poster, one panel of which was then reproduced in the Chicago Sun-Times on October 8, 2002 (next to my high school graduation photograph).

My mother's strength and independence became clearest to me as she nursed my father during his illness and as she has coped with and redefined her life after his death. My mother, now age 81, is impressively active and completely engaged in the events of the world. She regularly takes brisk walks for exercise and particularly enjoys walking at a nearby botanical garden. She is a bridge player, and loves the theater, the symphony, movies and art. She has long been politically active, and for some years has supported local and national actions of the National Council of Jewish Women. My mother has many friends, some of whom, albeit sadly a diminishing number, she has had for decades. My parents traveled together extensively, and my mother has continued her travels on her own, for example touring Australia, New Zealand and China. She also has visited Greece, Turkey and Costa Rica with my sister Carol. Their descriptions of the trip to Costa Rica, to see one of Carol's field stations in the tropical rainforest, were somewhat divergent.

Whereas Carol described the site as the Hilton of field stations, my mother emphasized the primitive conditions that she had experienced. Carol never told my mother that each night before going to sleep she would check my mother's bedding for deadly snakes and spiders. After my father died, my mother for 10 years taught English as a second language on a volunteer basis to immigrants from all over the world. She also has taken countless adult education courses, on topics as diverse as international relations, world religions, poetry, music appreciation and money management. She has very recently obtained and begun to learn to use a computer. My mother is highly organized, and both her home and her person are well-kempt. She has been described as "an elegant woman." She is constantly sending me newspaper clippings, sometimes about the world but more often with advice about foods, health, finance or child-rearing. Generally, the advice is very good.

My father and my mother were a team, in raising my sister and me and in life in general. Together they instilled in me a sense of responsibility, of commitment, of determination, of fairness, of pride, of ambition, of optimism and of love. People, particularly family, came first. Knowledge and learning were revered. We were all always expected to do our best. Prejudice was unacceptable and attributed to ignorance. Time together and family vacations were important, and we traveled together extensively within the U.S. Without doubt, my parents defined the priorities and many of the interests that came to be central in my life.

Growing up with my sister Carol was a special delight, except perhaps during a few of my teenage years when I regarded her at best as an intrusion. When little we would play together, and I would make up long stories of flying horses and imaginary friends. Carol voiced progressive views at an early age. When she was about five, she was listening to a conversation between my fa-

Me, with my sister Carol and our parents, Oscar and Mary Horvitz, on vacation in Biloxi, Mississippi (circa 1958).

ther and me. After he told me "Everything is made of atoms," she immediately cried in response, "That's not fair! Isn't anything made of Eves?" Carol and I were quite divergent in our interests, with her gravitating toward the artistic and poetic aspects of life, whereas I was more focused on math, science and schemes to make money. Carol also was more rebellious, first against our quite secular home – she adapted the orthodox Jewish custom of lighting candles and saying prayers on Friday nights and went to a Hebrew overnight camp – and later politically, leaving Barnard College (at Columbia University) with the plan of becoming a left-wing political activist. Our parents intervened, and she registered at highly liberal Antioch College in Ohio, which because of its workstudy program allowed her to be almost any place but Ohio. While she was a student at Antioch, Carol spent time in Colombia, South America, and developed her passion for botany, Spanish and Latin culture. During this period Carol was rescued rather dramatically and against her will by our parents from a nearly fatal disease and an ill-fated engagement to a Colombian merchant marine, after which she stayed briefly in Chicago. I invited her to come to Boston, and she moved in with me, living on a front porch that we insulated for her and working at the Harvard Herbarium. Carol then went back to school and obtained a Ph.D. degree in biology, just like me. It is amazing, given our backgrounds and interests, that we both became biologists. Carol married Randy Nutt, an artist also from Chicago, and the two of them now live in Miami, Florida, where Carol is on the faculty of

Me, with my sister Carol and our parents, Oscar and Mary Horvitz, at Carol's bat mitzvah (1963).

the University of Miami. Carol has spent much of her life in the tropical rainforests of Mexico, Central and South America. I believe we have always been close and mutually supportive, and I often think about the different view of life held by those who are not lucky enough to have siblings.

In addition to my family, there was another constant presence in my young life that I believe influenced me greatly. At age two and a half, I acquired asthma. For many years thereafter, nights (in particular) could plague me with a shortness of breath that made me feel (and occasionally wish) that I was dying. When I was small, my father would carry me in his arms through the night. Later I would just lie in bed, waiting for the dawn and hoping for the relief that often came with daybreak. I had tests, medications, and shots. The asthma continued and affected my life in many ways. Most sports seemed precluded. At age 10, I insisted upon going away to Camp Chi, an overnight camp in nearby Wisconsin, and spent almost the entire two weeks in the infirmary, with an adrenaline inhalator next to my bed. Despite these stresses, instead of becoming increasingly afraid of life, I think I became more and more determined to experience and conquer it. Nothing, I thought, could be worse than what I had already suffered. I became fearless, unafraid to go anywhere and try anything and determined to persevere, even when my shortness of breath made my physical suffering so great I could barely force myself to move.

In retrospect, I think back to growing up in Chicago with the impression that I had far more freedom and independence than I would allow a child today. After school, my friends and I wandered the neighborhood, which was assumed to be and probably was safe. For a while, Ira and I went house-to-house selling magazine subscriptions, hoping to earn all sorts of wonderful prizes. We rang every doorbell within the distance we could walk between school and dinner. We never sold a single subscription, but we did meet and talk with lots of people. For greater distances, my bicycle was my major source of transportation. Bicycle explorations went far beyond my local neighborhood and as often as possible into unknown areas. These adventures perhaps helped kindle the wanderlust that stays with me today.

In high school, the means of my explorations progressed from bicycles to cars. Before I could drive, I hitch-hiked through Chicago, simply going wherever the drivers were headed. Once the first of my friends turned age 16, we

My sister Carol Horvitz-Nutt and my brother-in-law Randy Nutt (1993).

had access to cars, and a much greater world was opened. We went north to the expansive Lake Michigan beaches of wealthy Chicago suburbs and once even to Wisconsin (discovering in the process how fast a station wagon could go). We went south into the heart of the Chicago slums. 14th and Peoria, reportedly the site of various houses of ill-repute (we never got out of the car, so I never ascertained if that was true), was a favorite target.

As a high school student I did well academically. My introduction to biology was in the ninth grade, when I did my *Drosophila* breeding experiments. The class consisted almost entirely of dissections of formaldehyde-preserved animal corpses. The teacher was allergic to formaldehyde (or so he told us), and left us on our own each day. Mostly, I played dots with my laboratory partner. Based upon this experience, I never would have guessed that I would later become a professional biologist. For a project for this class, I made an extensive insect collection, which the school kept and displayed. While collecting insects in Chicago's Humboldt Park, Ira and I were mugged by two older boys. They put their arms around our necks, and one said, "Your money or your lives." Ira and I laughed, which was the wrong response. They squeezed more tightly, and I relinquished my butterfly net and thirty-nine cents.

As a sophomore, I received my first B, which was for one quarter of a summer school World History class. It was traumatic. I had been encouraged at home to have high expectations. Later B's were less upsetting. I tended to work reasonably hard and developed a method of study that consisted of my trying to think of every possible question the teacher could ask followed by my writing both the question and the answer to the question on paper, which then allowed me to review both easily. This approach took some time and presumably reinforced the various facts I was supposed to be learning. I was the Assistant Editor-in-Chief of the school newspaper, The Nilehilite, which was my main extracurricular activity.

While I was in high school, my parents became convinced I would later become an entrepreneurial businessman. I had a variety of money-making plans. I published a magazine that I called Brigand (again, I simply liked the sound of the word), using a mimeograph machine for mass production. I sold advertisements for Brigand to local merchants and sold the magazines to friends, neighbors and relatives – mostly relatives. In another scheme, a shipment of sweatshirts had been damaged in one of the trucks from the trucking company at which my father worked. The trucking company had to buy the sweatshirts, which I purchased and resold at a profit. I also had a number of part-time jobs. For a while, I worked at an insurance company, looking at a handwritten number on a new insurance application and stamping that number onto the upper righthand corner of the first page the application. My most memorable job was at the large discount house E. J. Korvette's, where I was employed in the record department and was able to buy records at wholesale prices. My boss was a bit of an eccentric, and every day after the store officially closed he would play at maximal volume his favorite album, "The War Whoops of the North American Indians."

I did not have a clear idea about either where I wanted to go to college or what I wanted to study. I enjoyed science, particularly math and chemistry, but I also liked English and working on the school newspaper. It seemed likely that I, like most of my classmates who chose to continue their education, would attend the University of Illinois in Champaign-Urbana. Alternatively, some of the best students went a bit further away, to the University of Michigan in Ann Arbor. I applied to both schools. However, a guidance counselor insisted that I also apply to "some school in the East." I said that I would be happy to do so and that she should pick the school. She picked MIT, and I applied. I visited the University of Michigan the weekend of a Michigan vs. Michigan State football game, and the beer and partying convinced me this was not where I wanted to be. I was scheduled to visit the University of Illinois one weekend in November, but on Friday, November 22, President John F. Kennedy was assassinated, and my trip and much of the rest of what was happening in the U.S., were cancelled. The upshot was that when I was accepted to MIT, I saw no reason not to go there (except for the pleas from my English teacher, who warned me that my intellectual development would cease if I went to such a technical school).

The assassination of JFK for me, as for so many others, was an enormously deep shock. I was 16 years old and simply could not believe such things could happen in the country in which I lived. This event was one of a small number that no doubt are etched deeply in the minds of many, causing us all to remember precisely where we were at the time. In this case, I was in my high school cafeteria. Five years later, when JFK's brother Bobby was shot, I learned the news from the radio upon awakening in my apartment in Brighton, Massachusetts. Much more recently, on September 11, 2001, I was at a meeting of the Howard Hughes Medical Institute in Chevy Chase, Maryland, sitting next to the HHMI President, Tom Cech, when Tom was called outside to be told what was soon to be announced to the entire group about the attacks on New York City and Washington, D.C.

MIT, THE FIRST TIME

I flew from Chicago to Boston to begin life as an undergraduate student at MIT in September, 1964. At the freshman orientation, Dean Fred Fassett (whom I later got to know better after getting into some trouble) said "Look to your left. Look to your right. One of you won't be here when the rest of you graduate." My undergraduate years were a mix of academics and extracurricular activities, more of the latter than the former. I became Features and then Managing Editor of the MIT student newspaper, The Tech, and had the great excitement of scooping the Boston daily newspapers in getting a story and the paper out the morning after the great blackout in the northeast in 1965. I was very active in student government and was a member of a variety of committees and councils. When I became a candidate for student body president, I learned the names, hometowns, faces and majors of all approximately 3500 undergraduates. I spent full-time campaigning, met lots of people, and won the election. The next year I lived in a three-piece suit and had endless meetings with MIT administrators and, less often, with students. I talked with James Killian (President Eisenhower's Science Advisor), Jerry Wiesner (President Kennedy's Science Advisor), Jay Forrester (the inventor of magnetic core storage) and other MIT luminaries. Professor Forrester offered the view that everyone should change professions every seven years, a comment I have thought about many times since. I felt that I was grooming myself for a career in law and politics or possibly in business.

My coursework probably suffered, not in my grades, particularly, but in what I was learning. Mostly I would wait until the night before an examination, stay up all night, and learn what I needed to know to answer the questions the next day. For some courses I did nothing whatsoever until the week before the final examination. When one of my professors penalized me for turning in all of my problem sets at the end of the course, I thought he was being quite unreasonable. I majored in mathematics, with an emphasis on its theoretical as opposed to its applied aspects. However, I knew that in contrast to some of my classmates, I was not a mathematician in my soul. I had sufficient credits to graduate after three years, but did not want to do so, in part so that I could continue with my extracurricular activities and in part because I was simply having too much fun to want to leave. I decided to earn a second undergraduate degree, in economics. I then had the enormous good fortune to be able to write my undergraduate thesis under the tutelage of Bob Solow of the Economics Department. Bob was amazing. I would struggle for two weeks, get stuck, go talk with him for 30 minutes, and be set back on course for the next two weeks. The title of my thesis was "The Profit-Maximizing Utilization of Exhaustible Resources," and in short what it said was that left to their own devices with a profit motive as the only goal, businesses would deplete natural resources (such as oil, timber or minerals) as rapidly as possible, with no reason for or thoughts of conservation for the future. I enjoyed this project very much, and thought about becoming an economist. My minors, which were called something different, were computer science and psychology.

My computer background was enhanced by my undergraduate summers, when I worked for IBM. My first summer (1964) I spent wiring panels for accounting machines. A friend of my father's had arranged the job, and I was employed by the IBM Chicago Transportation Office. Weekly business meetings were held at 7:30 a.m. Monday mornings, and the office was over an hour's drive from my home in Skokie. Business attire was suit and white shirt, with socks and tie that matched each other in color. The second summer I wrote computer programs for an IBM 1440 computer. There was lots of spare time, and I wrote a program that would randomize a virtual deck of cards and deal bridge hands. (I had started playing in duplicate bridge tournaments with my friend Ira.) The summer of 1966 I taught computer programming in the language Autocoder to business executives at the IBM Chicago Education Office in downtown Chicago. I still wonder what these high-level executives felt when they entered their first class and saw a 19 year-old standing in the front of the room. One special aspect of these summers was driving between

Skokie and Chicago each day with my father. (The day I forgot to fill the gas tank and ran out of gas on the expressway during rush hour was particularly memorable.) My last summer in college, I stayed in Boston and worked at the IBM Boston Programming Center helping to develop CPS (Conversational Programming System), an early timesharing system that used the first computer language that looked more like English than algebra. CPS was never released, but I had a great time and many good lunches.

It was that last summer in Boston that I met Joe Schwarz, one of my closest friends. A year later, Joe and I became roommates, as I began graduate school and he continued his graduate studies of astrophysics. Despite the fact that Joe and his family lead complicated two-city lives – splitting their time between Munich, Germany, and Milan, Italy – we see each other often and have shared family vacations on Cape Cod a number of times.

During those summers in Chicago, my friends and I continued a tradition we had started in high school – Saturday night poker. A group of us would gather at Ira Zarov's house – Ira had been my closest friend since third grade – to play poker, whether Ira was there or not. (Often he was not.) My father's accounting approaches had taken root in me by then, and I kept a detailed log of my winnings each night, to the penny. (During this same period, I also kept a log of the precise time I went to bed and woke up each day, to the minute.) I say winnings, because there never was a night that I lost. I had studied the odds and poker strategy, and I was disciplined and cautious. In fact, after a year or two, I became obsessed with winning and distressed by the possibility of losing on a single night. I altered my play toward the end of evenings to ensure that I would be ahead when we finished. I realized that this state of mind was not healthy, and one day I decided that the next time I played I would lose, no matter what. I did so, and after that I sometimes won and sometimes lost when I played what I believe was both a better and a more enjoyable game of poker. Ira, as I recall, lost more often but may well have won more overall, and later became a successful professional poker player for a period of time. Marty Chalfie, long a friend and later a colleague in the *C. elegans* field, sometimes joined the game.

Typically during these games we would break for a while so that someone could pick up pizza. During one such break, those of us who had stayed behind started talking about how we were wasting time. Each Saturday we sat playing cards and trading our money back and forth. We should be able to come up with something more productive to do. We sat, and we thought. Then I had an idea. When our pizza-buying group returned, they were surprised to find the rest of us reading telephone directories. The idea was simple. The Chicago Daily News was running a contest called "The Wizard of Odds." Each day during the week the newspaper published a one-digit number. On Saturdays, the paper would publish three letters. Anyone with a telephone number that ended in four digits included in the five published numbers and with a last name that included all three of the published letters would share in that week's prize, which like a lottery grew during weeks when no prize was claimed. All we had to do was to read the phone books and find winners. So we did. We telephoned people and told them they had won a substantial cash prize and that we would tell them what it was if they agreed to give us one-third. People, not surprisingly, were exceedingly dubious. However, almost everyone agreed to meet with us and when they saw the newspaper description of the contest were convinced of our legitimacy. Only one person refused to pay us. Another invited me to tour the Chicago Mercantile Exchange and offered me a job there with him. One of my college roommates came to visit for a weekend and made a fair amount of money by joining us.

I was very unsure about what I wanted to do after graduation. Having worked each summer for IBM, I probably was expected either to join the company or to continue my education in the area of computer science. Graduate school in mathematics or economics as well as law school or business school also seemed to fit my undergraduate experiences. But the time was the late 1960's, and I wanted to do something "relevant," particularly if it did not involve wearing a three-piece suit. I began to think about medicine. However, I knew nothing about biology. A roommate, Al Singer, now a physician-scientist at the National Cancer Institute, convinced me that modern biology was more than formaldehyde-fixed specimens, and during the first term of my senior year I took the introductory course in biology. I loved it. As a text, we used Jim Watson's "Molecular Biology of the Gene." My interest in medicine became secondary to a growing intrigue with biology. Six weeks into the course I went to the professor in charge, Cy Levinthal. "Professor Levinthal," I said, "I am a senior, all I know about biology is what I have learned during the past six weeks from you, and I'm thinking about going to graduate school in biology. Am I crazy?" "I went to graduate school in physics," Professor Levinthal replied, "And I'm teaching your course. You're starting early." Biology it was. The next semester I took a course in Genetics from Maury Fox, and was captivated by Sturtevant and Beadle's "An Introduction to Genetics." I also took a laboratory course, which introduced me to neurobiology and to the electrophysiological methods used in the classic studies by Haldan Hartline of the eye of the horseshoe crab *Limulus*. I applied to Harvard, MIT and Stanford, was accepted at all three, and decided to simply move the mile up Massachusetts Ave. to the Harvard Department of Biology.

The summer after my graduation from MIT was an adventure. With three friends, I set out in June with a car, a tent and four sleeping bags. Our first stop was Chicago, where we earned some spending money doing office jobs for my father's trucking company. Our only subsequent obligation was a wedding in Omaha, Nebraska, in early August. We traveled across the Midwest, into the west, up to western Canada, down the west coast and back across the country. We met many people, and had a wonderful time. Even the night we spent in the Butte, Montana, jail was interesting. (We had befriended a boxer from Monterey, California, or so we thought. He was staying next to our campsite. When we returned from town, he and our sleeping bags had all disappeared. We went to the police station to report the theft, and the police kindly offered us a place to stay for the night.) Three of us had been math majors, and as we drove we studied the recently published book "Beat the Dealer," by Edward Thorp. From this book we learned how to play a winning game of blackjack and also how to spot a stacked deck of cards. We discovered such a stacked deck in a well-known Las Vegas casino and proudly told the dealer. He called the pit boss, who was very friendly and asked us accompany him to another room. There he became less friendly, and informed us, "If I ever see any of you in here again, no one else will ever see you after that." We left. I was not winning much money anyway. The other book I remember reading while traveling during the summer of 1968 was Jim Watson's autobiography, "The Double Helix."

At the end of that summer, I returned to Chicago in time to experience the Democratic National Convention, where the peace platform calling for an end to the Vietnam War was defeated and the pro-peace protestors were considered enemies of the country and in particular of Mayor Richard J. Daley's Chicago. Meeting Alan Ginsberg, standing face to face with the National Guardsmen, trying to help slow down the crowds running from the tear gas, watching the Chicago police beating innocent people, escaping by hiding in the back seat of a car driven by a resident of a nearby black ghetto through a neighborhood that on other occasions I might have feared entering – there are many images and memories engraved in my mind. These days still make me ashamed of my city and my country.

GRADUATE SCHOOL AT THE HARVARD BIO LABS

In September, 1968, I entered graduate school in the Harvard Department of Biology, located at The Biological Laboratories, or Bio Labs, in Cambridge. I felt like a fish out of water. Everyone else seemed vastly more prepared, more knowledgeable about biochemistry and biology and more engaged. Because of my limited background, I was assigned mostly to undergraduate classes. In a cell biology course taught by Keith Porter and Dick McIntosh, the first exam had a question involving ribosomes. I had no idea what a ribosome was. In Konrad Bloch's biochemistry course, one of the few graduate courses I was taking, we focused every day on how electrons moved in the series of biochemical reactions of the Krebs cycle. I had never heard of the Krebs cycle. In a genetics seminar course, I was simply lost. The experimental approach to problem solving was completely unlike the way I had been trained to think as a student of mathematics. I was a teaching assistant in an introductory biology course taught by Lynn Riddiford and Carroll Williams. For the dissection laboratory, the students had to choose between rats and lobsters. I encouraged them all to study lobsters, although I had no idea what was inside either a rat or a lobster, because after class the TAs were allowed to cook and eat the dissected lobsters. I decided that if at the end of the first year my understanding of biology had not improved substantially, I would leave and do something else. However, my comprehension increased, and I stayed.

My advisor was Matt Meselson, and I was supposed to begin working in his laboratory my first summer, in 1969. However, I had long planned and was very excited about spending that summer in Europe. The only foreign country I had ever visited was Canada – once during my trip the summer before and once during my freshman year at MIT when I had rather cold experience hitch-hiking to Montreal from Boston in January – and foreign travel appealed to me enormously. No one told me that in graduate school you were supposed to work in the laboratory over the summer. Matt was incredulous that I planned to be away during the period my professional training was supposed to begin. Still, I went, and had a wonderful time exploring new places, foods and peoples. The only downside was my continuing problem with asthma, which left me in bed for days at a time in Paris, Milan, Malaga and on the island of Skiathos in Greece, from which I had to be carried on a stretcher through the village to get to a boat back to the mainland and to a hospital in Volos, where I stayed for some time.

In the fall, I continued with my courses and began research in Matt's laboratory. I had no idea what I was doing, and had to ask a fellow graduate student, Patricia Foster, to show me how to use a pipette. Pat and I had gotten to know each other as teaching assistants in the introductory biology course, for which she and I had been exiled together to the cold room to grind spinach for chloroplast preparations. Pat very much educated me about and encouraged me to persevere in biology. We became a couple in 1970, and lived together for the next 13 years.

Matt Meselson is a great scientist, but he was more focused on the crucially important issues of chemical and biological warfare than on his laboratory. Matt was spending much of his time in Washington, D.C. as an advisor to the U.S. government. I had to be in an alternative environment to acquire the training I needed. I considered changing schools. I was intrigued about the idea of living in Europe and thought about a number of European laboratories. However, my cousin Ed Brody (eight years older than me and an older brother of Harvey, of Chicago little league baseball fame) convinced me that staying at Harvard was a better idea. It was one of a number of times Ed had

Me, at the Harvard Biological Laboratories (1973).

a major influence on my life. Ed is a physician-scientist who trained at the University of Chicago, did postdoctoral work in Geneva, Switzerland, and then ran a research laboratory in Paris, France for many years. It was Ed who obtained for me the fruit flies I used in my high school science fair project.

I talked with Wally Gilbert and Jim Watson about transferring from Matt's laboratory into theirs. Jim had recently become Director of the Cold Spring Harbor Laboratory on Long Island, New York, and he suggested I go to the annual Cold Spring Harbor Symposium to be held in June, 1970, to become exposed to the field of molecular biology in a professional way. The meeting was focused on transcription, and was very exciting, with cutting edge research being presented in a way that made me want to know what was coming next. At the end of the meeting, still unsure if I was going to be accepted to join Jim's laboratory, I asked him what I should do. Jim told me to return to Cambridge and talk with Klaus Weber, another Harvard professor who with Jim and Wally ran a joint laboratory, about a project. I gathered his answer was yes.

The Watson-Gilbert-Weber laboratory at Harvard offered an unparalleled training experience. The three of them were all exceptional scientists, and they were highly synergistic in their approaches and talents. Jim has a superb biological intuition, and it was he who defined the problems most of the students embarked upon. Wally has profound critical abilities, and could see the flaws in any experimental design or interpretation. Klaus has magic fingers, and could devise ways – old or new – to make any experiment work. It could not have been more stimulating, or more challenging.

My thesis project derived from Jim's interests in transcriptional regulation. He had written in his 1970 edition of "Molecular Biology of the Gene" about how development was likely to involve differential gene expression. With this in mind, Jim had focused a part of his laboratory on RNA polymerase, the enzyme responsible for synthesizing RNA from DNA. Jim's laboratory had shown that the enzyme consisted of two major components, a multi-protein complex that did the work of transcribing RNA and a more loosely associated protein that conferred specificity, i.e. that caused the enzyme to begin at specific sites along the DNA. This specificity factor they named sigma, and Jim's model for development was that a sequence of sigma factors defined a temporal series of distinct transcriptional products. The experimental basis for this model was some recent observations made in the laboratory concerning *E. coli* phage T4. These studies suggested that the bacterial RNA polymerase with a bacterial sigma factor transcribes a first set of T4-specifc RNAs, called "pre-early"; one of the pre-early genes encodes a viral-specific sigma factor, which then reads the "early" genes; and finally, a second T4 sigma factor encoded by one of the early genes reads the "late" genes. In this way, a series of three sigma factors defines three distinct developmental phases of T4 gene expression. However, no such factors had actually been isolated, and the evidence for this model was rather circumstantial. My project was to identify T4-specific sigma factors.

To begin, I purified *E. coli* RNA polymerase from bacterial cells infected with T4 and examined the subunit composition of the enzyme. I worked with the expert, constant and generous guidance of Klaus Weber and graduate student Chris Goff. (Chris was studying another change in the T4-modified *E. coli* RNA polymerase.) Indeed, the host RNA polymerase had a number of new subunits, as had been recently reported by Audrey Stevens at the Oak Ridge National Laboratory. In a cleverly designed experiment (the clever design was that of a postdoctoral scientist working next door, Jeff Roberts), I was able to show that one of these subunits was the direct product of T4 gene 33. Other studies had indicated that gene 33 regulated the transition from the early to the late stage of T4 development, so this result was quite exciting and led to my first publication, in *Nature New Biology*. I was very proud of this achievement. However, my enthusiasm was dampened a bit by a conversation with Larry Gold, a T4 researcher at the University of Colorado in Boulder. Larry knew how hard I had worked to prove that this small RNA polymerase-binding protein was the product of gene 33 and also knew the data that had suggested this was likely to be the case. Was it worth, he asked, working so hard to prove something that was clearly very likely to be true? Could I not have been better spending my time to find something that was unexpected, rather than something that was expected? Larry's questions were provocative – I'm still not sure I know the best answer – and I have thought about them often in designing further research projects.

Most of my days as a graduate student were spent in a cold room at 4°C, purifying RNA polymerase. After some time, I devised a rapid method for isolating small quantities of enzyme based upon precipitating RNA polymerase with antibody that had kindly been raised by Chris Goff. (Although I had tried, I proved to be too allergic to inoculate and bleed the rabbits myself.) Then I could work in my laboratory room at a normal temperature. To visualize changes in the enzyme, I used radioactivity, and routinely did bench-top experiments in the open laboratory involving 25 millicuries of radioactive phosphate or sulfate. I also worked with comparably high levels of radioactive iodine. Such experiments today would have to be done under special conditions, but during the early 1970s, scientists were much less cautious. After some years, I officially became Wally's student, as Jim left Harvard to spend full-time at the Cold Spring Harbor Laboratory.

I published four papers as a graduate student, all involving T4-induced modifications of the *E. coli* RNA polymerase. I was the sole author on all four publications. Both Jim and Wally put their names only on papers to which they felt they had made major and direct contributions. These papers were assembled into my Ph.D. thesis, which included an introductory chapter focused not on the biochemistry of RNA polymerase but rather on the biology of phage T4. I liked thinking about T4 as an organism, just as more recently I have enjoyed thinking about all aspects of the biology of the nematode *C. elegans*.

Life at Harvard was intense, which suited me well. Work started early in the morning and ended late at night. Students in the group were highly independent, relying more on other students and postdoctoral researchers than on faculty for input. "Sink-or-swim" seemed to be the prevailing attitude. If you managed to swim, you really learned to do science. However, there were a number of students who sank who I believed had outstanding potential. We had three group meetings a week, over lunch on Mondays, Wednesdays and Fridays. For each session, one student or postdoc presented his/her most recent findings. These were serious times, as an audience of Jim Watson, Wally Gilbert, Klaus Weber and often Mark Ptashne and David Dressler (two other Harvard faculty members) left little uncritiqued. Giving my first public talk, at a phage meeting at Cold Spring Harbor, was a very gentle experience by comparison. If you could survive a Harvard group meeting, you could survive anywhere. In preparing for these group meetings, I acquired a habit that I have continued to this day – writing a complete text of my presentation. I find this practice helps me organize my thoughts and time my talk, gives me an aide for those moments when I go blank on-stage and allows me to prepare a related talk at a later date very easily.

Some years into my graduate studies, our Harvard group meetings acquired a name. One of the underground Boston newspapers, either the Phoenix or the Real Paper, did a story about Wally Gilbert and entitled it "Stalking the Secret of Life." Thereafter, our group meetings were always labeled "Stalking the Secret of Life: Part 247," "Stalking the Secret of Life: Part 248," etc. I learned an enormous amount as a Ph.D. student, both about how to do and how not to do experimental biology. I came away with two beliefs that have driven my research ever since. First, do the do-able; working on an important but intractable problem would not suit me. Second, engraved in me from Jim Watson: since it is no harder to work on a problem that is important than on one that is not important, always choose the former.

Studying the bacterial virus T4 introduced me to a community of interactive and cooperative scientists. During one visit to Chicago to see my family, I went to the University of Chicago to talk with a fellow T4 researcher, who kindly gave me a set of mutant strains I needed. My mother supplied a small mayonnaise tub for me to use to transport the glass vials. I carried the tub in my hand as I boarded the plane at O'Hare Field for my return to Boston. The woman sitting to my right was curious, and asked what I was carrying. "Oh, nothing important," I said, "Just some viruses." This was an error. "VIRUSES!!" she yelled. After considerable subsequent conversation, the viruses flew to Boston in the cockpit, and the woman to my right said not another word to me the entire flight. On another occasion, I visited biochemist Ray Gesteland at the Cold Spring Harbor Laboratory to ask for his advice about an experiment I had planned. As I spoke with Ray, he was emptying rack after rack of test tubes from a large refrigerator and pouring the contents down the sink. "Ray," I finally said. "What are you doing?" "Oh," he replied, almost gleefully. "These are my last six months of experiments. They didn't work, so I'm going to start something new." To me, six months literally down the drain seemed like a disaster. To Ray, they offered an opportunity to explore some exciting new idea. Ray was a scientist, and I still had a lot to learn.

As a graduate student, I lived in north Cambridge, sharing the top two floors of a house with my friend Joe Schwarz and a cast of others. After a year, Pat Foster also moved in. Our roommates were interesting and varied and included graduate students in biology and astrophysics, people who worked at a great diversity of occupations and people who did not work at all. At one point, both the front and back porches were rented, someone was sleeping on the living room couch and a total of 14 people (including my sister Carol) called our five-bedroom house their home. For a while, we also had a German Shepherd and, briefly, her 11 pups. We took turns cooking. Because I had the only car, I did the shopping. We joined a food co-operative and when handed a form that asked for "Name," I wrote "Bob." After that, our house became known as a commune called "Bob."

I began to think about what I wanted to do after completing my Ph.D. I had a strong desire to spend some time in Europe, and visited a number of laboratories there. Jim Watson thought I should go to Stanford – there was a regular exchange of graduate students and postdocs between Harvard and Stanford at that time. Klaus Weber had an alternative thought and suggested that I talk with Sam Ward, who had just moved to the Harvard Medical School from Sydney Brenner's group in Cambridge, England. Sydney and Sam had been using the nematode *Caenorhabditis elegans*, an organism that had been little studied previously, to pursue problems in neurobiology. I had long been intrigued by the nervous system, in particular by complex issues, like the mechanistic basis of learning, memory and consciousness. As an undergraduate I had taken a course given by Jerry Lettvin entitled "The Biological Basis of Perception and Knowledge," which had whetted my appetite but supplied few answers. As a graduate student, for a course about protein synthesis taught by John Hershey, I had written a term paper entitled "A Research Proposal on Protein Synthesis and Learning." Sam told me about Sydney's efforts to reconstruct the worm's nervous system from serial section electron micrographs, promising to reveal the complete wiring diagram of a nervous system for the first time. *C. elegans* was eminently suited for genetic analysis, and I had become enamored with the power of genetics from my studies of T4. After a number of conversations with Sam, I wrote Sydney, and asked if I could join his laboratory. My expressed interests were developing methods for studying *C. elegans* at the molecular level and using genetics to analyze memory and learning. His response was "As far as I know, all attempts to show learning in nematodes have failed," but that I was free to choose my research project when I arrived. After some further correspondence, he wrote, "Go ahead and apply for fellowships." This meant "yes."

There were a number of other factors that influenced my decision to join Sydney's laboratory. First, Chris Goff, with whom I had worked closely, had gone there and raved about the place. Second, I wrote Ed Brody, in Paris, and asked him for his views of "Brenner and his nematodes," noting that the phrase "sounds like a new rock band." Ed was encouraging. Third, Pat was delighted at the prospect of living in England. Cambridge indeed was a great place to live, and the Laboratory of Molecular Biology a fantastic place to do science.

When I told Jim Watson that I planned to go to Cambridge, England, to study the neurobiology of *C. elegans* he asked me if I knew anything about neurobiology. I had to admit that I did not. Jim then suggested an immersion education – three consecutive Cold Spring Harbor summer courses in neurobiology. I enrolled in and took three such courses during the summer of 1974: An Introduction to Neurobiology, taught by John Nicholls; Experimental Methods in Electrophysiology, taught by Enrico Stefani and Dante Chiarandini; and The Neurobiology of *Drosophila*, taught by Bill Pak. Each course was intense and stimulating. I learned an enormous amount, and made many friends. Cold Spring Harbor courses have a reputation for being both exhilarating and exhausting, and I have been told that I am the only person who has ever taken three in a single summer.

ENGLAND AND WORMS

Sydney Brenner was my fourth official research supervisor, after Bob Solow, Jim Watson and Wally Gilbert. Amazingly, they, and I, now all have Nobel Prizes.

Pat and I arrived in England just before Guy Fawkes Day (November 5) in 1974. Chris and Eleanor Goff, friends from our Harvard days, kindly provided our initial lodging. My first assignment for laboratory space was a two-foot wide area of benchtop in a room in which biochemists on each side of me were labeling tRNAs using vast quantities of 32P-labelled phosphate. As I began my studies of *C. elegans*, I wondered whether the worms, or I, might acquire mutations as a consequence of the nearby radioactivity. Soon I moved down the hall, into a room that I shared for the rest of my stay with a number of scientists, including John Sulston, a young staff member in Sydney's group.

I had received a fellowship from the Muscular Dystrophy Association of America. The fellowship was for a much greater amount of money than Pat and I needed to live in England, and I deposited half of the funds I received into a Swiss bank account, believing this would guard us against an unstable British economy and also against currency fluctuations between Europe and the U.S. This account remains untouched. My fellowship application had focused on an analysis of the chemosensory nervous system of *C. elegans*. However, studies of nematode chemosensation had been begun by Sam Ward, and I decided I should do something different. My interests and my MDA support both drove me to want to study aspects of worm biology that might relate to human neuromuscular disorders. In particular, just before I left the U.S. I had a moving conversation with a close friend of my parents who was dying of ALS. I did not know anything about the disorder, but could see that it was terrible. Her last words to me were, "I know it is too late for me, but please try to do something for others who suffer from this horrible disease." Beyond studying the worm's musculature and/or nervous system, I had no idea how to try to do so, and in any event it was the basic biology that intrigued me.

Sydney suggested I study muscle cell growth, in particular that I try to determine how individual muscle cells add new myofilaments as they increase in size. The problem sounded interesting, and remains unsolved. However, I had begun conversations with John Sulston and found a topic that excited me more. John had recently found that he could directly observe cell divisions in living *C. elegans* larvae and in this way determine aspects of the worm's cell lineage, the pattern of cell divisions and cell fates that occurs as a multicellular organism develops from a single-celled egg. So far, he had examined only the development of part of the nervous system. John and I decided that together we would explore more of the worm's cell lineage, with the goal of determining the complete pattern of cell divisions that generates the adult animal. I began by examining the musculature, in part because of my MDA fellowship. John had already noted that the number of muscle cells in the main body musculature increases as the animal develops from a newly hatched larva. My first goal was to determine the precise number of body muscle cells in the young animal and the cell lineage responsible for adding additional muscle cells during larval development.

Counting muscle cells proved more challenging than I had anticipated, in part because using Nomarski optics one visualizes nuclei, and the difference between the nucleus of a muscle cell and that of certain non-muscle cells was not always obvious. More importantly, the worm was not designed the way I thought it should have been. The animal is shaped like a tube, with four quadrants of muscle: dorsal-right, dorsal-left, ventral-right and ventral-left. It never occurred to me that these four quadrants might contain differing numbers of muscle cells. So I counted, recounted, re-recounted and finally asked John to count, too. The upshot was that the superficially radially symmetric young animal contains 21 muscle cells in each dorsal quadrant, 20 in the ventral-right quadrant and 19 in the ventral-left quadrant. By tracing the postembryonic cell lineages involved in muscle development, I discovered that to this number is added an additional 14 muscle cells, three in each dorsal quadrant and four in each ventral quadrant. I learned that preconceived notions in biology can be very misleading. Only observation and experimentation can reveal biological truths.

Life as a postdoctoral researcher at the Laboratory of Molecular Biology was exceedingly stimulating, great fun and involved far more conversation than did life at the Harvard Bio Labs. Morning coffee and afternoon tea on the second floor, where the Division of Cell Biology was housed, were key, whether or not one drank coffee or tea. These breaks were opportunities to talk, to think, to listen and to learn. Sitting in the tea room with Sydney Brenner after lunch never failed to be both amusing and interesting. Sydney has an incredible wit, an amazing breadth of knowledge and always enjoys a good conversation. Latenight science – and sleep – were often endangered by Sydney. A number of times when I was working at 2 or 3 a.m. and desperate for a cup of tea, I would go to the tea room only to find myself soon joined by Sydney, attracted by the rattle of the spoon in the tea cup. In Sydney's (but not my) view, 2 a.m. was an excellent time to talk! I learned to stir silently.

Lunch was an especially good time for scientific interaction. Everyone in the building went upstairs to the top floor to a cafeteria run by Gisela Perutz, the wife of the famed structural biologist Max Perutz. Here one would pick up lunch – bangers, spuds, faggots, toads-in-the-hole or whatever (I quickly saw that despite the U.S. and England ostensibly sharing a common language, much of the English spoken in England was completely foreign to me) – and sit at whatever table had an open seat and talk about whatever science came to mind. I had the opportunity to meet many outstanding scientists. One of these lunches, with fellow worm researchers Jonathan Hodgkin and John White, led to an experiment by Jonathan that revealed a fundamentally new aspect of *C. elegans* sex determination. Monday nights were special. Dinner at the University of Cambridge's Clare Hall, hosted by Tim Hunt and often led in conversation by Francis Crick, was followed by a research seminar, in which Peter Lawrence and/or Michael Ashburner, two *Drosophila* developmental geneticists, would invariably ask the speaker why he or she had bothered to do whatever it was that he or she had done. Friday evenings involved a short walk to the Frank Lee Center, for a pint (or more) of England's best brew.

The worm group was very much the center of my scientific and much of my personal life. I worked most closely with John Sulston, who became a mentor, a friend and in some ways a hero. I had the impression that John could do anything. He was warm, personable, unassuming and completely dedicated. Often after his bicycle ride to the laboratory (John bicycled to reduce the pollution that results from driving cars), insects and other debris would fall from his long beard onto his samples. John argued passionately that research biologists should receive salaries much lower than those of trash collectors, because the former have so much more fun. I made many other friends, including Jonathan Hodgkin and John White. Toward the end of my stay, I had the opportunity to renew my high-school friendship with Marty Chalfie for the months that we overlapped. I developed scientific collaborations with Jonathan, John and Marty. The collaboration with Jonathan was particularly instructive to me. He had discovered a class of *C. elegans* mutants abnormal in the segregation of the sex chromosome during meiosis, and I had isolated a number of additional mutants of this sort. We decided to characterize our mutants together. We sat down, agreed upon a set of experiments, and assigned one of us to do each experiment. However, neither of us had ever collaborated in such a way before, and we each felt that we would trust our own data more than data from someone else. When we met again to examine our collected data, Jonathan had done all of his experiments as well as all of mine, and I had done all of mine as well as all of his. Luckily, our results agreed. The upshot was a paper in which we had counted precisely 183,001 worms for the published data alone. This experience made it easier for me later to trust the data of my students without having to repeat everything they did. (Even so, in the early days of my laboratory, there were some experiments I did repeat, privately, just to be sure.)

Pat and I lived first in Cambridge, not far from the train station, and later in the village of Great Shelford. Going to and from the lab meant driving past open farmland, which was highly refreshing after having lived my previous 27 years in Chicago and Boston. I found the whole area, with its charming villages, thatched cottages, beautiful flower gardens and comparatively slow pace, extremely appealing. The nearby village of Grantchester, with The Green Man pub and a lovely tea garden, was a particular favorite. One could punt to Grantchester from the center of Cambridge and stop for a pub lunch or just a pint of beer. For food, Indian cuisine seemed vastly better than in the U.S., and the Curry Queen in Cambridge was a common target. Pat, who had left Harvard to work in what was a very early version of a biotechnology company (she was the Biology Department), resumed her studies as a Ph.D. student at the University of Cambridge, which gave us both the opportunity to partake of life at Wolfson College. Pat and I used Cambridge as a base for travel, and visited most of Britain and much of Europe during my three years in Sydney's laboratory. I considered staying in England, but decided that a return to the U.S. would allow me greater flexibility in developing an independent research program.

My job at MIT came about through a misunderstanding. Sydney had visited Brandeis University, and he suggested that I apply for a faculty position there. I did not want to leave England yet, but I could not pass up such an opportunity. I sent my *curriculum vitae* and a statement of research interests, but no publications, as I had not yet published any of my postdoctoral studies. I decided that if there was a possibility that I might be offered a job, I should also apply to other places and see what the options might be. I came to the U.S. for a six-week visit and gave seminars at 11 institutions. My first visit was to Cornell University in Ithaca, N.Y. I was lucky to be offered a job while I was still there, just before I flew to Boston to visit Brandeis and MIT. This offer made my subsequent visits very relaxed. To my surprise, when I arrived at Brandeis I was told that although they wanted to meet me, no faculty position was currently available. Thus, I need not have started sending out job applications at all. Not long after my return to England I was contacted by Boris Magasanik, Chair of the MIT Department of Biology, and offered a faculty position. I accepted.

MY LABORATORY AT MIT

I moved back to Boston in January, 1978, between the two huge snowstorms that blanketed the northeastern U.S. that month. My friends Fred Ausubel and Stephanie Bird kindly offered me a spare room in their home on Irving St. in Cambridge, near the Harvard Bio Labs. They also let me use their cross country skies to get to and from MIT, skiing over barely detectable mounds of snow that had cars beneath them. Pat joined me in April, and we rented an apartment in north Cambridge.

Setting up the new laboratory was a busy time. I had arrived with more than 500 strains of *C. elegans*, the transport of which had proved slightly problematic: to import nematodes, I needed permission from the U.S. Department of Agriculture. However, the Department of Agriculture gave import approval only for parasitic nematodes, and *C. elegans* was not a parasite.

Therefore, they could not grant my request. It took a number of letters back and forth before they finally agreed that they could not be sure that *C. elegans* was not a parasite and thus could approve my importing a potential parasite. I also had arranged to avoid having to subject the living animals to X-rays at Heathrow Airport in London. The airline noted that this permission had been granted in my reservation information in their computer system. When I checked in, the woman behind the counter became wide-eyed as she looked at the computer screen, and said, "The computer says you have worms. Can I see them?" I explained that they were too small to be seen and that they were packaged in sealed boxes in any event. She let the opportunity pass.

Every day in my new laboratory I transferred worms, establishing and then freezing each of my many mutant strains. In April I was joined by Nancy Tsung, whom I hired as a technician. Nancy could not get over the fact that a professor could be so young, and for some years referred to me as her young boss. Nancy stayed in the laboratory for over 21 years, and she provided crucial technical and emotional support for projects and people, respectively. Soon thereafter, my first two graduate students joined the laboratory, Iva Greenwald and Chip Ferguson. Each proved to have a deep interest in biology and a natural flare for genetics, and the new lab was off to a very strong start in the areas of the genetic analysis of *C. elegans* development and behavior. Chip focused on the genetics of intercellular signaling, while Iva studied a muscle mutant with what then appeared to be unusual genetic and behavioral properties. Iva discovered that although this mutant had a severe muscular disorder, the inactivation of the gene responsible had no discernable effect whatsoever. This finding led to my first publication as a mentor, a paper with Iva in the journal Genetics in 1980.

Next, graduate students Paul Sternberg, Bill Fixsen and Carol Trent and postdoc Victor Ambros entered the lab. Paul started by studying the cell lineage of another nematodes species, *Panagrellus redivivus*, with an eye toward an analysis of the genetic basis of evolutionary change at the level of cell lineage. Bill studied the genetic control of cell migration. Carol analyzed the genetic basis of the behavior of egg laying. Victor pioneered studies of genes that control developmental timing.

The diversity of problems – intercellular signaling, muscle, the genetics of evolution, egg laying and developmental timing – seems unusual and perhaps dangerous for a beginning faculty member. "No focus," would be the criticism made by some today. However, these projects were highly coherent, in history, in philosophy and in experimentation. Each had derived directly from the studies of cell lineage and cell lineage genetics I had begun in collaboration with John Sulston in England. Most of my time examining the *C. elegans* cell lineage was spent observing the development of the musculature, including the muscles used for egg laying, and of the vulva, also used for egg laying. In large part for this reason, in seeking cell lineage mutants, I primarily sought those abnormal in egg laying. Amongst those mutants were all of those to be analyzed in our initial explorations into intercellular signaling, muscle, developmental timing and the behavior of egg laying. The comparative evolutionary project had a similar rationale – unlike *C. elegans, Panagrellus* does not lay eggs but rather releases hatched larvae; furthermore, the vulva of *Pangrellus* is displaced posteriorly compared to that of *C. elegans.* The cell lineages of the egg-laying systems of the two species seemed very likely to be different.

In addition, intellectually and experimentally, there was a single driving theme behind almost every project: use analytic genetics to define the genes and genetic pathways responsible for each biological phenomenon. How you design a mutant hunt, isolate additional mutations in a known gene, define a null phenotype, and order genes into a pathway is the same, whatever the problem being addressed. Each member of the laboratory could help each other member in thinking about such issues. There were two other factors that drove me to initiate so many distinct projects. First, given my experiences at Harvard and the LMB, I believed that one of the most important aspects of training a young scientist is to give that person the freedom to pursue his or her discoveries. Discovery leads to excitement, commitment, and fun. Elbow room is crucial. Training has always been one of my major goals, with the belief that training outstanding scientists will of necessity generate outstanding science. (The converse statement – that generating outstanding science will necessarily train outstanding scientists – in my view, is not necessarily the case.) In addition, the training environment is enhanced by exposure to a breadth of biological problems. A second factor that drove me to initiate so many distinct projects was, as I noted above, the fact that I am fascinated by the biology of the organism. The diversity of problems helped satisfy my personal curiosity.

My philosophy for the lab, then as now, was that we should be a community of scientists, with each graduate student and postdoctoral researcher focused on a biological problem of interest. Lab members should be independent, but highly interactive. Postdocs should be free to continue their projects independently after leaving the laboratory.

The next two young scientists to join my laboratory were graduate students Eun-chung (Joan) Park and Hilary Ellis. Again, I advised each to begin projects descended from but also distinct from projects already underway. Joan took a genetic approach that derived from the studies of Iva Greenwald to estimate the number of genes in *C. elegans.* Joan found that animals have a large number of genes that when inactivated have no obvious consequence to the organism. This finding was not broadly known and when rediscovered some years later in the field of mammalian genetics, as a consequence of the study of the first mouse "knock-out" mutants, was generally regarded as a great surprise. Hilary Ellis embarked on the study of the genetics of programmed cell death.

Postdoctoral researcher Gary Ruvkun entered the laboratory slightly later. Gary and I shared a driving desire to find a way to clone the many genes our laboratory was defining by methods of classical genetics. We devised more than a dozen distinct possible approaches, involving a variety of molecular techniques. Gary was a Junior Fellow at Harvard University and his Harvard sponsor was Wally Gilbert. Gary was able to do many more molecular experiments using Wally's funds than could have been supported by my rather limited budget. Gary, Wally and I published one paper together, my only publication with my official Ph.D. advisor. Gary introduced current methods of molecular biology to the laboratory. He succeeded in cloning one of the genes that Victor Ambros had shown to be key in the control of developmental timing.

The MIT lab grew in number, but not in physical size. Our only laboratory room, with seven lab benches, soon housed 11 scientists, three of whom worked at small desks where previously there had been a single –70°C freezer and one of whom sat at the end of a high bench with a cabinet door open so he would have some place to put his legs. I asked the Department Chair, Gene Brown, for more space, and he instantly provided it, moving his own laboratory, which was across the hall from mine, to do so. David Botstein, whose laboratory was located one floor upstairs, proved to be my major mentor on issues of science and non-science both. Lunches were a special treat, as varying groups of faculty would assemble in the faculty lounge at precisely noon (when Gene Brown began his lunch) and sit and talk. Collected wisdom concerning science and life was shared by all, and I had the good fortune to be able to learn from many of my senior colleagues, including Salvador Luria. Salva was a humanist, an independent thinker and a scholar, and I was struck by the fact that he offered the first-year graduate students the opportunity to get together with him once a week to discuss literature, primary fiction, because he believed that a single-minded focus on science was intellectually stifling.

OTHER PROFESSIONAL ACTIVITIES

My scientific pursuits have led to many opportunities and responsibilities beyond those of simply doing research. For example, as a beginning graduate student it never occurred to me that the life of a scientist could involve so much travel, something that I have always loved. In the course of presenting lectures, attending meetings, serving on advisory committees and teaching courses, I have visited Alaska, most European countries, Russia, Israel, Egypt, India, Japan, Korea, China and Tibet. (Out of interest, and without scientific excuse, I have also traveled to Peru and Senegal.) I have been an advisor to many national and international organizations, including the U.S. National Cancer Institute; the National Human Genome Research Institute (the U.S. sponsor of the Human Genome Project); the Sanger Centre (the British arm of the Human Genome Project); and the Tropical Disease Program of the World Health Organization, which amongst other interests is concerned with diseases caused by parasitic nematodes, relatives of *C. elegans.*

I have also been very involved with scientific societies. I served as the President of the Genetics Society of America in 1995 and have been a member of the Public Policy Committee of the American Society for Cell Biology and of the Joint Steering Committee for Public Policy, an organization that advocates for biomedical research funding from the U.S. government. I feel strongly that the support of biomedical research is an important investment for society as a whole and that both the Congress and the public have the right to know what is being done with their funds. We, as scientists, have a responsibility both to advocate and to educate. Toward this end, in 1997 I presented a briefing of the Congressional Biomedical Research Caucus. I was hosted by Massachusetts Representative Joe Kennedy, and my lecture was entitled "All Creatures Great and Small: The Universality of Genes." My theme, based in part upon findings of my research group, was that studies of simple non-human organisms –such as a yeast used for baking bread or making beer, a tiny fruit fly or a microscopic worm – can provide crucial breakthroughs important for the understanding, prevention and cure of human disease.

My laboratory and our studies of *C. elegans* have progressed in new directions and new dimensions. Our current interests include cell lineage and cell fate; programmed cell death; signal transduction; morphogenesis, micro RNAs; nervous system development; behavior; and the modulation of behavior by the environment and by experience. This last topic is essentially the one that drew me to *C. elegans* and was the interest that I wrote about when I first contacted Sydney Brenner – the genetic analysis of memory and learning. In addition, I have become increasingly interested in aspects of human disease. Much of the reason is that I have been repeatedly intrigued by the striking molecular genetic conservation between processes we have analyzed in *C. elegans* and those involved in human biology and human disease. The basic science has driven me toward human disease. My active involvement started in 1979, when Ed Kravitz, from the Department of Neurobiology at Harvard Medical School, invited me to a meeting about cell death. The meeting was sponsored by the Hereditary Disease Foundation, which is focused on Huntington's Disease, and was until very recently run by Nancy Wexler and Allan Tobin, both of whom became good friends. This meeting opened my eyes to the possibility that our studies of programmed cell death in *C. elegans* might prove relevant to a variety of human neurological disorders and also indicated that discussions of some of the basic principles of analytic genetics used to study simple organisms like *C. elegans* might be helpful to those working on human genetic diseases. My involvement with the Hereditary Disease Foundation continues to this day.

In 1986, when my father was diagnosed as having amyotrophic lateral sclerosis, or ALS, he came to Boston and was seen by MGH neurologist Bob Brown. Bob proved to be a highly supportive and sympathetic physician. Bob and I talked and soon agreed to establish a collaboration to try to better understand ALS though genetics. This collaboration has involved my having a small group of people working in Bob Brown's laboratory at MGH. This effort, in conjunction with the work of many others from around the world, led to the discovery in 1993 that one gene responsible for familial ALS encodes the enzyme copper-zinc superoxide dismutase.

During this period I began thinking that I might want to deepen my knowledge of medicine by attending medical school, but another friend, Mark

Fishman, who later became Chief of Cardiology at MGH and is now the Global Head of Research for the pharmaceutical company Novartis, persuaded me not to do so. He told me that what I would learn in medical school I could more efficiently learn by reading text books, and that to understand medicine I should simply follow him while he did rounds in the clinic. So for brief periods during two summers, I did so.

I have also been introduced to aspects of clinical medicine through my involvement with biotechnology companies. Although for some time I was adamant about being "pure," a conversation with friend and co-*C. elegans* researcher Jonathan Hodgkin changed my mind. When I told him I was about to turn down an invitation to become a company consultant, he said to me, "Don't you ever want to do anything useful?" I thought about his comment, and accepted the consultancy. I now feel that my three hats, under which I supervise basic research concerning *C. elegans* at MIT, help with medical research concerning ALS at MGH and advise companies how to apply basic knowledge of biology and medicine to drug development, together both foster my continuing education and synergistically enhance my efforts in all three arenas.

MY PERSONAL LIFE AND FAMILY

My scientific life would be empty were it not complemented by my personal life. Since becoming a research scientist, I have enjoyed the companionship

Me, with my wife-to-be, Martha Constantine Paton, in Nanjing, China (1991).

and support of a small number of women with whom I have had close, intense long-term relationships. All have been dedicated scientists, strongly independent and understanding of the demands of my professional obligations while working similar hours themselves to fulfill their own professional goals. They have provided me with enormous support and helped me to grow emotionally and intellectually.

Since 1991, I have shared my life with Martha Constantine Paton, now my wife. Martha studies the development of the nervous system. I first met Martha soon after we had both begun jobs as assistant professors – she was my host when I visited Princeton University to present a seminar in 1979. However, it was only years later, in 1991, that we became romantically involved. We have interests that are shared and interests that are complementary – all in all, a good match. In 1991, Martha was a professor at Yale University, and she had two sons, Joe and Chris, ages 13 and 17, respectively. Martha and I married, and I suddenly acquired a family, complete with a station wagon, two dogs, two teenage stepsons, and before long, a house in the suburbs. Joe and Chris are wonderful, and being close to them as they have grown into independent, intelligent and interesting adults has been a special pleasure for me. At present, Joe is immersed in neurobiology and is currently a graduate student at Columbia University, while Chris has had no interest in science whatsoever and spent some years working in the music industry before recently entering law school at St. John's University in New York City.

Our marriage is a partnership with love. While I do most of the cooking and laundry, Martha is responsible for preventing the house and garden from degenerating into chaos and for maintaining our extensive indoor foliage, in-

My family: from left-to-right, my wife Martha Constantine Paton, my stepson Christopher Paton, my daughter Alexandra Constantine Horvitz, and my stepson Joseph Paton (2002).

cluding our growing collection of orchids. Shopping we do together, which may not be efficient but preserves time with each other. We very much wanted to have a child together. Given our ages, we were very lucky. On September 2, 1993, when Martha and I both were 46 years old, our daughter was born. We had discussed extensively what to call her. We agreed that the last name of Horvitz was appropriate. Martha wanted her father's name, Constantine, to be continued, so we had a Constantine Horvitz. But what should her first name be? We needed something substantial – a one syllable name would not suffice if followed by Constantine Horvitz. I desired a name in memory of my father Oscar, whose Hebrew name was Asher. So we focused on the A's and O's and decided upon Alexandra: Alexandra Constantine Horvitz.

Alex's birth had a direct but little known consequence on the MIT Department of Biology. The Department was planning a new building, and I was chair of the building committee. The day we were selecting color schemes for the reading rooms, I was called out of the meeting for a phone call from Martha, who had just learned that she was pregnant with a healthy baby girl. I was elated! The reading rooms ended up with rather exuberant purple couches. Later, I telephoned my mother to tell her the news. "Mother," I said, "How would you like to be a grandmother?" After an exceedingly long pause, in which my mother was no doubt thinking about Martha and me, both 46, she replied, "How would that happen?" I could think of no answer other than the obvious, "In the usual way."

Alex has given me an unprecedented excitement and joy. Nothing delights me more than to spend time with her and to see her happy. She is a wonderful, loving, intelligent, engaging and strong-willed nine year-old girl. Alex enjoys math, puzzles, games and piano, like me (we play piano about equally well), and, like Martha, arts and crafts, swimming and animals. Alex also has talents that are not obviously derived from either of us, such as in music. Alex's and Martha's combined love for animals seems boundless, and were it not for my allergies our home would no doubt resemble a zoo. As it is, we have two dogs, a bearded dragon and two aquaria with tropical fish. Our family lived a hectic two-city life for six years, commuting between Boston and New Haven. I listened to and became a fan of books-on-tape. Our marriage and Alex's birth have completely changed my life, which is now much more chaotic, much more interesting and much more fulfilling. In September, 1999, Martha moved from Yale to MIT, and we now all live in the lovely Victorian house in Auburndale, Massachusetts, we had purchased in 1994 and then renovated in 1999 in preparation for new lives together.

THOUGHTS AND DREAMS

It is easy to look back over the years that have brought me to writing this autobiography. I have been very lucky, both professionally and personally. I have had a fantastic family, wonderful friends and the opportunity to explore and experience much of the world. I have had the pleasure of running a successful research laboratory and the greater pleasure of having helped train a large number of young scientists, many of whom will contribute to the discoveries of the future. I hope that I have helped them develop not only the technical and intellectual abilities to do research and the confidence to persevere but also an approach to research that will make their efforts fun. Messages that emerge from my life's experiences so far are less obvious. One, which I believe strongly, is that one should not be afraid to try something new: becoming an undergraduate at MIT instead of staying in the Midwest; studying biology as a graduate student after having earned degrees in mathematics and economics; and embarking upon the study of *C. elegans* at a point when nothing was published about the organism and many of the techniques of modern biology could not be applied to its analysis – all of these decisions proved to be good. What comes next for me I cannot know. I have three dreams. The first and most important is to be as good a father as I possibly can for Alexandra. The second is that some of the discoveries for which I can claim some credit, or perhaps my discoveries yet to come, will lead to applications in medicine that will alleviate human suffering. My third dream is that I will be able to use my new label of Nobel Laureate to contribute to the world in ways that will benefit society and mankind.

WORMS, LIFE AND DEATH

Nobel Lecture, December 8, 2002

by

H. ROBERT HORVITZ

Howard Hughes Medical Institute, McGovern Institute for Brain Research, Department of Biology, Massachusetts Institute of Technology, 77 Massachusetts Ave., Cambridge, MA 02139, U.S.A.

I never expected to spend most of my life studying worms. However, when it came time for me to choose an area for my postdoctoral research, I was intrigued both with the problems of neurobiology and with the approaches of genetics. Having heard that a new "genetic organism" with a remarkably simple nervous system was being explored by Sydney Brenner – the microscopic soil nematode *Caenorhabditis elegans* – I decided to join Sydney in his efforts.

THE CELL LINEAGE

After arriving at the Medical Research Council Laboratory of Molecular Biology (the "LMB") in Cambridge, England, in November, 1974, I began my studies of *C. elegans* (Figure 1) as a collaboration with John Sulston. John,

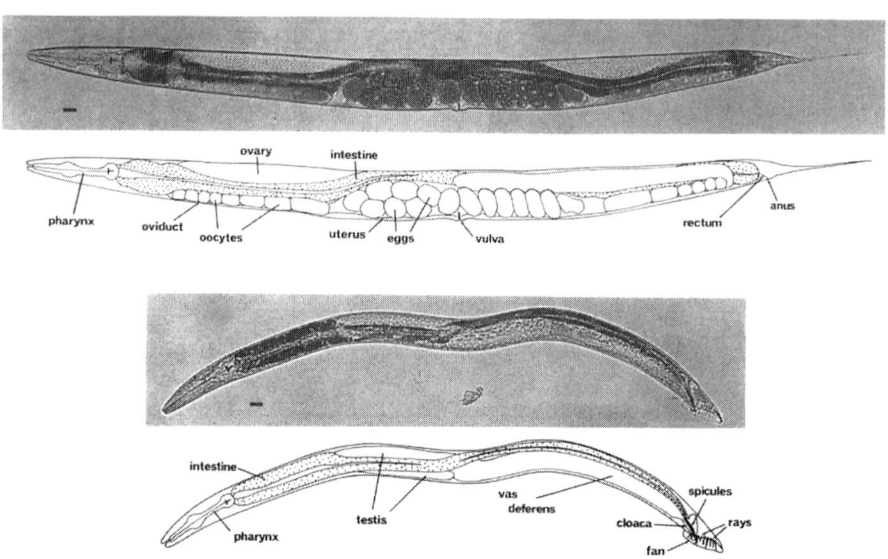

Figure 1. Caenorhabditis elegans adults. Hermaphrodite above, male below. John Sulston took these photographs, and I drew the diagrams. Bar, 20 microns. From (2).

Figur 2. John White.

trained as an organic chemist, had become a Staff Scientist in Sydney's group five years earlier. John's aim was to use his chemistry background to analyze the neurochemistry of the nematode. By the time I arrived, John had turned his attention to the problem of cell lineage, the pattern of cell divisions and cell fates that occurs as a fertilized egg generates a complex multicellular organism. John could place a newly hatched *C. elegans* larva on a glass microscope slide dabbed with a sample of the bacterium *Escherichia coli* (nematode food) and, using Nomarski differential interference contrast optics, observe individual cells within the living animal. In this way, he could follow cells as they migrated, divided and, in certain cases, died. That cells died as a normal aspect of animal development had been known by developmental biologists and neurobiologists for many years, and in 1964 Richard Lockshin and Carroll Williams had published a paper (1) in which they referred to such naturally-occurring cell death as "programmed cell death." The study of this phenomenon was later to engage a substantial proportion of my scientific efforts.

John's initial analyses of the *C. elegans* cell lineage were focused on the developing larval ventral nervous system. I found his discoveries about the relationship between cell lineage and nerve cell fate very exciting: 12 neuronal precursor cells underwent the same pattern of cell division, and descendant cells with equivalent cell lineage histories in general differentiated into the same nerve cell type. For example, the anterior daughter of the posterior daughter of each of 12 neuroblasts became motor neurons of a class called "AS" by John White (Figure 2), who with Sydney was then defining the complete anatomy and connectivity of the *C. elegans* nervous system. I asked John Sulston if I could join him in the examination of other aspects of the *C. elegans* cell lineage, and to my delight, he said "Yes."

John and I decided that it should be possible to track every cell division that occurred during larval development. With the added efforts of Judith Kimble, who as a graduate student with David Hirsh at the University of Colorado in Boulder was interested in the development of the *C. elegans* gonad, we did just that. John and I published a description of the complete nongonadal larval cell lineages in 1977 (2). The authorship of that paper was a topic of debate between John and me: we each insisted on being second author. John was rather unassuming and wanted to be second author to give me more credit and visibility. I, knowing that John had done the vast majority of the work, believed strongly that he should be first author. We were at a standstill until I had an idea. I knew that John hated writing manuscripts. I told him that I would write the paper on the condition that he would be first author. We had a deal.

Two years later, Judith Kimble and David Hirsh described the larval cell lineages of the gonad (3). John then completed the picture by tracking the pattern of cell divisions between the single-celled fertilized egg and the newly hatched larva. This achievement was far more difficult than what John, Judith and I had done previously, in part because the process of embryonic morphogenesis involves a major cellular rearrangement to generate a worm-shaped larva from what was previously a ball of cells. A simple analogy is to imagine that you are watching a bowl with hundreds of grapes, trying to keep your eye on each grape as it and many others move. John succeeded in following all 558 nuclei, and this effort (with input from Einhard Schierenberg, John White and Nichol Thomson) led to the description of the embryonic cell lineage of *C. elegans* (4). Together, these studies defined the first, and to date only, completely known cell lineage of an animal (Figure 3).

The *C. elegans* cell lineage, which is essentially invariant among individuals, presents many of the problems of developmental biology at the level of resolution of single cells. The issue now was how to proceed from description to mechanism. We discussed two general approaches. The first was based upon

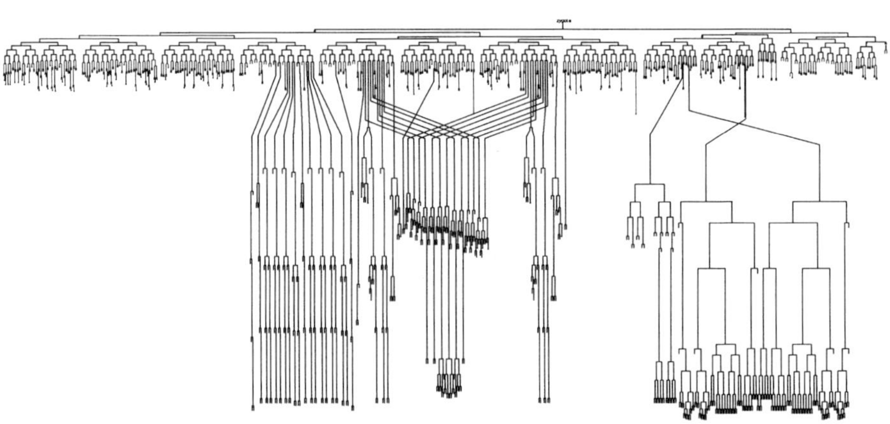

Figure 3. The complete cell lineage of *C. elegans.*

the classical methods of experimental embryology, involving the removal and/or transplantation of particular bits of developing animals. With this direction in mind, John White began pursuing a modern approach to cell removal: laser microsurgery. By focusing a laser beam though a microscope equipped for Nomarski optics, he could visualize and kill single cells. In this way, it would be possible to determine the functions of individual cells. Killing a cell in the developing embryo or larva should reveal if that cell influenced the developmental fate of another cell. The two Johns and Judith Kimble applied this technology to eliminate specific cells and analyze development (5, 6). They discovered that cell interactions play a substantial role in *C. elegans* development and that the invariance of the *C. elegans* cell lineage reflects in part the invariance of cell interactions.

CELL LINEAGE GENETICS

The second approach we considered for the analysis of the mechanisms responsible for the *C. elegans* cell lineage was genetics. Sydney had established *C. elegans* as a genetic system (7). However, we had no idea if there existed genes that had specific roles in controlling cell lineage or that controlled specific cell lineages. It was possible that mutating any gene that affected one cell division would also affect so many other divisions as to lead to an uninterpretable lethality. Even mutations that affected a single cell division could well prove to be leaky alleles of genes that act more broadly, e.g., a weak mutation in any gene involved in cell division presumably would cause a defect in the divisions of those cells most sensitive to decreases in the activity of that gene. I expressed this concern more optimistically on March 19, 1976, in my first presentation of the idea of using genetics to study the worm cell lineage, at a combined worm-fly group meeting ("Tea Talk") at the LMB (Figure 4). I

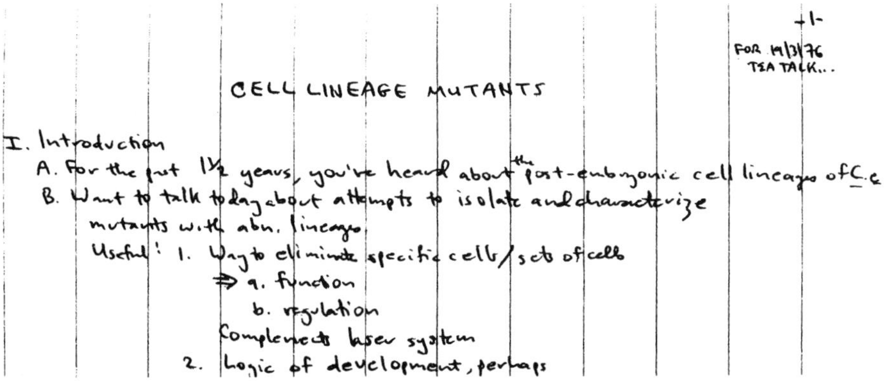

Figure 4. My notes for the beginning of a "Tea Talk" I presented on March 19, 1976, to an informal weekly gathering of the *C. elegans* and *Drosophila* researchers at the Cell Biology Division of the Medical Research Council Laboratory of Molecular Biology in Cambridge, England.

suggested that mutants with abnormal lineages could prove useful in two ways: first, they would provide a means of eliminating specific cells or sets of cells, thus "complement{ing the} laser system"; second, they could reveal aspects of the logic of development, "perhaps."

My studies of worm cell-lineage genetics began as another collaboration with John Sulston. We looked for mutant animals abnormal in specific aspects of behavior and/or morphology and then examined those mutant animals at the single-cell level for defects in cellular anatomy. If the number of cells was abnormal, we reasoned, an abnormality in the cell lineage might well be responsible. We then directly examined cell lineages in such mutants, using the same techniques we had used to define the cell lineage of the wild-type animal. Because John and I first determined larval as opposed to embryonic cell lineages, we began by seeking mutants defective in the larval cell lineages. To do so, we considered what biological features the larval cell lineages add to the newly hatched animal. A young worm and an adult worm are in general very similar. Most of the larval cell lineages are involved in sexual maturation, i.e., with the development of the gonad and of the neurons, muscles and vulval cells used for egg laying. For this reason, one approach we took was to seek mutants defective in egg laying and ask if these mutants were defective in cell lineage.

In our initial study (8, 9), John and I characterized a set of 24 cell-lineage mutants. Five of these mutants had been isolated some years earlier by Sydney Brenner in his pioneering screens for mutant worms of any sort. Over the years, my laboratory has isolated over 4,000 mutants, many of which are cell lineage mutants. In many cases the genes defined by such mutants have proved to have specific and interesting effects on the worm cell lineage.

Heterochronic mutants and the control of developmental timing

Some *C. elegans* cell lineage mutants perturb the developmental timing of specific aspects of the cell lineage. We called such mutants "heterochronic" and regard them as temporal counterparts of spatial homeotic mutants, since they transform cell fates in time rather than in space (10). The first heterochronic mutant, which defined the gene *lin-4* (*lin*, cell lineage abnormal), was isolated in Sydney Brenner's laboratory as a morphologically abnormal animal by P. Babu (personal communication). *lin-4* was characterized collaboratively by Marty Chalfie, John Sulston and me (11). (I knew Marty in high school, and after a chance encounter in which I told him about *C. elegans*, he, too, joined Sydney's laboratory as a postdoctoral researcher. Marty has focused his continuing studies of *C. elegans* on the animal's nervous system and introduced the jellyfish green fluorescent protein (GFP) as a reporter for gene expression and protein localization; GFP is now widely used in many fields of biology (12).) More detailed analyses of heterochronic genes have been performed by Victor Ambros and Gary Ruvkun, first as postdoctoral fellows in my laboratory (10, 13, 14) and later as independent researchers. Heterochronic mutations are generally of two classes, as exemplified by their effects on cell lineage: retarded mutations cause early developmental events to occur at late developmental times, and precocious mutations cause late developmental events to occur at early developmental times (Figure 5). The heterochronic genes are major regulators of developmental timing.

The Ambros and Ruvkun laboratories discovered that two of the heterochronic genes, *lin-4* and *let-7*, do not encode protein products but rather encode small (21–22 nt) RNAs (15, 16). These RNAs are the founding members of a family of RNAs, now called microRNAs, which have proved to be widespread in biology, with large families not only in *C. elegans* but also in insects, plants and mammals (17–21). MicroRNAs are currently the subject of intensive study in a rapidly growing and very exciting field.

Genes that control the generation of cell diversity

Other cell lineage mutants blocked the generation of cell diversity at specific cell divisions. For example, although the cell lineage diagram (Figure 6) looks very complicated, the logical decisions that occur in this lineage can be depicted much more simply (Figure 6A). In short, we can regard every cell in the cell lineage as having a fate. That fate may be to differentiate into a specific cell type or to divide in a specific pattern to generate a specific complement of descendant cells. At each cell division, a cell with one fate A can be said to divide to produce two daughter cells with fates B and C, where in general B and C differ both from each other and also from A. Some of the cell lineage mutants we found cause sister cells to be identical instead of different or daughter cells to be like their mothers instead of acquiring new fates (Figure 6B). If the mutation involved eliminated the activity of a gene, we could conclude that the gene is necessary to make sister cells different from each other or to make daughter cells different from mother cells, i.e., that the gene functions to generate cell diversity during development.

We have characterized a variety of such cell lineage genes genetically, developmentally and molecularly (reviewed in (22, 23)). One of the most inter-

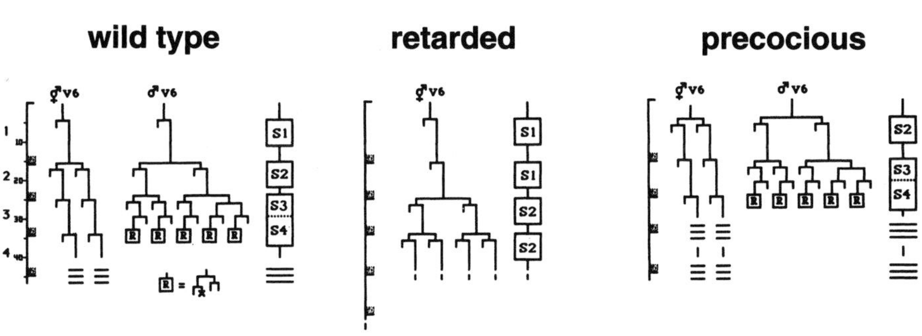

Figure 5. Heterochronic mutations cause perturbations in the timing of specific developmental events. 1, 2, 3, and 4 at the left indicate the first through fourth larval stages (L1-L4) of *C. elegans* development, respectively. S1, S2, S3 and S4 indicate stage-specific lineage patterns normally expressed at the L1-L4 stages, respectively. In retarded mutants, early events occur late, e.g., an S1 pattern can be seen at the L2 stage. Conversely, in precocious mutants, late events occur early, e.g,. an S2 pattern can be seen at the L1 stage. From (13).

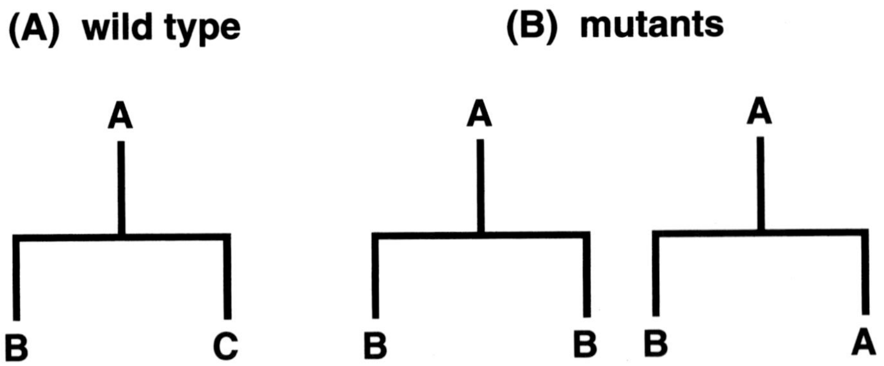

Figure 6. Every cell can be considered to have a fate, where that fate is either to express a particular differentiated state (programmed cell death can be regarded as one such fate) or to divide in a particular pattern and generate a particular complement of descendant cells. (A) Every cell that divides can be said to have a fate, A, which is to divide to generate two daughter cells that have fates B and C, which in general differ from each other and from A. (B) In certain cell lineage mutants, mother cells of fate A instead of generating daughters with fates B and C generate daughters with fates B and B or fates B and A. Such mutants can define genes that function in the generation of cell diversity.

esting and, I think, most significant findings to emerge from these and our other studies of the genetics of the *C. elegans* cell lineage is that most of these genes have counterparts in other organisms, including humans. For example, among the worm cell lineage genes we identified to be involved in the generation of cell diversity are founding members of the now well known and well studied POU (24) and LIM (25) families of transcription factors.

IDCGs, lin-12 and lin-14

As we were beginning our studies of genes that affect cell lineage, one issue that we faced was how to distinguish genes with direct as opposed to indirect roles in controlling a specific cell division or cell fate. In other words, our goal was to analyze "important developmental control genes," which we referred to (with amusement) as "IDCGs." How might we identify genes that directly control specific cell fates? We knew that in certain cases these cell fates were regulated by cell interactions, i.e. whether a particular cell expressed fate A or fate B was determined by the presence or absence, respectively, of a signal from another cell. In such cases, any gene needed for the generation, differentiation or functioning of the signaling cell would be required for the expression of fate A, but could be controlling that cell fate only indirectly.

To address this issue, we decided (e.g., see a subsequent discussion by graduate student Paul Sternberg and me (26)) that one appropriate focus would be genes for which opposite classes of mutations – those that eliminate gene activity and those that elevate gene activity – have opposite developmental effects. The paradigm indicating that genes with such characteristics are *bona fide* developmental genes had been established by Ed Lewis in his pioneering studies of the *Drosophila* Bithorax gene complex (27, 28). Thus, for the case noted above, we would be particularly interested in genes for which too little gene activity caused a cell that normally expresses fate A instead to express fate B, while too much gene activity caused a cell that normally expresses fate B instead to express fate A. Such genes could not simply be needed for some aspect of a process upstream of the cell-fate decision of interest. Rather, such genes could be considered to be not only necessary but also sufficient (in certain cellular contexts) for the expression of a particular cell fate.

The first two genes we studied that satisfied this IDCG criterion and, in fact, led us to this way of thinking were *lin-12* and *lin-14*. The *lin-12* gene was analyzed by graduate students Iva Greenwald, Paul Sternberg and Chip Ferguson (29–31). Opposite classes of *lin-12* mutations result in opposite homeotic transformations in cell fates precisely as outlined above. *lin-12* indeed proved to be an important developmental control gene. Molecularly characterized by Iva Greenwald after she left my laboratory (32), *lin-12* was a founding member of the LIN-12/Notch family, since shown to control intercellular signaling in many organisms and to be involved in human cancer (33–35). The LIN-12 protein showed regions of sequence similarity to the then recently characterized mammalian EGF protein. This finding was one of the first indicating that developmental genes from simple animals were strikingly conserved with human genes. In addition, continuing studies from Iva's laboratory established general principles of LIN-12/Notch signaling (36–38) and revealed that this family of proteins interacts with presenilins, making the LIN-12/Notch family of substantial interest to investigators studying Alzheimer's Disease (39).

lin-14, which was the second heterochronic gene we identified and which was characterized at about the same time as *lin-12*, also fulfilled our IDCG criterion of having opposite classes of mutations cause opposite biological consequences: mutations that reduce *lin-14* activity lead to precocious development, whereas mutations that elevate *lin-14* activity lead to retarded development (10). *lin-14* proved to play a pivotal role in the control of developmental timing (13). All of the other genes we have studied that fulfill our IDCG criterion also have proved to be interesting. For example, *let-60* encodes a Ras protein and acts as a binary switch in a signal transduction pathway of vulval development (40, 41) (see below); *egl-10* was a founding member of a key class of G protein regulators known as RGS proteins (42); and *ced-9* plays a crucial regulatory role in programmed cell death (43) (see below).

With hindsight, I think our rationale for choosing candidate IDCGs was perfectly reasonable. However, we know now (and suspected then) that many genes that do not fulfill this particular IDCG criterion also are of interest and importance. It may well be that the detailed and analytic study of most genes that affect development in specific ways will be instructive.

A genetic pathway for organ development

By studying groups of genes with related effects on the same aspect of the worm cell lineage, we have been able to define not only single genes involved in specific developmental steps but also extensive gene pathways. For example, we have studied a set of genes involved in organ development – specifically in the development of the vulva of the *C. elegans* hermaphrodite. The vulva defines the opening between the uterus and the external environment and is needed for egg laying and for mating with males. Mutants abnormal in vulval development were first isolated by Sydney Brenner, and the genes defined by these and similar mutants were initially characterized collaboratively by John Sulston and me. Vulval developmental mutants are of two basic classes (Figure 7) (8, 9). Some mutants lack a vulva, and are called "vulvaless." Because *C. elegans* is an internally self-fertilizing hermaphrodite, a vulvaless mutant generates fertilized eggs but these eggs develop and hatch *in utero*, after which the progeny devour the animal that was both mother and father to them. The young larvae are released into the environment, so vulvaless mutations are not lethal, at least in the genetic sense of causing the inviability of a strain. Animals in the second class of vulval developmental mutants have multiple ectopic vulva-like structures. Such mutants are called "multivulva."

Many of the genes defined by vulvaless and multivulva mutants proved to be involved in the cell interactions that specify worm vulval development (e.g., (29, 44)). Studies of these genes by our laboratory and by other laboratories, particularly that of my ex-graduate student Paul Sternberg, not only established the molecular genetic basis of vulva development but also helped elucidate the Ras pathway for signal transduction (40, 41, 45) (reviewed in

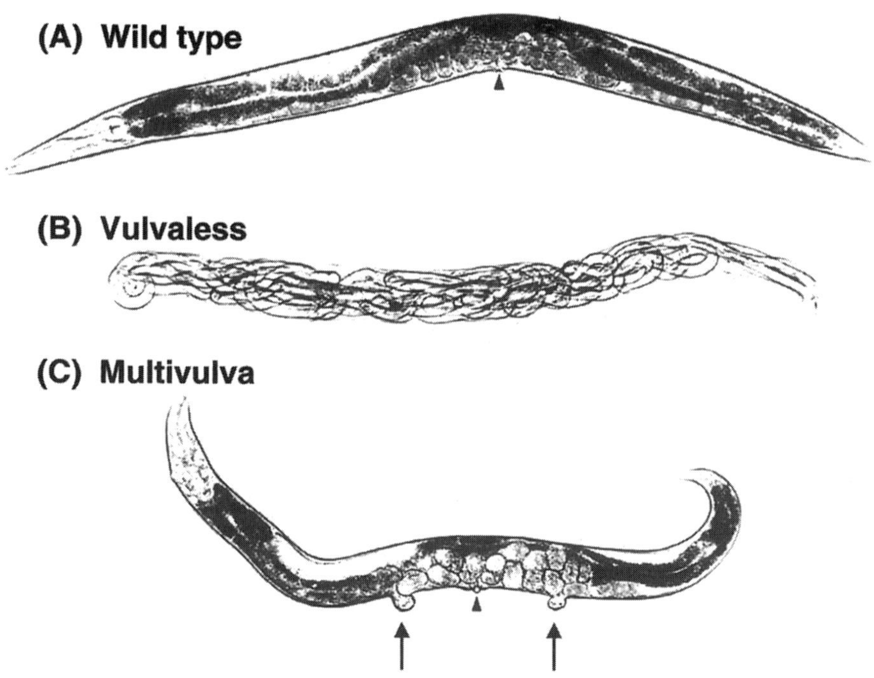

Figure 7. Mutants abnormal in vulva development define two basic classes. (A) Wild type. (B) A vulvaless mutant. (C) A multivulva mutant. Arrowhead indicates position of the vulva. Arrows indicate extra vulva-like structures.

(46)). Ras had been discovered as a human proto-oncogene (47), and an understanding of the normal function of Ras genes as well as of the pathway in which Ras genes act has proved fundamental to the fields of developmental biology and oncology. Our more recent studies of genes involved in vulval development (48, 49) are revealing how a Retinoblastoma (Rb) tumor suppressor gene (50) pathway acts to antagonize this Ras oncogene pathway.

PROGRAMMED CELL DEATH

As I and members of my new laboratory at MIT were beginning genetic studies of the *C. elegans* cell lineage, one aspect of the cell lineage particularly caught my attention: in addition to the 959 cells generated during worm development and found in the adult, another 131 cells are generated but are not present in the adult (2, 4). These cells are absent because they undergo programmed cell death. This phenomenon of naturally-occurring cell death had long been observed as a feature of animal development (e.g., (1, 51, 52)) and seemed a fundamental but essentially unexplored area of developmental biology. Furthermore, programmed cell death is a striking feature of nervous system development (e.g., 105 of the 131 programmed cell deaths in *C. elegans* are in the nervous system), and it was neurobiology that had first attracted me to *C. elegans.* In addition, Kerr, Wyllie and Curie in 1972 (53) had suggested, based upon ultrastructural studies of dying cells, that the mechanisms responsible for naturally-occurring developmental cell death might also be involved in the cell deaths seen during tissue homeostasis as well as in untreated malignant neoplasms and in some cases of therapeutically-induced tumor regression. They noted that the process of cell death in all of these cases is characterized by a series of specific structural changes, and they named this process "apoptosis," a term that has become widely used and is often considered to be synonymous with programmed cell death.

It seemed likely that we could apply the approaches we were taking to study the *C. elegans* cell lineage to analyze programmed cell death. In particular, from the cell lineage we knew that specific cells with diverse developmental origins undergo programmed cell death at specific times during development and that programmed cell death is characterized by a series of specific morphological changes. Thus, we could think of programmed cell death as a cell fate, much like other cell fates, such as differentiating into a muscle cell or a serotonergic neuron. If so, we reasoned, there should be genes that control both the decision to express that fate and the execution (so to speak) of the fate itself.

The first cell-death gene, nuc-1

The first gene to be identified that affects *C. elegans* programmed cell death was discovered by John Sulston in his initial search for mutants defective in the cell lineages of the ventral nervous system, the first cell lineages he studied (54). John was screening for animals abnormal in the number of cells in the ventral nervous system. To visualize these cells, he used Feulgen DNA staining. John observed that in one mutant individual DNA-positive pycnotic bodies were located precisely in positions where there should have been prior programmed cell deaths. He showed that this mutant is defective in the degradation of DNA in cells undergoing programmed cell death and named the gene defined by this mutant *nuc-1* (nuclease abnormal), because it controls the activity of a DNA endonuclease.

ced-1 and ced-2, two genes needed for the engulfment of cell corpses during programmed cell death

Next, Ed Hedgecock, then a postdoctoral researcher at the LMB, identified two genes important for the engulfment of cell corpses by neighboring cells during programmed cell death (55). This process of engulfment, or phagocytosis, normally removes dying cells from the body of the animal. Ed named these two genes *ced-1* and *ced-2*, for cell death abnormal.

The killer gene ced-3

In my laboratory, we were interested in identifying genes responsible either for causing cells to die during programmed cell death or for deciding which cells are to live and which are to die by programmed cell death. As an approach, we decided to seek mutants abnormal either in the presence or in the pattern of programmed cell deaths. However, using Nomarski optics to visualize programmed cell deaths was problematic – dying cells are rapidly engulfed, so that at any given developmental stage very few if any cell deaths can be seen. Following by direct observation the process of cell death in living larvae or embryos is too slow to allow for an efficient mutant hunt. Instead, beginning soon after my 1978 arrival at MIT, we performed mutant hunts using the *nuc-1* mutant to allow the visualization of cell deaths. We obtained a variety of mutants with abnormalities in patterns of cells deaths, but in each case the abnormalities proved to reflect more general defects in cell lineage. For example, in *lin-22* mutants five rather than one programmed cell death occur along each side of the second stage larva; the sources of these extra cell deaths proved to be the lateral blast cells V1, V2, V3 and V4, which normally do not generate dying cells but in *lin-22* mutants are all transformed to express the fate of V5, which normally generates one cell that undergoes programmed cell death (56). Thus, the fundamental role of *lin-22* seemed to be in specifying the fates of V1-4, not in determining which cells live and which die. It is worth noting that this understanding of *lin-22* would have been difficult to attain in the absence of the ability to study at single-cell resolution the *C. elegans* cell lineage. No mutants specific for defects in programmed cell death were isolated from our *nuc-1* screens.

The *ced-1* and *ced-2* mutants of Hedgecock *et al.* (55) offered us a more powerful approach. In *ced-1* and *ced-2* mutants, dying cells initiate the process of programmed cell death but are not engulfed, so that cell corpses persist in an intermediate stage that is easily visualized in living individuals using Nomarski optics. Visualizing programmed cell deaths in living individuals is far more efficient for a genetic screen than examining fixed and DNA-stained specimens, as we were doing in our screens with *nuc-1* mutants. Specifically, our *nuc-1* screens required establishing lines of animals, so that if a fixed and stained mutant animal of interest was identified the mutation responsible could be recovered in a living sibling of the mutant animal. By contrast, if a mutation were identified in a *ced-1* mutant background using Nomarski optics to visualize programmed cell deaths, the individual carrying that mutation could be picked and used to establish a mutant strain. We could screen many more *ced-1* animals than *nuc-1* clones of animals. Thus, our idea was to mutagenize *ced-1* animals and look for mutants with abnormalities in the presence or in the pattern of programmed cell deaths as seen with Nomarski optics. In this way, we hoped to be able to identify mutants in which the process of programmed cell death had not been initiated or in which the pattern of programmed cell deaths was altered. We were encouraged to suspect that such mutants would exist based upon our prior success in isolating cell-lineage mutants with highly specific defects.

Hilary Ellis (Figure 8), a graduate student in the laboratory, undertook this mutant hunt. Although *ced-1* and *ced-2* had not yet been described in any publication, Ed Hedgecock generously sent Hilary a *ced-1* mutant to use in her screen. Such an open sharing of both information (we knew all about Ed's results prior to publication) and resources (he sent us his unpublished mutant) has been a characteristic of the *C. elegans* field since its beginning and has, I believe, allowed the field to blossom in a highly efficient, rapid and enjoyable fashion. Hilary mutagenized *ced-1* animals and found a mutant in which no cell corpses could be seen (Figure 9). We named the gene defined by this mutant *ced- 3*. In a series of experiments, Hilary demonstrated that if *ced-3* activity is reduced or eliminated by mutation, essentially all 131 cells that normally die instead survive (57, 58). Leon Avery, a postdoctoral researcher in my laboratory, later introduced the term "undead" to refer to such surviving cells

Figure 8. Hilary Ellis and her daughter Alina.

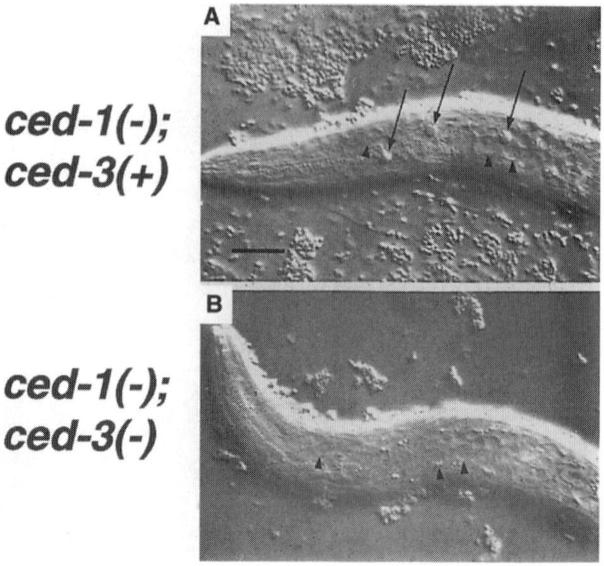

Figure 9. Programmed cell death does not occur in a *ced-3* mutant. (A). A *ced-1* mutant, defective in the engulfment of dying cells, contains persisting cell corpses (arrows). (B) A *ced-1* and *ced-3* double mutant, in which there is no programmed cell death, does not contain cell corpses. The triangles indicate marker cells present in both animals. Bar, 10 microns. From (58).

and showed that at least one undead cell is sufficiently normal to be able to function as a motor neuron and act in feeding behavior (59).

These findings indicated that the activity of the gene *ced-3* is required for cells to die by programmed cell death. This discovery was key, as it demonstrated that programmed cell death requires the function of a specific gene and hence that programmed cell death is an active biological process, analogous to other fundamental biological processes, such as cell division, cell migration and cell differentiation. Based upon our discovery of *ced-3,* we proposed our first draft of a genetic pathway for programmed cell death: *ced-3* acts to trigger programmed cell death upstream of *ced-1* and *ced-2,* which control the engulfment of cell corpses and themselves act upstream of *nuc-1,* which degrades the DNA in dying cells (57).

The second killer gene, ced-4

In what I thought to be a completely unrelated line of study, graduate student Carol Trent (Figure 10A) in my laboratory was studying an aspect of *C. elegans* behavior, egg-laying. (Our studies of the behavior of egg-laying were begun as a consequence of my having isolated egg-laying defective mutants while seeking cell lineage mutants and finding that some of the mutants were abnormal in egg-laying despite being normal in cell lineage and having a normal complement of the cells used for egg-laying.) Carol and technician Nancy Tsung (Figure 10B) isolated and characterized a set of 145 mutants defective in egg-laying and found that one of these mutants is egg-laying defective because it lacks the HSN motor neurons, which innervate the vulval muscles and drive egg-laying (60). We named the gene defined by this mutant *egl-1,* for egg-laying abnormal.

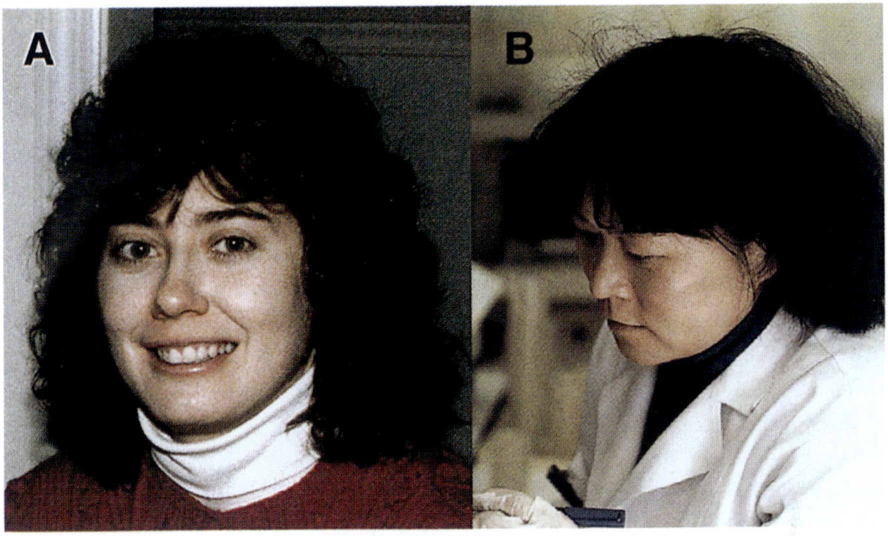

Figure 10. (A) Carol Trend. (B) Nancy Tsung.

Why are the HSN neurons missing in *egl-1* mutants? We considered two possibilities. First, perhaps the HSN neurons are never generated. Second, perhaps they are generated but once generated they die. This latter alternative struck us as plausible, because in his studies of the *C. elegans* cell lineage John Sulston had found that in males the cells homologous to the HSN neurons undergo programmed cell death (4). We reasoned that if in fact HSN neurons were dying by programmed cell death in *egl-1* hermaphrodites, then a *ced-3* mutation, which blocks programmed cell death, should block this HSN death and restore both HSN neurons and egg laying to an *egl-1* mutant. Hilary tested this possibility by constructing a *ced-3; egl-1* double mutant. She found that indeed a mutation in *ced-3* suppresses the effects of the *egl-1* mutation (Table 1). Thus, the absence of HSN neurons in *egl-1* mutant animals seemed likely to reflect a defect in which the HSN neurons instead of surviving undergo programmed cell death. John Sulston directly confirmed that this was the case by examining the embryonic cell lineage of *egl-1* mutants.

Table 1. A *ced-3* mutation, which prevents programmed cell death, suppresses the effects of an *egl-1* (gain-of-function) mutation. Adapted from (58).

	HSN present?	Egg-laying defective?
wild type	yes	no
egl-1	no	yes
ced-3; egl-1	yes	no

This observation proved very useful, as it suggested a very efficient approach for the isolation of more *ced-3*-like mutants: look for mutations that suppress the egg-laying defect of *egl-1* mutants. At least some such mutations should act by preventing programmed cell death. Both Hilary Ellis and graduate student Chand Desai (Figure 11) sought *egl-1* suppressors. Chand's goal was different from Hilary's, as Chand wanted mutations that allowed egg laying in the absence of HSN neurons and thus could define genes responsible for aspects of neuronal wiring. Chand isolated a mutant that restored HSN neurons and proved to be defective in a new gene with properties that Hilary showed to be essentially identical to those of *ced-3*. We named this gene *ced-4* and added it to the genetic pathway for programmed cell death (58).

Programmed cell death involves a process of cellular suicide

Graduate student Junying Yuan (Figure 12A) then asked whether the *ced-3* and *ced-4* killer genes act within the cells that die or elsewhere in the body of the animal, for example to control humoral factors. To address this issue, she performed genetic mosaic analyses, in which in single animals some cells are mutant (e.g., lack *ced-3* function) while others are wild-type (e.g., have *ced-3* function), and asked whether the genetic identity of a cell that normally dies (*ced- 3(+)* vs. *ced-3(–)*) determines its fate (programmed cell death vs. survival). Junying's observations suggested that both *ced-3* and *ced-4* act within the dying cells themselves (61). These findings indicated that, to this extent at least, programmed cell death is a process of cellular suicide.

Both CED-3 and CED-4 have human counterparts that function in programmed cell death

Junying Yuan cloned the *ced-4* gene and discovered that the CED-4 protein was novel, i.e. in its sequence unlike any other protein known at the time. We

Figure 11. Chand Desai.

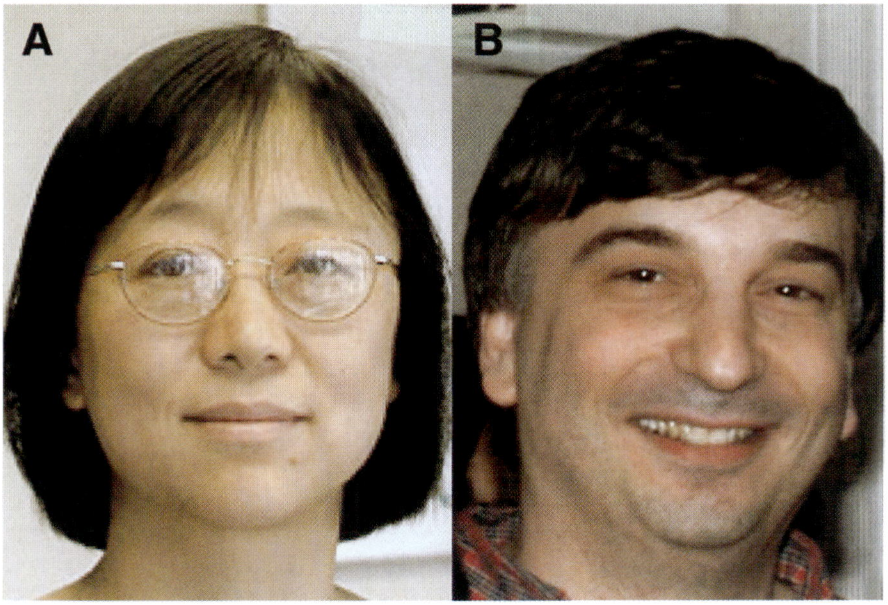

Figure 12. (A) Junying Yuan. (B) Shai Shaham.

published a paper entitled "The *Caenorhabditis elegans* cell death gene *ced-4* encodes a novel protein and is expressed during the period of extensive programmed cell death" (62). Some years later, in 1997, a protein similar to CED-4 was identified. The laboratory of Xiaodong Wang had characterized biochemically an *in vitro* system for cell death and identified a factor, which they called Apaf-1 (apoptotic protease activating factor), with a domain with significant similarity to CED-4 (63). Thus, Apaf-1 is a pro-apoptotic human protein similar in both sequence and function to the *C. elegans* programmed cell death killer gene CED-4.

Just after cloning *ced-4* and five years before Apaf-1 was identified, Junying Yuan also cloned the *ced-3* gene and characterized it molecularly in collaboration with graduate student Shai Shaham (Figure 12B). *ced-3*, too, was novel, not matching any sequence in the databases at that time. However, we could not see publishing a paper with the title "Another *C. elegans* cell death gene encodes a novel protein." So we did not. Instead, we waited. Each day or two for nearly two years, Shai searched the database, until finally there was a hit (Figure 13) (64). The CED-3 protein is similar in sequence to an enzyme that had been purified biochemically (65, 66) by two pharmaceutical companies interested in human inflammatory disease. This enzyme is the protease interleukin-1-beta converting enzyme, or ICE, which converts the pro-form of the cytokine interleukin-1-beta into the active molecule. CED-3 and ICE proved to be the founding members of a family of cysteine proteases now known as caspases (67).

Based upon this finding and upon additional observations (68) made by

```
Query=   TITLE CED-3_PROTEIN.TXT;2
         (503 residues)

Database:   Non-redundant SwissProt+PIR+GenPept+GPUpdate, 4:54 AM EDT Apr 27,
            1992
            65,370 sequences; 18,314,537 total residues.
Searching...............................................done

                                                                  Smallest
                                                                   Poisson
                                                           High  Probability
Sequences producing High-scoring Segment Pairs:            Score  P(N)       N

GPU:HUMIL1BCE_1 Homo sapien interleukin-1 beta convertase...  97  1.4e-14    2
SP:VE2_HPV47     E2 PROTEIN. >PIR:W2WL47 E2 protein - Huma...  56  0.019      2
PIR:A35419       *Neutrophil protein - Pig (fragment) >GP:...  58  0.021      2
SP:PDXJ_ECOLI    PYRIDOXAL PHOSPHATE BIOSYNTHETIC PROTEIN ...  66  0.28       1
GP:CHKPR264_1    G.gallus PR264 mRNA >GPU:CHKPR264_1 G.gal...  62  0.72       1
GP:HUMPR264_1    H.sapiens PR264 mRNA >GPU:HUMPR264_1 H.sa...  62  0.72       1
GPU:HUMSC35A_1   Human splicing factor SC35 mRNA, complete...  62  0.72       1
SP:POLG_SVDVH    GENOME POLYPROTEIN (COAT PROTEINS VP1 TO ...  56  1.0        1
SP:POLG_SVDVU    GENOME POLYPROTEIN (COAT PROTEINS VP1 TO ...  56  1.0        1
GP:SVDSVDV_11    Swine vesicular disease virus complete ge...  56  1.0        1


>GPU:HUMIL1BCE_1 Homo sapien interleukin-1 beta convertase (IL1BCE) mRNA,
            complete cds.
            Length = 404
```

Figure 13. Printout from the database search done by Shai Shaham on April 27, 1992, revealing that CED-3 is similar to the human protein interleukin-1-beta converting enzyme. The search was done using the BLAST network service of the U.S. National Center for Biotechnology Information.

Ding Xue, a postdoctoral researcher in our laboratory, we concluded that CED-3 functions to kill cells during programmed cell death in *C. elegans* by acting as a cysteine protease. We suggested that, given the existence of ICE, ICE or some other CED-3/ICE-like cysteine proteases likely function in programmed cell death in mammals. Junying Yuan, in research performed in her new independent laboratory at the Massachusetts General Hospital, provided support for this hypothesis in a paper (69) published back-to-back with our joint paper reporting the similarity between CED-3 and ICE. Extensive further support for the hypothesis that caspases act in programmed cell death in mammals has been provided by studies from many laboratories, which have identified at least 14 mammalian caspases and shown that many function in programmed cell death, e.g., (67, 70).

ced-9 *protects cells against programmed cell death*

To identify additional genes involved in *C. elegans* programmed cell death, graduate student Ron Ellis (Figure 14A) focused on one particular cell that undergoes programmed cell death, the sister of the serotonergic NSM neuron in the pharynx, the feeding organ. Ron used Nomarski optics to seek mutants in which the NSM sister cell survives. He obtained two classes of mutants: (1) mutants in which the NSM sister survives but other dying cells still die, and (2) mutants in which not only the NSM sisters but also all other dying cells survive. The former set of mutants helped us begin to analyze how specific cells decide whether to live or die (71). Five of the six mutants in the latter set were defective in *ced-3* or *ced-4*. However, one mutant in which there

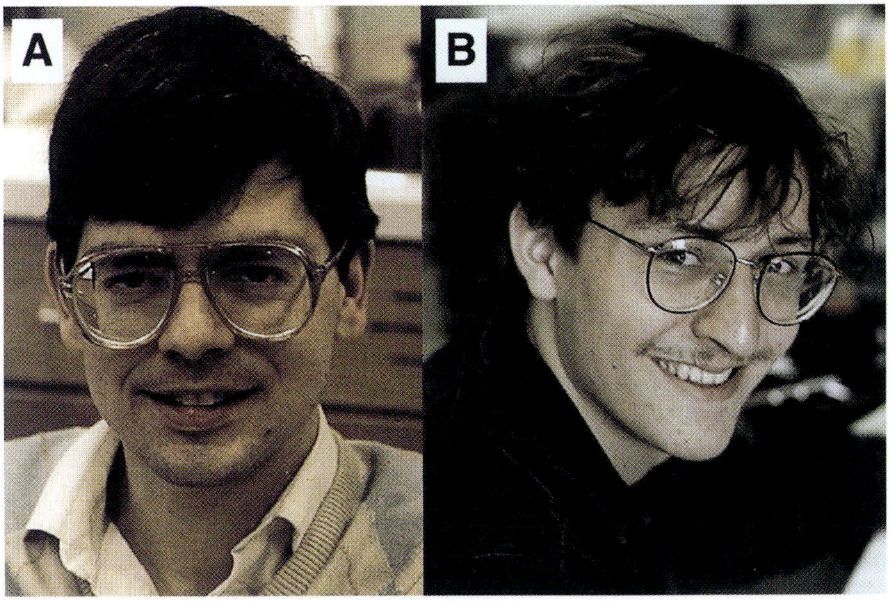

Figure 14. (A) Ron Ellis. (B) Michael Hengartner.

was no programmed cell death was different, and defined a new gene. We named this gene *ced-9* (43).

Genetic studies of *ced-9* by Ron Ellis and subsequently by Michael Hengartner (Figure 14B), also a graduate student in my laboratory, revealed that our original *ced-9* mutation caused a gain rather than a loss of *ced-9* function and, furthermore, that a loss of *ced-9* function had the opposite effect (43). Specifically, whereas a mutation that results in a gain of *ced-9* function causes cells that should die by programmed cell death instead to live, a mutation that results in a loss of *ced-9* function causes cells that should live to die by programmed cell death. These studies indicated that *ced-9* controls the decision between cell survival and programmed cell death (Figure 15) and established that unlike *ced-3* and *ced-4*, which promote programmed cell death, the gene *ced-9* protects cells against programmed cell death.

Michael Hengartner cloned *ced-9* and discovered (72) that it encodes a

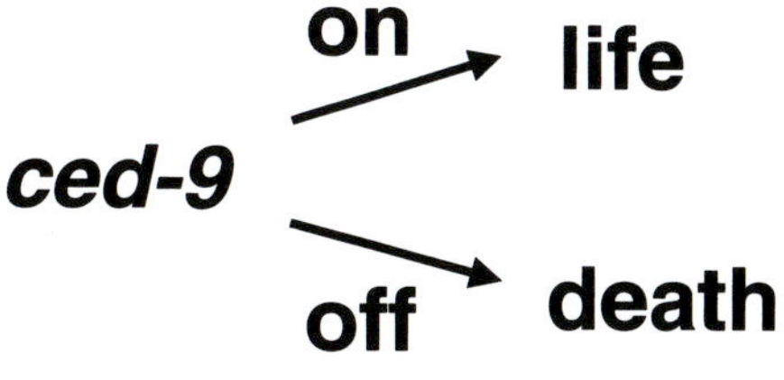

Figure 15. ced-9 controls the life vs. death decisions of cells in *C. elegans.*

protein with similarities to the product of the human proto-oncogene Bcl-2 (B cell lymphoma), the misexpression of which had been shown to cause follicular lymphoma (73–75). Prior studies of Bcl-2 had indicated that Bcl-2 could protect cells in the mammalian immune system from undergoing apoptosis (e.g., (76, 77)). Thus, the molecularly similar genes *ced-9* and Bcl-2 can protect cells against programmed cell death and apoptosis, respectively, supporting the idea that the morphological changes of apoptosis are effected by the same mechanisms as those responsible for programmed cell death. Subsequent studies have revealed a family of CED-9/Bcl-2-like proteins involved in programmed cell death in mammals (e.g., (78, 79)).

The thrill of discovery and the path to scientific visibility

People sometimes ask, when does a scientist feel that "aha!" thrill of discovery? In the case of our studies of programmed cell death, my biggest thrill was probably on February 12, 1992 (Figure 16). That was the day that Michael Hengartner obtained the CED-9 sequence, searched the database and found Bcl-2 at the top of the similarity list. We immediately realized that the pathway of cell-death genes we were studying in *C. elegans* was very likely to be similar to the pathway that controls apoptosis/programmed cell death in humans.

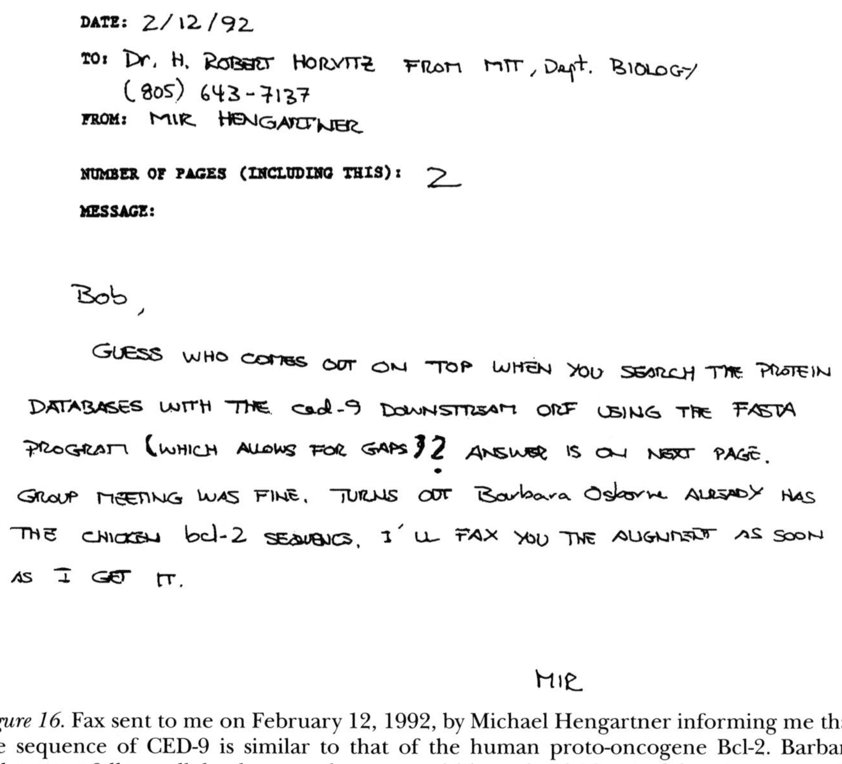

DATE: 2/12/92

TO: Dr. H. ROBERT HORVITZ FROM MIT, Dept. BIOLOGY
(805) 643-7137

FROM: MIR HENGARTNER

NUMBER OF PAGES (INCLUDING THIS): 2

MESSAGE:

Bob,

GUESS WHO COMES OUT ON TOP WHEN YOU SEARCH THE PROTEIN DATABASES WITH THE ced-9 DOWNSTREAM ORF USING THE FASTA PROGRAM (WHICH ALLOWS FOR GAPS)? ANSWER IS ON NEXT PAGE.
GROUP MEETING WAS FINE. TURNS OUT Barbara Osborne ALREADY HAS THE CHICKEN bcl-2 SEQUENCE. I'LL FAX YOU THE ALIGNMENT AS SOON AS I GET IT.

MIR

Figure 16. Fax sent to me on February 12, 1992, by Michael Hengartner informing me that the sequence of CED-9 is similar to that of the human proto-oncogene Bcl-2. Barbara Osborne, a fellow cell-death researcher, was a visiting scientist in our laboratory on a sabbatical leave.

Publication dates can belie scientific history. We published the similarity between CED-3 and ICE before we published the similarity between CED-9 and Bcl-2. Yet our discovery of the similarity between CED-3 and ICE was the later discovery, made on April 27, 1992. After the discovery of the similarity between CED-9 and Bcl-2, I immediately and very broadly discussed this finding. Bcl-2 was already of great interest in the field of oncology. I believe that the fact that Bcl-2 proved to look like a worm protein that antagonized programmed cell death helped convince researchers that the function of Bcl-2 was to antagonize the cell death process. I also believe that this similarity made the worm cell-death pathway suddenly a topic of major interest in the biomedical community, as this pathway was no longer simply an abstract formalism derived from complicated genetic studies of a microscopic soil-dwelling roundworm but rather a framework for a process fundamental to human biology and human disease.

The back-to-back publication of the discoveries that CED-3 looks like a human protease (64) and that CED-3 expressed in mammalian cells could induce those cells to undergo programmed cell death (69) revealed for the first time a mechanistic basis of the process of programmed cell death. This work also strongly suggested that mammalian cells contain the machinery for cellular suicide. The day these two papers were published (November 19, 1993), I received telephone calls from scientists at five pharmaceutical companies wanting to know how these findings could best be used to help them develop novel drugs. There was no question that the biomedical community had become interested in *C. elegans* programmed cell death.

The core molecular genetic pathway for programmed cell death

At about the time that we found the similarity between CED-9 and Bcl-2, David Vaux, working at Stanford University with Irv Weissman and Stuart Kim, demonstrated that when expressed as a transgene in *C. elegans* human Bcl-2 can protect against programmed cell death in worms (80). Michael Hengartner in my laboratory confirmed this finding and further established that human Bcl-2 can substitute for worm *ced-9* in a *ced-9* mutant (72). These observations strongly indicated that *ced-9* and Bcl-2 act to prevent programmed cell death by similar mechanisms and, furthermore, since Bcl-2 can substitute for *ced-9* in worms, that *ced-9* and Bcl-2 act in similar molecular genetic pathways.

By analyzing the genetic interactions among *ced-9*, *ced-3* and *ced-4*, Michael Hengartner and Shai Shaham helped define the genetic pathway for the core, killing step of programmed cell death in *C. elegans*: *ced-3* kills; *ced-4* kills by promoting the killing activity of *ced-3*; and *ced-9* protects by preventing *ced-4* from promoting the killing activity of *ced-3* (43, 81). This pathway raised the question, what regulates *ced-9*? The answer proved to be the gene *egl-1*, which I discussed above as the gene that can mutate to cause the HSN motor neurons to undergo programmed cell death. In a series of genetic studies, postdoctoral fellow Barbara Conradt (Figure 17) discovered that the normal function of *egl-1* is not specific to the HSN neurons but rather is needed for

Figure 17. Barbara Conradt.

essentially all programmed cell deaths, like *ced-3* and *ced-4* (82). However, whereas *ced-3* and *ced-4* act downstream of *ced-9,* the *egl-1* gene acts upstream of *ced-9.* These findings suggest a pathway in which *egl-1* kills by preventing *ced-9* from preventing *ced-4* from promoting the killing activity of *ced-3.* Barbara cloned *egl- 1,* found that it encodes a small protein of 91 amino acids with similarity to the so-called BH3-only members of the Bcl-2 family. She showed that the EGL-1 and CED-9 proteins interact physically. A variety of studies, mostly by others, indicated that the CED-9 and CED-4 proteins also interact physically, as do the CED-4 and CED-3 proteins. Taken together, these observations defined a molecular genetic pathway for programmed cell death involving a cascade of protein interactions and a predicted conservation with mammals (Figure 18).

The overall molecular genetic pathway for programmed cell death in C. elegans

In further studies using similar approaches, we and others have analyzed events both upstream and downstream of this core pathway. Our current picture of the overall molecular genetic pathway for programmed cell death in

EGL-1 —| CED-9 —| CED-4 → CED-3 → Cell Death

BH3-only	**Bcl-2-like**	**Apaf-1-like**	**caspase**

Figure 18. Core molecular genetic pathway for programmed cell death in *C. elegans* and the correspondence of the *C. elegans* gene products with mammalian counterparts. →, positive interaction; –|, negative interaction.

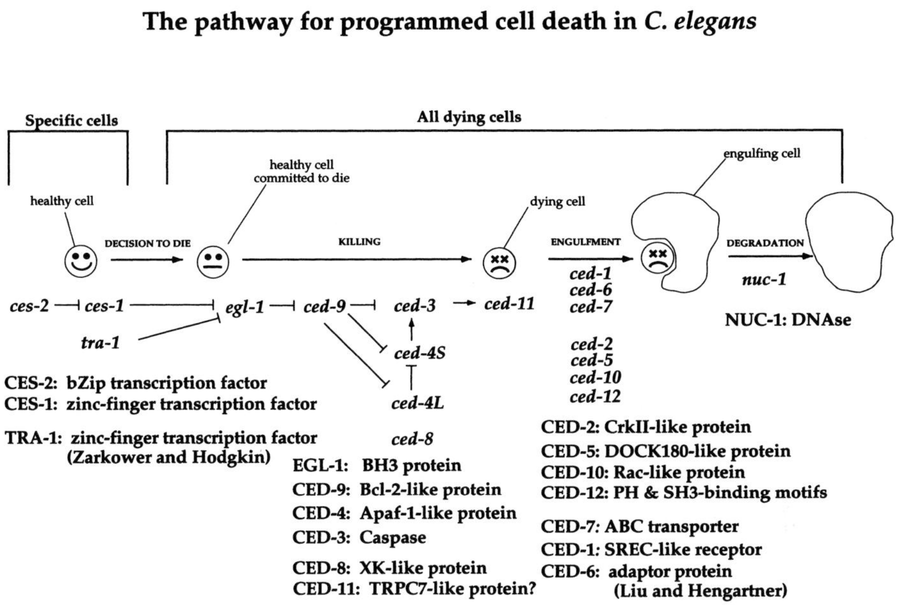

Figure 19. The overall molecular genetic pathway for programmed cell death in *C. elegans.*

C. elegans is shown in Figure 19. First, every cell in the animal must decide whether it is to live or die by programmed cell death. We identified two of the three genes known to be involved in this decision in a mutant hunt related to our studies of the behavior of egg laying. We knew that the neuromodulator serotonin was involved in egg laying (83), and graduate student Carol Trent and technician Nancy Tsung sought mutants abnormal in the only serotonergic neurons we knew about at that time, the NSM neurons in the pharynx. Carol and Nancy identified two mutants that appeared to have extra NSM neurons, and graduate student Ron Ellis found that the extra NSM neurons in these mutants were surviving NSM sister cells (71). These mutants defined the genes *ces-1* and *ces-2.* Ron Ellis found two additional *ces-1* mutants in the screen described above in which he looked directly for mutants defective in the deaths of the NSM sister cells (71). *ces-2* encodes a transcription factor similar to the product of the human proto-oncogene HLF (hepatic leukemia factor), which has been implicated in acute lymphoblastic leukemia (84). *ces-2* directly represses the transcription of *ces-1,* which encodes a transcription factor of the zinc finger family (85). Based on our studies of *ces-2* and *ces-1,* both the oncogenic form of HLF and the human CES-1 counterpart, SLUG, have been shown to regulate programmed cell death in mammalian cells, suggesting that these proteins act in acute lymphoblastic leukemia through effects on programmed cell death (86, 87).

The third gene we have shown to regulate cell-type specific programmed cell death in *C. elegans* is *tra-1* (88). The *tra-1* gene, like the *ces-1* gene, encodes a Zn finger transcription factor; the TRA-1 protein is similar to members of the human GLI protein family (89), which has been implicated in glioblastoma (90) and also in the developmental disorder Grieg cephalopolysyndactyly syndrome (91). *tra-1* controls sexual identity in *C. elegans*; the activity of *tra-1* is regulated by the ratio of X chromosomes to autosomes, so that it is active in hermaphrodites (XX) and inactive in males (X0) (e.g., (92)). Barbara Conradt discovered that *tra-1* regulates the sexually dimorphic programmed cell deaths of the HSN neurons (which survive in hermaphrodites and die in males) by repressing the transcripton of the *egl-1* BH3-only killer gene (88). Her findings suggest that if *egl-1* is expressed (as in males, which have low TRA-1 repressor activity), the resulting EGL-1 protein binds to the CED-9 protein, causing the activation of CED-4 and the subsequent activation of CED-3, resulting in cell death. By contrast, if *egl-1* is repressed by TRA-1 (as in hermaphrodites), HSN neurons survive. Our original *egl-1* mutations, which result in a gain rather than a loss of *egl-1* gene activity, disrupt the TRA-1 binding site of *egl-1*, causing *egl-1* to be inappropriately expressed and thus the HSN neurons to die in hermaphrodites.

Overall, our studies of the cell-type specific control of programmed cell death indicate that the decision about whether a given cell is to live or die is controlled by the actions of specific transcription factors. This finding is consistent with our hypothesis from many years ago that programmed cell death can be regarded as a cell fate, since the expression of many cell fates is specified by the actions of and interactions among particular transcription factors.

The engulfment step of programmed cell death has proved to involve at least seven genes that define two parallel and partially redundant signal transduction pathways responsible for communication between the cell that will die and the engulfing cell (Figure 20) (55, 93–99). These two pathways act not only in the recognition by engulfing cells of cells that are to die and in controlling the cytoarchitectural changes necessary for the process of phagocytosis but also in the killing process itself (100).

Finally, as shown by my graduate students Yi-Chun Wu and Gillian Stanfield (101), the DNA in cells undergoing programmed cell death is degraded (in part) by the direct protein product of the first cell-death gene ever identified, *nuc-1*.

As indicated in Figure 12, most and possibly all of the genes involved in programmed cell death in *C. elegans* have human counterparts, most of which have been implicated in programmed cell death in humans and some of which have been implicated in human disease. This pathway is clearly incomplete. Unpublished studies by us and by others have identified a number of additional genes with roles in programmed cell death, and there are many aspects of the pathway that have not been explored fully. There is still much to learn about both the genetic pathway and the molecular mechanisms of programmed cell death in *C. elegans*.

Programmed cell death and human disease

As the understanding of programmed cell death in *C. elegans* and in other organisms has progressed, more and more human disorders have been shown

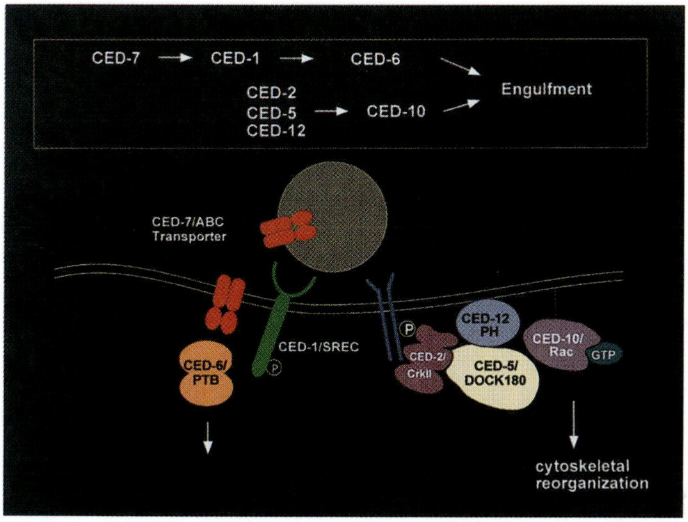

Figure 20. Model for the two signaling pathways that lead to cell-corpse engulfment. (A) Genetic pathways. (B) Diagram of proposed molecular pathways. See text for references.

to be, or at least strongly suspected to be, caused by abnormalities in programmed cell death. The relationship between programmed cell death and human disease has been reviewed numerous times (e.g., (102–105)), and here I will note only that either too much or too little cell death can cause disease. For example, the neurodegenerative diseases – such as Alzheimer's Disease, Parkinson's Disease, Huntington's Disease and amyotrophic lateral sclerosis or ALS – all involve neuronal cell death. In each case, specific classes of nerve cells die, leading to the particular clinical features of each of these neurologic disorders. One major hypothesis today is that the nerve cell deaths in some of these disorders are essentially ectopic programmed cell deaths, i.e., deaths that use the same mechanisms as those that occur during normal development but that for some reason are caused to occur at the wrong time, in the wrong place or affecting the wrong cell type. The evidence supporting this hypothesis varies among the neurodegenerative disorders, and at present is probably strongest for certain retinal degenerations (e.g., (106)). Other disorders characterized by too much cell death and that may well involve ectopic programmed cell death include cerebral stroke, traumatic brain injury, AIDS, myocardial infarction, congestive heart failure, acute liver injury and aplastic anemia.

Conversely, some human disorders involve too little cell death. The number of cells in our bodies is defined by an equilibrium of opposing forces: mitosis adds cells, while programmed cell death removes them. Just as too much cell division can lead to a pathological increase in cell number, so can too little cell death. Certain cancers, including follicular lymphoma, which can be caused by the misexpression in B cells of the CED-9-like proto-oncogene Bcl-2 (77, 107), are clearly a consequence of too little programmed cell death. It may well be that most cancers involve a misregulation of programmed cell death. Similarly, the clinical features of certain autoimmune diseases and viral infections are consequences of too little programmed cell death.

Because of the relationship between programmed cell death and human disease, the identification of the genes and proteins that function in the process of programmed cell death has provided new targets for possible intervention in a broad diversity of disorders. My dream is that the pathway we have identified for programmed cell death in *C. elegans* will help lead to better diagnostics, treatments and cures for some of these diseases responsible for so much human suffering.

The principle of biological universality

One point that emerges from the studies of programmed cell death in *C. elegans* and other organisms is the striking similarity of genes and gene pathways among organisms that are as superficially distinct as worms and humans. Many studies over the past 10 or so years involving *C. elegans*, yeast, the fruit fly *Drosophila melanogaster* and other simple organisms have repeatedly led to analogous findings concerning evolutionary conservation and have established one of the most striking themes of modern molecular biology. I like to refer to this theme as "The principle of biological universality," and it underlies my strong conviction that the rigorous, detailed and analytic study of the biology of any organism is likely to lead to findings of importance in the understanding of other organisms, including ourselves.

ACKNOWLEDGMENTS

First, I thank Sydney Brenner and John Sulston, who are my ex-colleagues, my mentors and my good friends, and with whom I have had the enormous pleasure of sharing the 2002 Nobel Prize in Physiology or Medicine. I am grateful to Jim Watson, Wally Gilbert and Klaus Weber, in whose combined laboratory I worked as a graduate student and who taught me not only how but also why to do science. I thank my colleagues at MIT, who have provided a spectacularly supportive environment that fostered my growth as an independent scientist. In particular, my department chairs – Boris Magasanik, Gene Brown, Maury Fox, Richard Hynes, Phil Sharp, and Bob Sauer – all helped me in many ways over the years. Amongst my many wonderful colleagues, Salva Luria, David Baltimore and David Botstein particularly stimulated me to think, to grow and to appreciate that it is important as a scientist to fulfill a responsibility to a greater community.

I thank my family. My wife Martha Constantine Paton has given me love and support, while ensuring that I maintain a perspective on what is important. Martha's two sons, Chris and Joe Paton, introduced me to the special pleasures of family life. My daughter Alexandra Constantine Horvitz constantly brings me amusement, wisdom and joy, as well as an appreciation for biology that goes well beyond what I have ever observed through a microscope looking at worms. My sister Carol Horvitz-Nutt and her husband Randy

Nutt have shared their love and support for many years. Most deeply, I thank my parents, Mary and Oscar Horvitz, who raised me to ask "Why?" and "How?" and who sustained a belief in me that gave me the confidence to do science and more generally to explore all that life has to offer.

I thank the members of and visitors to my laboratory at MIT, some of whom were directly responsible for the research described above and all of whom together generated the stimulating and enjoyable environment that made such work possible. It is very much their efforts, their insights, and their accomplishments that have been recognized. These researchers include Namiko Abe, Gwen Acton, Mark Alkema, Ezequiel Alvarez-Saavedra, Victor Ambros, Na An, Erik Andersen, Colleen Asbury, Leon Avery, Cori Bargmann, Michael Basson, Greg Beitel, Andrew Bellinger, Ala Berdichevsky, Barbara Biber, Kris Bieker- Brady, Laird Bloom, Ned Buttner, Scott Cameron, Gordon Campbell, Beth Castor, Craig Ceol, Wendy Champness, Fangli Chen, Andrew Chisholm, Chantal Christ, Scott Clark, Barbara Conradt, Brian Davies, Ewa Davison, Lisa Delissio, Chand Desai, Elizabeth DeStasio, John Doll, Bob Edgar, Mary Kaye Edwards, Hilary Ellis, Ron Ellis, Scott Federhen, Chip Ferguson, Mike Finney, Bill Fixsen, Brendan Galvin, Gian Garriga, Elad Gil, Dorit Ginsberg, Susie Glass, Iva Greenwald, Melissa Harrison, Erika Hartwieg, Anna Heineman, Andrew Hellman, Michael Hengartner, Tory Herman, Michael Herman, Brad Hersh, Emma Jean Hess, George Hess, Megan Higginbotham, Dennis Hom, Michael Huang, Brigitte Huber, Melissa Hunter-Ensor, Mike Hurwitz, Ho-Yon Hwang, Yishi Jin, Carl Johnson, Erik Jorgensen, Josh Kaplan, Sam Katz, Saechin Kim, Stuart Kim, Michael Koelle, Kerry Kornfeld, Michel Labouesse, Eric Lander, Joshua Levin, Chris Li, Leslie Lobel, David Lombard, Xiaowei Lu, Andras Madi, Steve McIntire, Marianne McPherson, Mark Metzstein, Eric Miska, Mireya Nadal-Vicens, Dan Omura, Barbara Osborne, Andrea Page, Joan Park, Dianne Parry, Erica Pascal, Ignacio Perez de la Cruz, Rajesh Ranganathan, Peter Reddien, Niels Ringstad, Denise Roberts, Marsha Rosner, Richard Russell, Gary Ruvkun, Hitoshi Sawa, Beth Sawin, Hillel Schwartz, Shai Shaham, Liz Speliotes, Gillian Stanfield, Kristen Stephens, Michael Stern, Paul Sternberg, Jeff Thomas, Jim Thomas, Carol Trent, Nancy Tsung, Nadine Urann, Sander Van den Heuvel, Clem Wang, Eve Wolinsky, Yi-Chun Wu, Fang Xie, Ding Xue, Junying Yuan and Zheng Zhou.

I thank the members of the *C. elegans* research community for establishing and maintaining a field that has been friendly, supportive, interactive and open in the sharing of data, reagents, techniques and even ideas. In particular, Bob Edgar played a pivotal role by founding The Worm Breeder's Gazette, in which unpublished observations and methods have been broadly distributed, sometimes long before publication.

I thank the funding agencies that have supported my research. The Muscular Dystrophy Association and the UK Medical Research Council funded my initial studies of *C. elegans* while I was a postdoctoral researcher in Sydney Brenner's laboratory in Cambridge, England. After my move to MIT in 1978, my research was funded by the U.S. National Institutes of Health and, beginning in 1988, by the Howard Hughes Medical Institute. I am very appreciative of this financial support, which I believe illustrates an important point. These agencies funded me to do very basic research at a time where neither the generality nor the application of our efforts was at all clear. *C. elegans* was an obscure organism. Genetic studies are often highly abstract, and there was no existing technology available to link genetic observations of *C. elegans* with the molecular and biochemical findings being made in studies of other organisms. My research did not target any disease, nor did I know if the biology we uncovered would be relevant to any organism not closely related to *C. elegans*. Nonetheless, our studies established mechanisms that appear to be universal among animals, and our findings may help provide the basis for new treatments for a broad variety of human diseases.

Some important points emerge from these considerations. First, basic research can lead in unexpected ways to insights of substantial practical importance. Time and time again, truly basic studies of simple experimental organisms have proved directly relevant to human biology and human disease. An investment in such basic studies is an effective investment indeed. However, this investment must be made outside of the private sector, by governments and foundations, because only such organizations can act based on the fact that discoveries from basic research have a high likelihood of benefiting humanity, but in ways that cannot be predicted and hence cannot constitute a robust "business plan." I thank those responsible for my early support who, either for these reasons or perhaps more simply because they believed that the pursuit of knowledge is a vital aspect of human culture, saw fit to support our studies of the genetics of cell lineage and programmed cell death in a microscopic worm.

I dedicate this lecture to my mother, Mary Horvitz, and to the memory of my father, Oscar Horvitz.

REFERENCES

1. Lockshin, R. A., and Williams, C.M. **1964.** Programmed cell death. II. Endocrine potentiation of the breakdown of the intersegmental muscles of silkmoths. *J. Insect Physiol.* 10: 643–649.
2. Sulston, J. E. & Horvitz, H. R. **1977.** Post-embryonic cell lineages of the nematode, *Caenorhabditis elegans. Dev. Biol.* 56: 110–156.
3. Kimble, J. & Hirsh, D. **1979.** The postembryonic cell lineages of the hermaphrodite and male gonads in *Caenorhabditis elegans. Dev. Biol.* 70: 396– 417.
4. Sulston, J. E., Schierenberg, E., White, J. G. & Thomson, J. N. **1983.** The embryonic cell lineage of the nematode *Caenorhabditis elegans. Dev. Biol.* 100: 64–119.
5. Kimble, J., Sulston, J. & White, J. **1979** *Regulative development in the postembryonic lineages of Caenorhabditis elegans,* in Cell lineage, stem cells and cell determination, ed. LeDouarin, N. (Elsevier, New York), pp. 59–68.
6. Sulston, J. E. & White, J. G. **1980.** Regulation and cell autonomy during postembryonic development of *Caenorhabditis elegans. Dev. Biol.* 78: 577–597.
7. Brenner, S. **1974.** The genetics of *Caenorhabditis elegans. Genetics* 77: 71–94.
8. Horvitz, H. R. & Sulston, J. E. **1980.** Isolation and genetic characterization of cell-lineage mutants of the nematode *Caenorhabditis elegans. Genetics* 96: 435– 54.

9. Sulston, J. E. & Horvitz, H. R. **1981.** Abnormal cell lineages in mutants of the nematode *Caenorhabditis elegans. Dev. Biol.* 82: 41–55.
10. Ambros, V. & Horvitz, H. R. **1984.** Heterochronic mutants of the nematode *Caenorhabditis elegans. Science* 226: 409–16.
11. Chalfie, M., Horvitz, H. R. & Sulston, J. E. **1981.** Mutations that lead to reiterations in the cell lineages of *C. elegans. Cell* 24: 59–69.
12. Chalfie, M., Tu, Y., Euskirchen, G., Ward, W. W. & Prasher, D. C. **1994.** Green fluorescent protein as a marker for gene expression. *Science* 263: 802–805.
13. Ambros, V. & Horvitz, H. R. **1987.** The *lin-14* locus of *Caenorhabditis elegans* controls the time of expression of specific postembryonic developmental events. *Genes Dev* 1: 398–414.
14. Ruvkun, G., Ambros, V., Coulson, A., Waterston, R., Sulston, J. & Horvitz, H. R. **1989.** Molecular genetics of the *Caenorhabditis elegans* heterochronic gene *lin-14. Genetics* 121: 501–16.
15. Lee, R. C., Feinbaum, R. L. & Ambros, V. **1993.** The *C. elegans* heterochronic gene *lin-4* encodes small RNAs with antisense complementarity to *lin-14. Cell* 75: 843–54.
16. Reinhart, B. J., Slack, F. J., Basson, M., Pasquinelli, A. E., Bettinger, J. C., Rougvie, A. E., Horvitz, H. R. & Ruvkun, G. **2000.** The 21-nucleotide *let-7* RNA regulates developmental timing in *Caenorhabditis elegans. Nature* 403: 901–6.
17. Pasquinelli, A. E., Reinhart, B. J., Slack, F., Martindale, M. Q., Kuroda, M. I., Maller, B., Hayward, D. C., Ball, E. E., Degnan, B., Muller, P., Spring, J., Srinivasan, A., Fishman, M., Finnerty, J., Corbo, J., Levine, M., Leahy, P., Davidson, E. & Ruvkun, G. **2000.** Conservation of the sequence and temporal expression of *let-7* heterochronic regulatory RNA. *Nature* 408: 86–9.
18. Lee, R. C. & Ambros, V. **2001.** An extensive class of small RNAs in *Caenorhabditis elegans. Science* 294: 862–4.
19. Lau, N. C., Lim, L. P., Weinstein, E. G. & Bartel, D. P. **2001.** An abundant class of tiny RNAs with probable regulatory roles in *Caenorhabditis elegans. Science* 294: 858–62.
20. Lagos-Quintana, M., Rauhut, R., Lendeckel, W. & Tuschl, T. **2001.** Identification of novel genes coding for small expressed RNAs. *Science* 294: 853–8.
21. Reinhart, B. J., Weinstein, E. G., Rhoades, M. W., Bartel, B. & Bartel, D. P. **2002.** MicroRNAs in plants. *Genes Dev* 16: 1616–26.
22. Horvitz, H. R. **1990.** *Genetic control of* Caenorhabditis elegans *cell lineage,* in The Harvey Lectures,Wiley-Liss, Inc., Vol. 84, pp. 65–77.
23. Horvitz, H. R. & Herskowitz, I. **1992.** Mechanisms of asymmetric cell division: two Bs or not two Bs, that is the question. *Cell* 68: 237–55.
24. Finney, M., Ruvkun, G. & Horvitz, H. R. **1988.** The *C. elegans* cell lineage and differentiation gene *unc-86* encodes a protein with a homeodomain and extended similarity to transcription factors. *Cell* 55: 757–69.
25. Freyd, G., Kim, S. K. & Horvitz, H. R. **1990.** Novel cysteine-rich motif and homeodomain in the product of the *Caenorhabditis elegans* cell lineage gene *lin- 11. Nature* 344: 876–879.
26. Sternberg, P. W. & Horvitz, H. R. **1984.** The genetic control of cell lineage during nematode development. *Annu Rev Genet* 18: 489–524.
27. Lewis, E. B. **1963.** Genes and developmental pathways. *Am. Zool.* 3: 33–56.
28. Lewis, E. B. **1978.** A gene complex controlling segmentation in *Drosophila. Nature* 276: 565–70.
29. Greenwald, I. S., Sternberg, P. W. & Horvitz, H. R. **1983.** The *lin-12* locus specifies cell fates in *Caenorhabditis elegans. Cell* 34: 435–44.
30. Ferguson, E. L. & Horvitz, H. R. **1985.** Identification and characterization of 22 genes that affect the vulval cell lineages of the nematode *Caenorhabditis elegans. Genetics* 110: 17–72.
31. Ferguson, E. L., Sternberg, P. W. & Horvitz, H. R. **1987.** A genetic pathway for the specification of the vulval cell lineages of *Caenorhabditis elegans. Nature* 326: 259–67.

32. Greenwald, I. **1985.** *lin-12,* a nematode homeotic gene, is homologous to a set of mammalian proteins that includes epidermal growth factor. *Cell* 43: 583–90.
33. Greenwald, I. **1998.** LIN-12/Notch signaling: lessons from worms and flies. *Genes Dev* 12: 1751–62.
34. Frisen, J. & Lendahl, U. **2001.** Oh no, Notch again! *Bioessays* 23: 3–7.
35. Allenspach, E. J., Maillard, I., Aster, J. C. & Pear, W. S. **2002.** Notch signaling in cancer. *Cancer Biol Ther* 1: 466–76.
36. Seydoux, G. & Greenwald, I. **1989.** Cell autonomy of *lin-12* function in a cell fate decision in *C. elegans. Cell* 57: 1237–45.
37. Wilkinson, H. A., Fitzgerald, K. & Greenwald, I. **1994.** Reciprocal changes in expression of the receptor *lin-12* and its ligand *lag-2* prior to commitment in a *C. elegans* cell fate decision. *Cell* 79: 1187–98.
38. Struhl, G., Fitzgerald, K. & Greenwald, I. **1993.** Intrinsic activity of the Lin-12 and Notch intracellular domains *in vivo. Cell* 74: 331–45.
39. Levitan, D. & Greenwald, I. **1995.** Facilitation of *lin-12*-mediated signalling by *sel-12,* a *Caenorhabditis elegans* S182 Alzheimer's disease gene. *Nature* 377: 351–4.
40. Beitel, G. J., Clark, S. G. & Horvitz, H. R. **1990.** *Caenorhabditis elegans ras* gene *let-60* acts as a switch in the pathway of vulval induction. *Nature* 348: 503–509.
41. Han, M. & Sternberg, P. W. **1990.** *let-60,* a gene that specifies cell fates during *C. elegans* vulval induction, encodes a ras protein. *Cell* 63: 921–31.
42. Koelle, M. R. & Horvitz, H. R. **1996.** EGL-10 regulates G protein signaling in the *C. elegans* nervous system and shares a conserved domain with many mammalian proteins. *Cell* 84: 115–25.
43. Hengartner, M. O., Ellis, R. E. & Horvitz, H. R. **1992.** *Caenorhabditis elegans* gene *ced-9* protects cells from programmed cell death. *Nature* 356: 494–499.
44. Sternberg, P. W. & Horvitz, H. R. **1989.** The combined action of two intercellular signaling pathways specifies three cell fates during vulval induction in *C. elegans. Cell* 58: 679–93.
45. Clark, S. G., Stern, M. J. & Horvitz, H. R. **1992.** *C. elegans* cell-signalling gene *sem-5* encodes a protein with SH2 and SH3 domains. *Nature* 356: 340–4.
46. Chang, C. & Sternberg, P. W. **1999.** *C. elegans* vulval development as a model system to study the cancer biology of EGFR signaling. *Cancer Metastasis Rev* 18: 203–13.
47. Parada, L. F., Tabin, C. J., Shih, C. & Weinberg, R. A. **1982.** Human EJ bladder carcinoma oncogene is homologue of Harvey sarcoma virus ras gene. *Nature* 297: 474–8.
48. Lu, X. & Horvitz, H. R. **1998.** *lin-35* and *lin-53,* two genes that antagonize a *C. elegans* Ras pathway, encode proteins similar to Rb and its binding protein RbAp48. *Cell* 95: 981–91.
49. Ceol, C. J. & Horvitz, H. R. **2001.** *dpl-1* DP and *efl-1* E2F act with *lin-35* Rb to antagonize Ras signaling in *C. elegans* vulval development. *Mol Cell* 7: 461–73.
50. Friend, S. H., Bernards, R., Rogelj, S., Weinberg, R. A., Rapaport, J. M., Albert, D. M. & Dryja, T. P. **1986.** A human DNA segment with properties of the gene that predisposes to retinoblastoma and osteosarcoma. *Nature* 323: 643–6.
51. Glücksmann, A. **1951.** Cell deaths in normal vertebrate ontogeny. *Biol. Rev.* 26: 59–86.
52. Saunders, J. **1966.** Death in embryonic systems. *Science* 154: 604–612.
53. Kerr, J. F., Wyllie, A. H. & Currie, A. R. **1972.** Apoptosis: a basic biological phenomenon with wide-ranging implications in tissue kinetics. *Br J Cancer* 26: 239–57.
54. Sulston, J. E. **1976.** Post-embryonic development in the ventral cord of *Caenorhabditis elegans. Philos. Trans. R. Soc. London Ser. B* 275: 287–298.
55. Hedgecock, E. M., Sulston, J. E. & Thomson, J. N. **1983.** Mutations affecting programmed cell deaths in the nematode *Caenorhabditis elegans. Science* 220: 1277–1279.
56. Horvitz, H. R., Sternberg, P. W., Greenwald, I. S., Fixsen, W. & Ellis, H. M. **1983.** Mutations that affect neural cell lineages and cell fates during the development of the nematode *Caenorhabditis elegans. Cold Spring Harb Symp Quant Biol* 2: 453–63.
57. Horvitz, H. R., Ellis, H. M. & Sternberg, P. **1982.** Programmed cell death in nematode development. *Neuroscience Commentaries* 1: 56–65.

58. Ellis, H. M. & Horvitz, H. R. **1986.** Genetic control of programmed cell death in the nematode *C. elegans. Cell* 44: 817–829.
59. Avery, L. & Horvitz, H. R. **1987.** A cell that dies during wild-type *C. elegans* development can function as a neuron in a *ced-3* mutant. *Cell* 51: 1071–8.
60. Trent, C., Tsung, N. & Horvitz, H. R. **1983.** Egg-laying defective mutants of the nematode *Caenorhabditis elegans. Genetics* 104: 619–647.
61. Yuan, J. Y. & Horvitz, H. R. **1990.** The *Caenorhabditis elegans* genes *ced-3* and *ced-4* act cell autonomously to cause programmed cell death. *Dev Biol* 138: 33– 41.
62. Yuan, J. & Horvitz, H. R. **1992.** The *Caenorhabditis elegans* cell death gene *ced- 4* encodes a novel protein and is expressed during the period of extensive programmed cell death. *Development* 116: 309–320.
63. Zou, H., Henzel, W. J., Liu, X., Lutschg, A. & Wang, X. **1997.** Apaf-1, a human protein homologous to *C. elegans* CED-4, participates in cytochrome c-dependent activation of caspase-3. *Cell* 90: 405–13.
64. Yuan, J., Shaham, S., Ledoux, S., Ellis, H. M. & Horvitz, H. R. **1993.** The *C. elegans* cell death gene *ced-3* encodes a protein similar to mammalian interleukin- 1 beta-converting enzyme. *Cell* 75: 641–652.
65. Thornberry, N. A., Bull, H. G., Calaycay, J. R., Chapman, K. T., Howard, A. D., Kostura, M. J., Miller, D. K., Molineaux, S. M., Weidner, J. R., Aunins, J., Elliston, K., Ayala, J., Casano, F., Chin, J., Ding, G., Egger, L., Gaffney, E., Limjuco, G., Palyha, O., Raju, S., *et al.* **1992.** A novel heterodimeric cysteine protease is required for interleukin-1 beta processing in monocytes. *Nature* 356: 768–774.
66. Cerretti, D. P., Kozlosky, C. J., Mosley, B., Nelson, N., Ness, N., Greenstreet, T. A., March, C. J., Kronheim, S. R., Druk, T., Cannizzaro, L. A., Huebner, K. & Black, R. A. **1992.** Molecular cloning of the interleukin-1b converting enzyme. *Science* 256: 97–100.
67. Nicholson, D. W. **1999.** Caspase structure, proteolytic substrates, and function during apoptotic cell death. *Cell Death Differ* 6: 1028–42.
68. Xue, D., Shaham, S. & Horvitz, H. R. **1996.** The *Caenorhabditis elegans* cell-death protein CED-3 is a cysteine protease with substrate specificities similar to those of the human CPP32 protease. *Genes Dev.* 10: 1073–1083.
69. Miura, M., Zhu, H., Rotello, R., Hartwieg, E. A. & Yuan, J. **1993.** Induction of apoptosis in fibroblasts by IL-1 beta-converting enzyme, a mammalian homolog of the *C. elegans* cell death gene *ced-3. Cell* 75: 653–60.
70. Shi, Y. **2002.** Mechanisms of caspase activation and inhibition during apoptosis. *Mol Cell* 9: 459–70.
71. Ellis, R. E. & Horvitz, H. R. **1991.** Two *C. elegans* genes control the programmed deaths of specific cells in the pharynx. *Development* 112: 591–603.
72. Hengartner, M. O. & Horvitz, H. R. **1994.** *C. elegans* cell survival gene *ced-9* encodes a functional homolog of the mammalian proto-oncogene bcl-2. *Cell* 76: 665–676.
73. Bakhshi, A., Jensen, J. P., Goldman, P., Wright, J. J., McBride, O. W., Epstein, A. L. & Korsmeyer, S. J. **1985.** Cloning the chromosomal breakpoint of t(14;18) human lymphomas: clustering around JH on chromosome 14 and near a transcriptional unit on 18. *Cell* 41: 899–906.
74. Cleary, M. L., Smith, S. D. & Sklar, J. **1986.** Cloning and structural analysis of cDNAs for bcl-2 and a hybrid bcl- 2/immunoglobulin transcript resulting from the t(14;18) translocation. *Cell* 47: 19–28.
75. Tsujimoto, Y. & Croce, C. M. **1986.** Analysis of the structure, transcripts, and protein products of bcl-2, the gene involved in human follicular lymphoma. *Proc Natl Acad Sci U S A* 83: 5214–8.
76. Vaux, D. L., Cory, S. & Adams, J. M. **1988.** *bcl-2* gene promotes haemopoietic cell survival and cooperates with c-*myc* to immortalize pre-B cells. *Nature* 335: 440–442.
77. Hockenbery, D., Nunez, G., Milliman, C., Schreiber, R. D. & Korsmeyer, S. J. **1990.** Bcl-2 is an inner mitochondrial membrane protein that blocks programmed cell death. *Nature* 348: 334–336.

78. Gross, A., McDonnell, J. M. & Korsmeyer, S. J. **1999.** BCL-2 family members and the mitochondria in apoptosis. *Genes Dev* 13: 1899–911.
79. Cory, S. & Adams, J. M. **2002.** The Bcl2 family: regulators of the cellular life-or-death switch. *Nat Rev Cancer* 2: 647–56.
80. Vaux, D. L., Weissman, I. L. & Kim, S. K. **1992.** Prevention of programmed cell death in *Caenorhabditis elegans* by human *bcl-2*. *Science* 258: 1955–1957.
81. Shaham, S. & Horvitz, H. R. **1996.** Developing *Caenorhabditis elegans* neurons may contain both cell-death protective and killer activities. *Genes Dev.* 10: 578– 591.
82. Conradt, B. & Horvitz, H. R. **1998.** The *C. elegans* protein EGL-1 is required for programmed cell death and interacts with the Bcl-2-like protein CED-9. *Cell* 93: 519–529.
83. Horvitz, H. R., Chalfie, M., Trent, C., Sulston, J. E. & Evans, P. D. **1982.** Serotonin and octopamine in the nematode *Caenorhabditis elegans*. *Science* 216: 1012–4.
84. Metzstein, M. M., Hengartner, M. O., Tsung, N., Ellis, R. E. & Horvitz, H. R. **1996.** Transcriptional regulator of programmed cell death encoded by *Caenorhabditis elegans* gene *ces-2*. *Nature* 382: 545–547.
85. Metzstein, M. M. & Horvitz, H. R. **1999.** The *C. elegans* cell death specification gene *ces-1* encodes a snail family zinc finger protein. *Mol Cell* 4: 309–19.
86. Inaba, T., Inukai, T., Yoshihara, T., Seyschab, H., Ashmun, R. A., Canman, C. E., Laken, S. J., Kastan, M. B. & Look, A. T. **1996.** Reversal of apoptosis by the leukaemia-associated E2A-HLF chimaeric transcription factor. *Nature* 382: 541– 544.
87. Inukai, T., Inoue, A., Kurosawa, H., Goi, K., Shinjyo, T., Ozawa, K., Mao, M., Inaba, T. & Look, A. T. **1999.** SLUG, a *ces-1*-related zinc finger transcription factor gene with antiapoptotic activity, is a downstream target of the E2A-HLF oncoprotein. *Mol Cell* 4: 343–52.
88. Conradt, B. & Horvitz, H. R. **1999.** The TRA-1A sex determination protein of *C. elegans* regulates sexually dimorphic cell deaths by repressing the *egl-1* cell death activator gene. *Cell* 98: 317–327.
89. Zarkower, D. & Hodgkin, J. **1992.** Molecular analysis of the *C. elegans* sex-determining gene *tra-1*: a gene encoding two zinc finger proteins. *Cell* 70: 237– 49.
90. Kinzler, K. W., Ruppert, J. M., Bigner, S. H. & Vogelstein, B. **1988.** The GLI gene is a member of the Kruppel family of zinc finger proteins. *Nature* 332: 371– 4.
91. Vortkamp, A., Gessler, M. & Grzeschik, K. H. **1991.** GLI3 zinc-finger gene interrupted by translocations in Greig syndrome families. *Nature* 352: 539–40.
92. Hodgkin, J. **1988.** Developmental genetics. Everything you always wanted to know about sex. *Nature* 331: 300–1.
93. Ellis, R. E., Jacobson, D. M. & Horvitz, H. R. **1991.** Genes required for the engulfment of cell corpses during programmed cell death in *Caenorhabditis elegans*. *Genetics* 129: 79–94.
94. Wu, Y. C. & Horvitz, H. R. **1998.** *C. elegans* phagocytosis and cell-migration protein CED-5 is similar to human DOCK180. *Nature* 392: 501–504.
95. Wu, Y. C. & Horvitz, H. R. **1998.** The *C. elegans* cell corpse engulfment gene *ced-7* encodes a protein similar to ABC transporters. *Cell* 93: 951–960.
96. Liu, Q. A. & Hengartner, M. O. **1998.** Candidate adaptor protein CED-6 promotes the engulfment of apoptotic cells in *C. elegans*. *Cell* 93: 961–972.
97. Reddien, P. W. & Horvitz, H. R. **2000.** CED-2/CrkII and CED-10/Rac control phagocytosis and cell migration in *Caenorhabditis elegans*. *Nat Cell Biol* 2: 131–136.
98. Zhou, Z., Hartwieg, E. & Horvitz, H. R. **2001.** CED-1 is a transmembrane receptor that mediates cell corpse engulfment in *C. elegans*. *Cell* 104: 43–56.
99. Zhou, Z., Caron, E., Hartwieg, E., Hall, A. & Horvitz, H. R. **2001.** The *C. elegans* PH domain protein CED-12 regulates cytoskeletal reorganization via a Rho/Rac GTPase signaling pathway. *Dev Cell* 1: 477–89.
100. Reddien, P. W., Cameron, S. & Horvitz, H. R. **2001.** Phagocytosis promotes programmed cell death in *C. elegans*. *Nature* 412: 198–202.

101. Wu, Y. C., Stanfield, G. M. & Horvitz, H. R. **2000.** NUC-1, a *Caenorhabditis elegans* DNase II homolog, functions in an intermediate step of DNA degradation during apoptosis. *Genes Dev* 14: 536–548.
102. Rudin, C. M. & Thompson, C. B. **1997.** Apoptosis and disease: regulation and clinical relevance of programmed cell death. *Annu Rev Med* 48: 267–81.
103. Nicholson, D. W. **2000.** From bench to clinic with apoptosis-based therapeutic agents. *Nature* 407: 810–6.
104. Mullauer, L., Gruber, P., Sebinger, D., Buch, J., Wohlfart, S. & Chott, A. **2001.** Mutations in apoptosis genes: a pathogenetic factor for human disease. *Mutat Res* 488: 211–31.
105. Robertson, J. D., Fadeel, B., Zhivotovsky, B. & Orrenius, S. **2002.** 'Centennial' Nobel Conference on apoptosis and human disease. *Cell Death Differ* 9: 468–75.
106. Chader, G. J. **2002.** Animal models in research on retinal degenerations: past progress and future hope. *Vision Res* 42: 393–9.
107. McDonnell, T. J., Deane, N., Platt, F. M., Nunez, G., Jaeger, U., McKearn, J. P. & Korsmeyer, S. J. **1989.** bcl-2–immunoglobulin transgenic mice demonstrate extended B cell survival and follicular lymphoproliferation. *Cell* 57: 79–88.

JOHN E. SULSTON

On my mother's side I come from Midlands engineers and on my father's from tenant farmers near Oxford. As far back as I remember, and earlier, I was an artisan, a maker and doer. Mechanically minded, my parents said.

Ted, my father, was an Anglican priest; after serving as an army chaplain in the second world war he joined the missionary society SPG (later USPG) and spent his life as an administrator, though he was active in the local parish at weekends. He had a great interest in the natural world, and was a keen gardener. When young I was mainly interested in the physical world, especially anything to do with electricity, but I think something of his love for living things may have rubbed off on me. He brought me up as a Christian, and it was a source of distress to him that I lost my faith, as they say, during my adolescence. That was a hard struggle, one of the hardest I've had. When I tried to talk to my fellow students about it at Cambridge I found them uncomprehending, not seeing it as very important in the scheme of things: but I had

Ted and Muriel Sulston, mid 50's.

had to choose between my judgement and my father. It was a slight worry to me that our children were raised faithless – not prohibited, just not encouraged – in case the religious upbringing was essential to their moral development. Great relief that they've got on fine! Daphne said don't be silly, and of course she was right.

Muriel, my mother, was my main confidant. She was a teacher of English at Watford grammar school, but took a break while my sister Madeleine and I were children. She held court in the kitchen, and we talked about everything. Questions, help with homework. She was more subtle than my father, and it was only after her death that I opened a letter from her and realised that she too desperately wanted me to have continued in the Christian faith.

From them both I gained a sense that there is not, or need not be, any clear distinction between work and play, and that one has a duty both to serve others and to do the best one can in everything.

They believed in private education, and at five I was sent to the local preparatory school. It was just round the corner. The classroom work was easy enough and I got on well with that, but I absolutely loathed games at which I was hopeless. Whether it was eyesight, reflexes or just daydreaming I don't know, but it just didn't turn out well. Cricket was the worst because it went on so long, on beautiful summer afternoons when there were so many better things to do. Once we had catching practice and the ball hit me in the forehead because I was thinking about something else to pass the time. It knocked me over and there was some concern about safety – rather embarrassing for my teacher.

From there I got a scholarship to Merchant Taylors at Northwood, which was a financial relief for my parents, who were not particularly well off. I could have gone to a grammar school like Madeleine, but I'm ashamed to say that there was the sense that the boy had to have the "best". One can't blame them too much – that was the tenor of the age. School gradually became

Madeleine Sulston, late 50's.

more interesting because I was able to specialise in science. Games continued to be a nightmare for a while, but that gradually faded as I realised that one didn't have to take them seriously and began to discover the joy of hill walking on my own.

When it came to choice of subjects, science was obvious – since I was uninterested in anything else – but a decision that caused consternation in some eyes was my demand to take biology for A-level. I was told that this was not sensible at all and would lower my chances, upon which what had initially been more or less a whim on my part hardened into determination. So biology, physics and chemistry it was, with maths abandoned. Not sensible at all, but I had tremendous fun dissecting animals and sectioning plants and wondering about their workings. This was the late 50's, and the molecular biology revolution was getting under way with reports from the front starting to trickle back to us. And we had great support: my enthusiastic zoology teacher, Richard Stokes, wrote to me regularly until his death last year.

In 1960 I arrived in Cambridge, with a scholarship to read Natural Sciences at Pembroke College. The first year was easy because those of us with scholarships had a head start, but the second year was a grind. I wasn't enjoying my course work much, because biology wasn't fulfilling its promise, and anyway I'm not a books person but a hands person, but mainly because by then I had become distracted by other activities, especially theatre lighting at the ADC. Meredith Dewey, my tutor, warned me that those who went into the theatre seldom did well at anything else, but of course that was of no great concern to me. I also became fairly depressed at my lack of social graces and everything seemed pretty pointless. It culminated in a night of drunken disorder that ended in the police station, and poor Meredith was hauled out to retrieve me. Covered in shame, I felt nobody would ever speak to me again. Of course my misdemeanour was hardly noticed, but it did have the effect of motivating me to make some sort of effort in my final year. I plumped for organic chemistry,

Colin Reese [personal photo].

reasoning that at least I could get a job with it, found it interesting because of my sparky supervisor Ian Fleming, and ended up with a 2.1, though without tremendous enthusiasm.

My only attempt at job-seeking had been an application to VSO, and I was actually expecting to join a scheme that summer. However it fell through at the last moment, and so with my 2.1 in hand I went along to see Alexander Todd at the chemistry department in Lensfield Road, on the off chance of being taken on as a research student. To my amazement I was taken on immediately and handed over to Colin Reese, who put me to work on his research into oligonucleotide synthesis.

And that was the beginning of my scientific career, if you can call it that. No more text books, just my own lab books, and the toys, the lovely toys, to play with. I was put in a large lab, sharing a long bench with Mike Tanner, swarthy and amiable. Life was simple, revolving largely around the lab and the Panton Arms where Iris Ambrose dispensed beer, egg on toast with beans (or sometimes as a variant beans on toast with egg), sympathy, and cashed our cheques.

That first year my lodging was in a bedsit on the other side of town. For the second year I arranged with another research student Henry Chan to share a flat. We advertised for a third, and a geophysicist Bob Grasty replied. Most Tuesdays we shared a meal, to which Bob brought two women from his group – Monica Dirac and Daphne Bate. We were all very cheerful together, and when the year ended we seemed to disperse.

But that autumn we went to London for Bob's wedding to his long standing girl friend Jenny. I sat behind Daphne, and she turned and smiled at me. I asked her if she wanted a lift home, and we haven't really been apart since. And Mike, Henry and Bob are our lifelong friends.

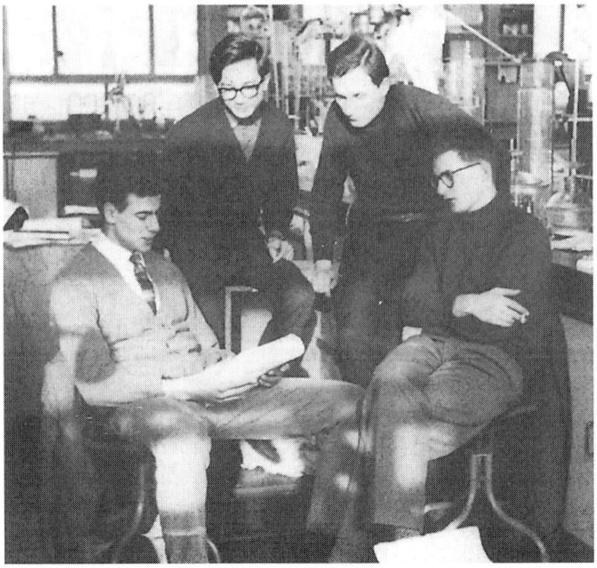

Research students in the large laboratory of the University Chemistry Department, Lensfield Road, Cambridge, ca 1964. Left to right: Mike Tanner, Henry Chan, Peter Schulz, John Sulston.

Meanwhile, thanks to Colin's efficient organisation and my love of playing with the toys, my labwork was trundling along nicely, and already it was time to be writing up and moving on. Colin suggested that I should take a post-doc position with Leslie Orgel at the Salk Institute, and Daphne agreed to get married. And so we whooshed away to California, where Daphne intended to get a job too but then found she was pregnant. So Ingrid was born there, and though at first we feared destitution it turned out to be idyllic. The lab was a great place, and we spent holidays travelling around the western states; shores, deserts, forests and mountains – an earthly utopia.

With Leslie's prebiotic chemistry I felt I was beginning to get back towards biology. It was through the discussions there that I really grasped the power of evolution for the first time. And Leslie spoiled me terribly, frequently inviting us to dinner when he had visitors, and making me feel pretty important. I fear that I became rather obnoxious as a result. In my second year there Leslie introduced me to Francis Crick, who interviewed me on behalf of Sydney Brenner. All I really knew at the time was that Sydney had picked a small animal to study neurobiology. There were lots of jokes about Sydney's worm, and general scepticism about its chances of coming to anything. This seemed a pretty good recommendation to me: there's little point in doing what everybody else is doing. In fact, Leslie suggested that I could just go away for a year and then apply for a junior fellowship at the Salk Institute. The buzz word was neurobiology, and the place was full of meetings and discussion about this next step in molecular biology. One of the events was a summer programme run by Steve Kuffler and Ed Furshpan from Harvard Medical School, and I was invited to join in. They put me on to studying the formaldehyde-induced-fluorescence of catecholamines, a method for revealing these neurotransmit-

Leslie Orgel [personal photo].

ters in sections of frozen tissue. Although I was not deeply involved in their objectives, learning this technique was to make a difference in my life.

In the spring of 1969 the three of us took our last holiday, swinging around Mexico and then departed on an extended break, travelling overland to the east coast via Florida and Canada. After the open spaces of America, southern England felt pretty cramped, but in no time we were settled again in Cambridge, I joined Sydney's group, and Adrian was born.

The group was housed in the Cell Biology Division of the Medical Research Council's Laboratory of Molecular Biology in Cambridge, where so many of the basic mechanisms of molecular biology had been worked out: a rather awe-inspiring place to arrive as a junior staff member. Space was at a premium, as it is in any successful institute, and I ended up in the "big lab". Hugh Robertson, Sidney Altman and Mary Osborn were all there. Next door was Mike Wilcox working on *Anabaena*, later switching to *Drosophila*. Our families became great friends, and we often shared weekend trips as the children grew up. His death from cancer in 1992 was a great loss.

My first years there were devoted to exploring all manner of things to do with the worm: it was a largely virgin field, and so one didn't have to know much in advance. I worked on neurotransmitters, and, because of my background, I spent some time on the DNA using the global methods of hybridisation that were all we had available at the time.

But as a side project I tried out the FIF method, and got it to work on the tiny cells of the worm. The neurons I discovered have not proved especially important in themselves, but they led me to start looking at the cellular anatomy as a whole. And so, quite by chance, I was the one who began to watch the cell divisions unfold. For a decade I became essentially a pure zoologist, and with several colleagues worked out the entire cell lineage of the animal.

My key colleagues at this time were Bob Horvitz and John White, and after Bob returned to the US for a tenure track post at MIT I felt distinctly bereft.

Adrian, Daphne and Ingrid on the North Norfolk Railway, ca 1974.

But there was Judith Kimble as my first post-doc, and Marty Chalfie with Sydney, and Jonathan Hodgkin as Sydney's research student, so we continued as a vigorous little group. It suited me very well not to be in charge, so that I was free to go on playing. The LMB gave me a junior staff position, for which I am immensely grateful. To set up a formal research programme would have been impossible for me then.

The last and most difficult stage of the cell lineage to be completed, that of the embryo, was an important accomplishment for me. It quickly secured my election to the Royal Society, proposed by Sydney, and is now a factor in bringing me to Stockholm. Later, better methods of observation were devised, which to some extent are superseding direct observation. Of course one has to begin somehow, and that beginning then stimulates the development of new approaches, so that what had been difficult becomes routine. I was lucky to have the chance to start it off.

Another feature of the community that was instigated by Bob Edgar was the series of international worm meetings. The first was at Woods Hole in 1977, then it moved to Cold Spring Harbor, then to Madison. The meetings played a vital role in holding together and nurturing the growing *C. elegans* community, and for me they were particularly important when I moved on in 1982 to map and eventually sequence the worm's DNA. This was no longer to be a solitary endeavour for me. It began with new partnerships – Alan Coulson and Bob Waterston – and continued with an increasingly large group.

The worm genome work was funded throughout by the Medical Research Council, but in 1992 the Wellcome Trust accepted a proposal for tackling the human genome and under Michael Morgan's management built a new laboratory, the Sanger Centre, to house worm, human and other projects. From then until I stepped down in 2000 I found myself to my astonishment being an actual director. Thanks to a close knit group of colleagues (Alan Coulson, Jane Rogers, Richard Durbin, David Bentley, Bart Barrell and Murray Cairns)

Daphne during a trip along the river Nene, 1989.

we succeeded in our aims, and it all seems to have worked out well enough, but it was a strange time.

The human sequence is being brought to completion this year through the international consortium of the human genome project. More than half the effort is in the US; the central management is there – initially under Jim Watson, then under Francis Collins from1992 – as are other leading figures including Bob Waterston, Eric Lander, Maynard Olson and Phil Green. Outside the US are several groups, including Jean Weissenbach in France, and others in Germany, Japan and China. Michael Morgan is a key organiser of the international aspects.

It all took a lot of work, organisation, and of course money. Despite helping to set up the UK side of it, I didn't expect to be playing a visible role myself. But in 1999 we were drawn into defending our position against a vigorous bid, by Celera Genomics, to take over the project for profit, and it fell to me to be a major UK spokesman.

Having to deal with this extraordinary dispute, at what turned out to be a moderately unpleasant level of public acrimony, was quite a shock. It seemed to me self-evident that on both moral and practical grounds the human genome itself (as opposed to inventions that may be made from a knowledge of it) is an inappropriate subject for commercial investment and ownership. It was not just the commercial bid itself that shocked, what was worse was that it gained support from all sorts of people for whom I'd previously had respect. I still don't exactly know why, but part of the reason seems to be a business-style way in science nowadays of following bandwagons and avoiding controversy in case things turn out politically against you.

During these events, I thought all the time that I would return to my lab bench once we'd got the episode over with. It had never been my ambition to run a large enterprise, and the thought of returning to play with the toys buoyed me along. Well, get over it we did – Celera collapsed, and the human,

Board of management of the Sanger Centre, ca 2000. Left to right: back row: Bart Barrell, Murray Cairns, Alan Coulson, Mike Stratton; front row: Jane Rogers, John Sulston, David Bentley, Richard Durbin [photo Richard Summers].

mouse and other genomes are firmly in the public domain. It's immensely exciting to reflect that the plan has worked out, and that all these sequences are quietly providing a new and firm foundation for biological research. But I found myself impelled further away from practical science rather than returning.

First, I felt it important to write an account of what had happened. Other books had already been published, and more were to come, but all were written by journalists who drew on the press accounts, largely from the US where for no solid reason the balance of opinion had been in favour of the private effort. So I thought someone should provide an accurate record of events. Bob Waterston and I discussed writing something, but we were neither expert nor free enough to get it done in a reasonable time. So at Daphne's suggestion I got together with Georgina Ferry, a very accomplished science writer. I found it an enormously rewarding undertaking, and a wholly novel experience. Georgina drew on my email records, interviewed many people, and did the greater part of the actual writing, but I wrote some parts and we finished it off jointly, merging our ideas as we went (Sulston and Ferry, 2002).

The exercise had a great influence on me, because it forced me to examine some of the premises under which I had been working all my life. Like many scientists I had felt that my job was to get on with my work, and leave the world to be run by politicians who have the skills and experience to do it properly. Indeed while writing these words I've had a conversation with a colleague who argued exactly that. She feels that to stray from one's expertise is to become second rate.

However, it's become apparent to me that the problems that we encountered over the human genome are much more widespread than I had realised. The researchers at Oxfam, in particular, have educated me in the global consequences of ignoring common goods in the quest for short term profit. These issues are not simple, but I do think we would benefit from a

Paul, Micah and Ingrid at Setauket, Long Island, 2002 [photo Theo Pavlidis].

greater and genuine involvement by us non-professionals in deciding the sort of world we want. Many people say: "Yes, I agree with you, but what can I do?". For me, it's been like climbing up from a valley and reaching a col: suddenly you can see new territories, stretching away into the distance, and you wonder.

At present, of course, I'm indeed second-rate in these matters – an amateur. But one should not be afraid of being an amateur if one is willing also to be a student. Has my head been turned by the Nobel prize? No, this happened to me two years ago. But getting the prize provides us all with a higher platform from which (if we don't fall off through excessive gesticulation) we can have more influence than we did before. So it changes our lives.

It remains for me to say thank you to my family for putting up with me, and for all the talking. Daphne retires this year from her duties as librarian at DAMTP. Through her I've had a lot of interaction with the staff there, and shall miss seeing them at parties. After taking her PhD at Berkeley Ingrid went into interactive museums. She lives in New York with her husband Paul Pavlidis, and son Micah. Adrian is in Edinburgh, and works in software. Madeleine worked for many years in the Gray Lab at Mount Vernon Hospital; now retired, she lives with her husband John Harvey near Lewes. We all met in Stockholm on 7 December, and had a marvellous week together.

REFERENCE

Sulston, J. and Ferry, G. (2002) The Common Thread – a story of science, politics, ethics and the human genome. Bantam Press, February 2002; Joseph Henry Press, October 2002

C. elegans: THE CELL LINEAGE AND BEYOND

Nobel Lecture, December 8, 2002

by

JOHN E. SULSTON

Medical Research Council Laboratory of Molecular Biology and Wellcome Trust Sanger Institute, Cambridge, CB10, 1SA, United Kingdom.

Thank you so very much for inviting me to be here. It gives me a mingled sense – of humility at how much I owe to others, and of joy that the collective work on the worm has been recognised in this way.

Among the first of my many mentors was my PhD supervisor, Colin Reese, who was developing non-aqueous methods for oligonucleotide synthesis (e.g. Fromageot *et al.*, 1968). Colin passed me on to my postdoctoral supervisor Leslie Orgel at the Salk Institute, to work on prebiotic chemistry. The plan was to see to what extent we could copy RNA chains without enzymes. The products were in low yield, and the challenge was to work out the sequences that had been produced. We used cutting with different ribonucleases, and chromatography to separate the fragments (e.g. Sulston *et al.*, 1968).

My introduction to *C. elegans* came in 1969 with my move, at Leslie's suggestion, to Sydney Brenner's group at the Medical Research Council's Laboratory of Molecular Biology (Figure 1). Sydney was reputed to be setting

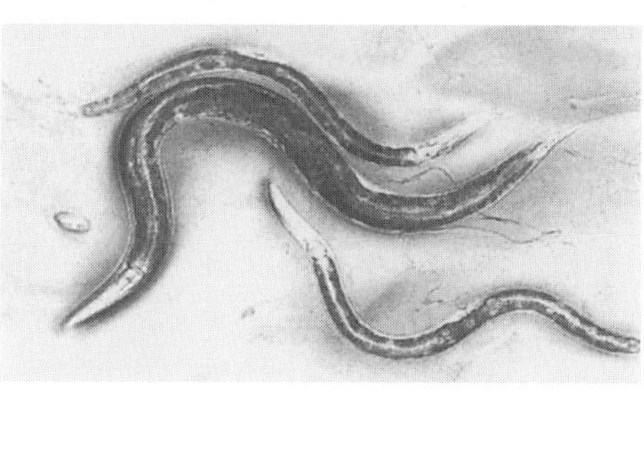

Figure 1. Sydney Brenner, laboratory and *Caernorhabditis elegans.* Photo: MRC Laboratory of Molecular Biology.

up a group to work on the nervous system of a nematode, though at that point nobody knew much about it (Brenner, 1973).

Sydney had already collected numerous mutations that affected the behaviour of the worms (Brenner,1974), and one of my first projects was to investigate the level of glutamate decarboxylase in them, for GABA was believed to function as an inhibitory transmitter. I continued to work on GABA for a while, with a view to finding mutations that affected its production and the behaviour of the worms, but didn't find anything useful – entirely my fault, since plenty of such mutations were found later on.

We also had a look at the DNA content of the worm genome, using the newly developed technique of renaturation analysis (Britten and Smith, 1969; Kohne and Britten, 1971), and came up with an estimate of 20X that of *E. coli* (Sulston and Brenner, 1974). At that time we were not certain of the ploidy of the cells, so even such a simple thing as the comparison of estimates from annealing analysis and chemical analysis was of interest. We counted the stable RNA genes, and attempted to estimate the proportion of the genome that was transcribed. This also led me into a brief collaboration with Gerry Rubin, then a research student with Andrew Travers; we demonstrated linkage between the 5S and larger ribosomal genes of yeast (Rubin and Sulston, 1973).

Meanwhile I played with the technique of formaldehyde induced fluorescence (FIF) (reviewed by Fuxe and Jonsson, 1973) of catecholamines, purely because I knew about it from working with Steve Kuffler and Ed Furshpan when they had run a summer school at the Salk Institute. After some fiddling, necessary on account of the small size of the nematode neurons, I was able to get nice results (Figure 2). The fluorescence spectra indicated that the transmitter was dopamine. Marilyn Dew and I mutagenised some worms and looked for mutants, and found several genes that were required, some for the dopamine to be visible at all, others for loading it into vesicles (Sulston, Dew and Brenner, 1975). We tried to find a function for the dopaminergic cells, reasoning from their appearance in electron micrographs that they would be involved in mechanosensation. Our search was unsuccessful, but it did reveal the first of another series of mutants – the *mecs* – that were insensitive to gentle touch and were adopted by Marty Chalfie in his research programme (Chalfie and Sulston, 1981; Chalfie and Au, 1989).

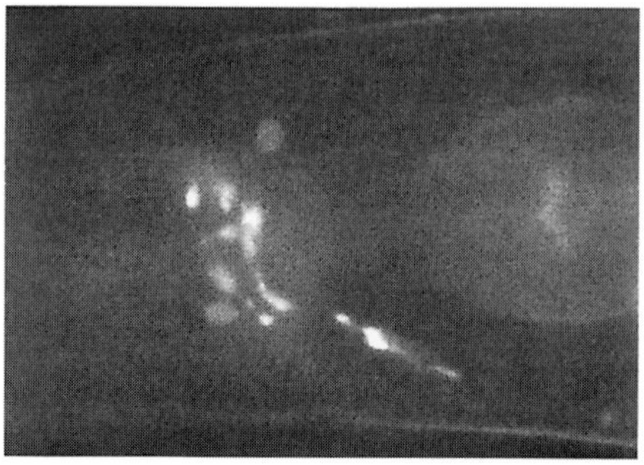

Figure 2. Formaldehyde-induced fluorescence of dopamine in the nerve ring and retrovesicular ganglion of the worm.

But another aspect of the FIF had now diverted me. There were half a dozen FIF-positive neurons in the head of the newly hatched larva, but curiously an extra pair appeared later on. Curiously because the review literature at that time indicated that the complement of neurons is complete at hatching. However, when I started to stain worms by Feulgen in order to discover which were the catecholamine containing ones, it became clear that numerous neurons in the ventral nerve cord were formed some time after hatching (Figure 3). Perusal of the primary literature showed that this observation was not actually novel (Wessing, 1953), but it sufficed to get us started in earnest on the cell lineage.

The significance was that although Sydney had wanted from the beginning of the project to follow cell lineages, it was presumed that the most interesting development happened in the egg. So this was the only stage that had been studied, but it was proving very difficult to see anything. It meant, however, that there were already Nomarski DIC (differential interference contrast) microscopes in the lab. DIC provides an image of a thin optical section of a specimen, with the rate of variation in refractivity from point to point, in a given direction, being represented by intensity. To the observer the effect is as though the refractivity were modelled in relief and then shadowed in the given direction: a powerful input to the eye. Since there is no preparation, the specimen can remain alive. Unlike phase contrast, the objective and condenser work at full aperture, so interference from regions above and below the focal plane is minimised.

Thus I started looking at the larvae. At first this seemed even more difficult then the egg, because all attempts to hold them down for viewing killed them or at least stopped them from developing. Then I realised that I shouldn't hold them down but let them go free, crawling between the top of a thin but perfectly flat layer of agar and the cover slip. By scraping a very thin coating of bacteria onto the centre of the cover slip I got them to stay in the area, browsing gently along at a pace that allowed me to watch and draw the cells. Seeing my first cell division was an exciting moment, because it implied that determining the larval lineage was possible (Figure 4).

And indeed so it proved. The first area that I looked at seriously was the ventral nerve cord (Figure 5), because John White was studying it and we

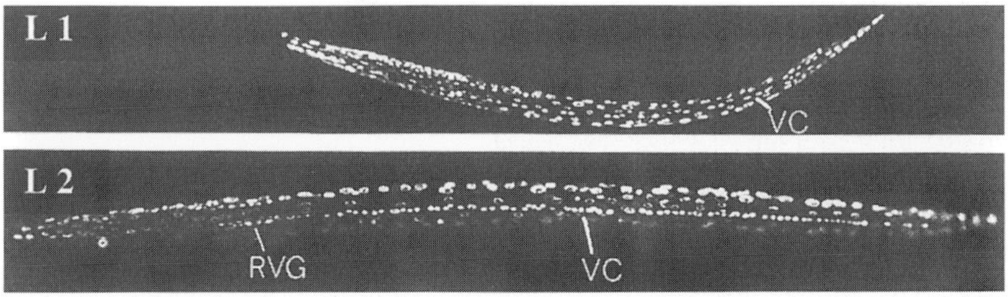

Figure 3. At hatching the ventral cord (VC) has 15 cells; a few hours later there are 57. DAPI stained specimens.

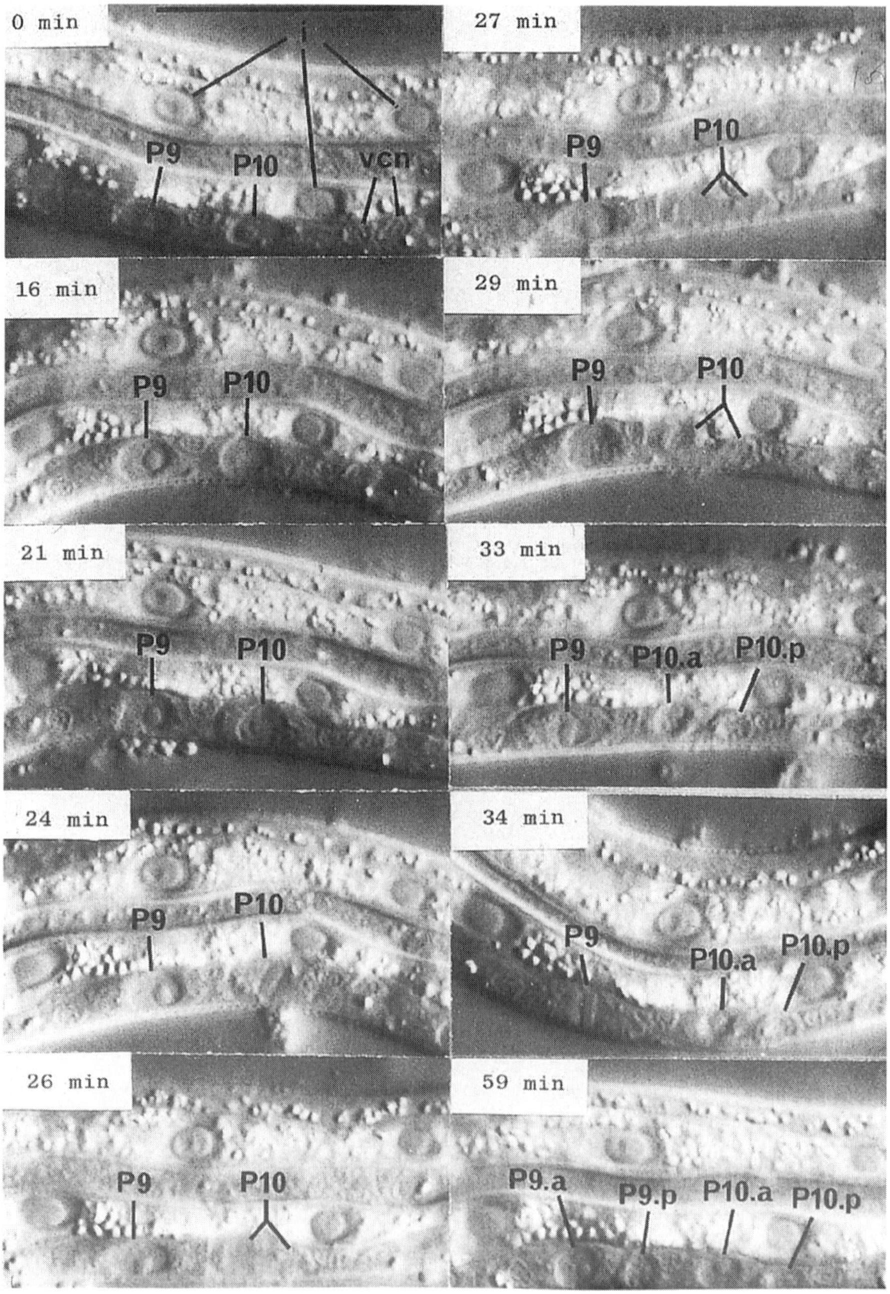

Figure 4. The first division of two of the ventral cord neuroblasts, viewed by Nomarski optics in a living animal. [Sulston and Horvitz, 1977].

were curious to find out how the various classes of cells were derived. John had joined the lab at the same time as me (White, 2000). His initial role was to work on automation of the reconstruction of the neuroanatomy of the worm, from serial section electron micrographs skilfully produced by Nichol

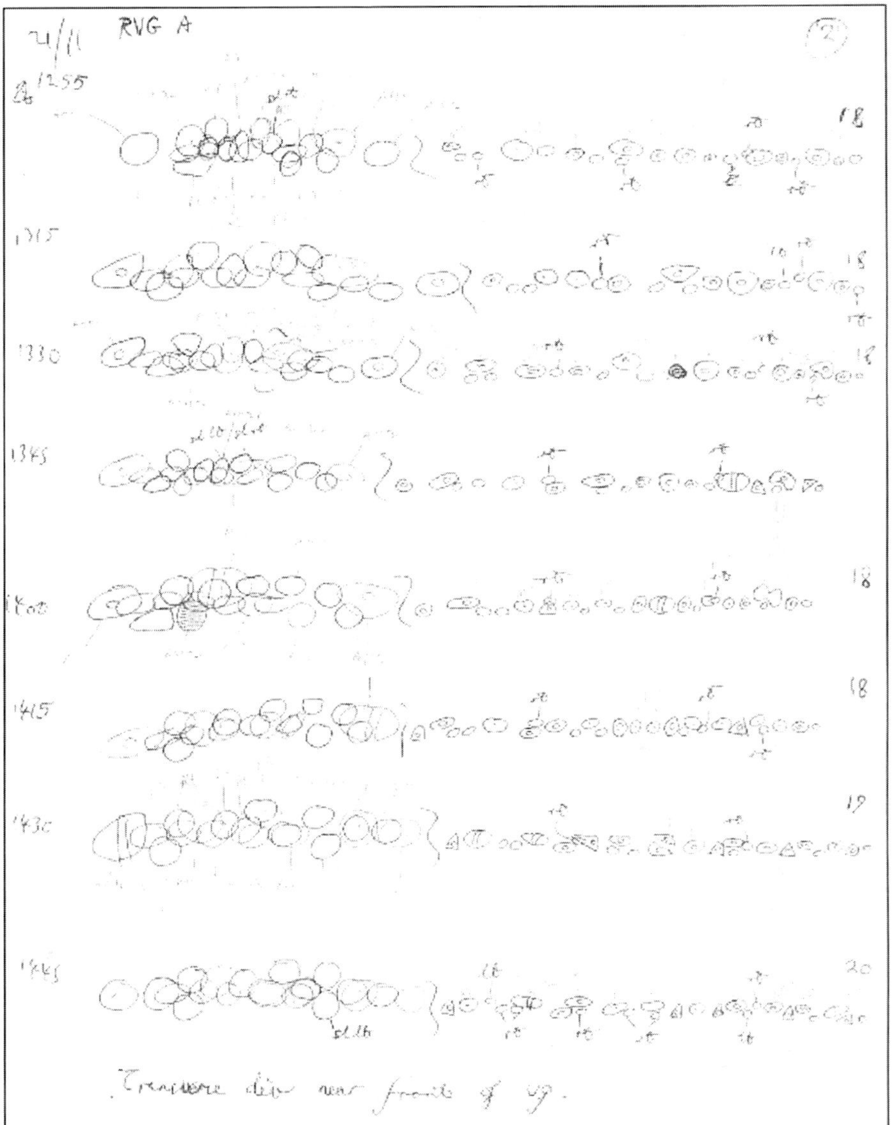

Figure 5. Drawings recording neuroblast divisions over a two hour period in the anterior ventral cord and retrovesicular ganglion, 1974.

Thomson (Figure 6). As it turned out, the computers of the time were not powerful enough to allow full automation, but, characteristically undaunted, John recruited Eileen Southgate to assist him and performed the reconstruction in short order by hand (White *et al.*, 1976).

So, if I could determine the cell lineages, John could correlate the division pattern with the fates of the cells. The answer was immediate: the cells were formed from a stereotyped sublineage repeated along the length of the cord, and each class of neuron arose from a particular branch of the sublineage (Figure 7; Sulston, 1976).

Figure 6. Nichol Thomson and John White. Photo: MRC Laboratory of Molecular Biology.

At the same time, it was obvious that certain cells died and disappeared, in a completely invariant but sex-specific pattern. Such a cell would become more refractive, over a period of half an hour, until it resembled a raised disc in the DIC image; then it would rapidly shrink and gradually disappear (Figure 8). Programmed cell death was already known from other systems (Saunders, 1966; Kerr, Wyllie and Currie, 1972), but now in the worm it was both visible and predictable.

At this point Bob Horvitz joined the group. He was accustomed to collecting information in a very different way – through carefully designed biochemistry, radioactive labels, and so forth – and was at first bemused by the idea of sitting and watching cells divide (Horvitz and Sulston, 1990). But in a very short time he became a tremendous enthusiast, and it was he who urged that we continue to determine the entire larval lineage of the hermaphrodite (Sulston and Horvitz, 1977). By now Judith Kimble (Figure 10) was in the group as my postdoc, but while in David Hirsh's lab she had already determined the gonadal cell lineages (Kimble and Hirsh, 1979). After Bob returned to the US, I went on to complete the male cell lineage, which fed into the neuroanatomy being determined by Donna Albertson (Figure 11) and the genetics of male determination being studied by Jonathan Hodgkin (Figure 12) (Sulston, Albertson and Thomson, 1980; Hodgkin and Brenner, 1977). So now we had all the larval lineages worked out.

Cell assignments were a fine thing, but the aim was to find out how they, and the cell divisions that preceded them, worked. A direct approach to learning about interactions between the cells, and about their functions, would be to kill them selectively. John White devised a system with a pulsed laser beam that projected through the objective of the microscope (Berns, 1972; White and Horvitz, 1979). It worked wonderfully well, because the

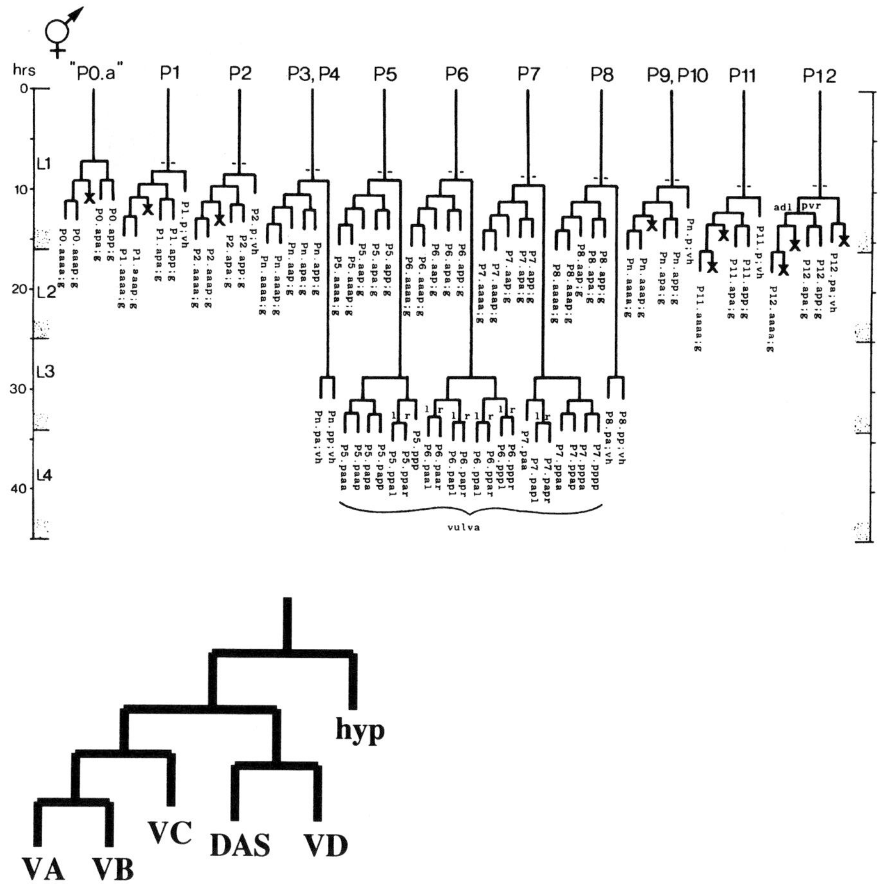

Figure 7. Cell lineages of the ventral cord. Between 6 and 15 hours after hatching each blast cell goes through a stereotyped set of divisions (with small variations at the ends of the cord), and in most places the fates of the progeny are correlated with their positions on the lineage as shown in the insert. Variations at the end of the cord include a defined pattern of programmed cell death, shown by X's in the main picture. [Sulston and Horvitz, 1977].

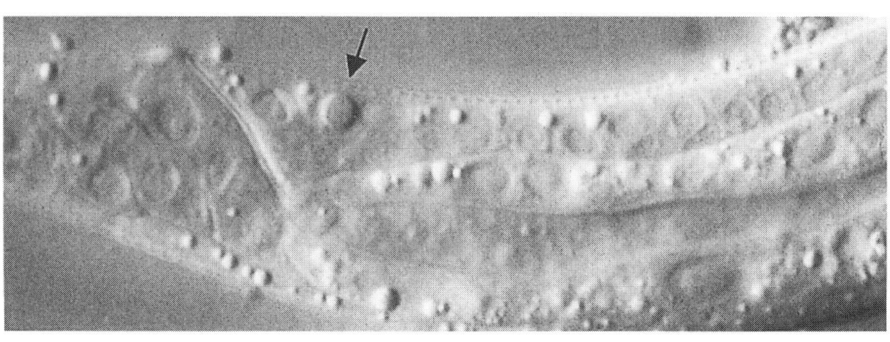

Figure 8. The arrow points to a programmed cell death as it passes through its peak refractivity; posterior end of the cord, Nomarski optics.

Figure 9. Bob Horvitz, mid 70's. Photo: MRC Laboratory of Molecular Biology.

Figure 10. Judith Kimble, late 70's. Photo: MRC Laboratory of Molecular Biology.

Figure 11. Donna Albertson, 70's. Photo: MRC Laboratory of Molecular Biology.

Figure 12. Jonathan Hodgkin, 70's. Photo: MRC Laboratory of Molecular Biology.

large aperture of DIC optics meant that the beam reached high intensity only in the cell of interest. In most cases killing a cell simply removed that cell, and any progeny, from the eventual structure of the animal and so allowed us to discern its function. For example killing the two HSN neurons that lie on either side of the gonad (HSN stands for hermaphrodite specific neuron; later found to undergo programmed cell death in the male) blocked egg laying. But in certain cases a killed cell was replaced by another (Figure 13), and so we were able to map out the regulatory possibilities in development (Sulston and White, 1980).

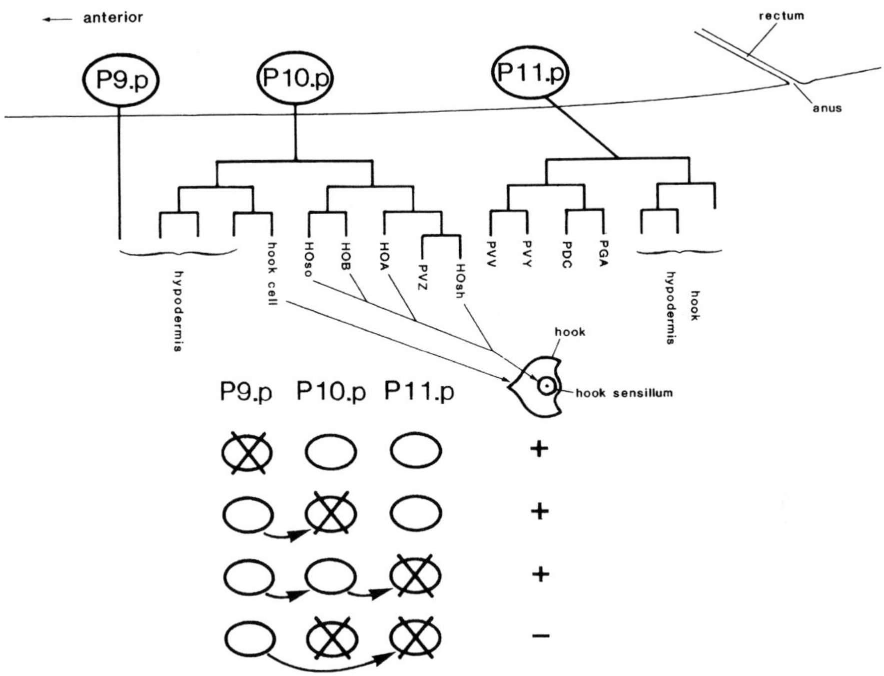

Figure 13. An example of fate regulation in the male tail. In the group of three precursor cells, any one can be killed without causing a phenotypic defect, because of replacement regulation. Killing two cells results in a defect, because further proliferation does not occur.

Among the cell interactions revealed by the laser experiments were two cases of cell death by murder (Sulston, Albertson and Thomson, 1980). In each of the cell pairs concerned, prior removal of the engulfing cell allowed survival of the cell programmed to die. We speculated that the majority of cell deaths were suicides, i.e. that engulfment was not a necessary trigger, and this subsequently proved to be the case.

But of course the real aim was to go further and discover the genes that control the programme. We began to hunt for mutations, but it was not easy to decide what to look for. Nevertheless we gradually accumulated a few mutations that affected the cell lineage, some by analysis of the existing collection, others isolated by predicting phenotypes (for example Bob Horvitz's continued foray into egg-laying defects: Ferguson and Horvitz, 1985), and others by looking at specimens taken randomly from mutagenised stocks (Horvitz and Sulston, 1980; Sulston and Horvitz, 1981). The animals were examined by DNA staining, FIF and Nomarski to determine whether the lineage was altered. The possibilities were endless, but by now the news of the cell lineage had been passed around the community, and many groups were engaged in mutation and laser experiments to study different functions and body parts of the worm. Sharing information in this way has always been a feature of the worm community, and was greatly fostered by Bob Edgar's initiative in starting the newsletter known as the Worm Breeders' Gazette (Figure

Figure 14. Bob Edgar, recent photo, with the first issue of the Worm Breeder's Gazette.

14). The newsletter continues to appear, but in recent years has been hugely extended by electronic resources (elegans.swmed.edu; www.wormbase.org; biosci.umn.edu/CGC).

As for genes involved in programmed cell death, the deoxyribonuclease *nuc-1* was known from an early stage as a result of the DNA stain screens, but it was only a scavenging enzyme and did not have to do with initiating cell death (Figure 15). The first step into the pathway itself was taken by Ed Hedgecock, who did what the rest of us thought was too difficult and screened live mutagenised worms by DIC optics. By this time we knew that the

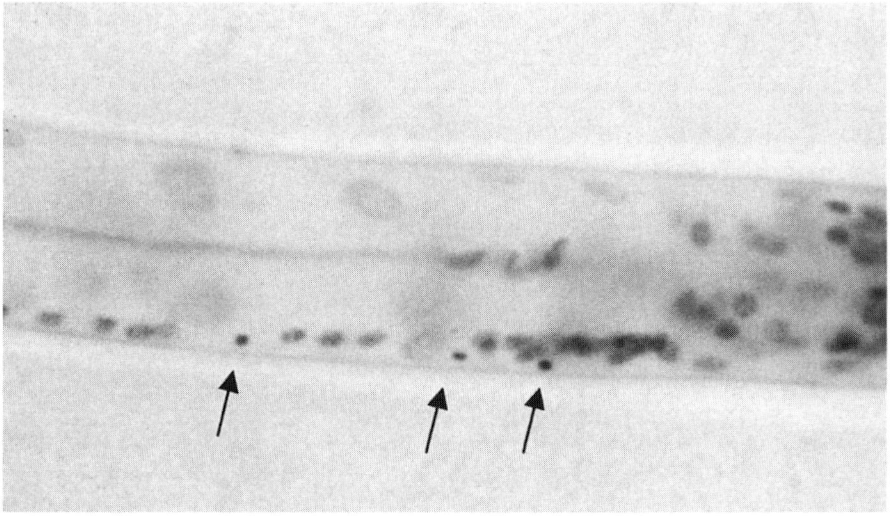

Figure 15. Three cell deaths (arrows) at the anterior end of the cord, revealed in a *nuc-1* mutant by persistence of their DNA in small clumps. Feulgen stain.

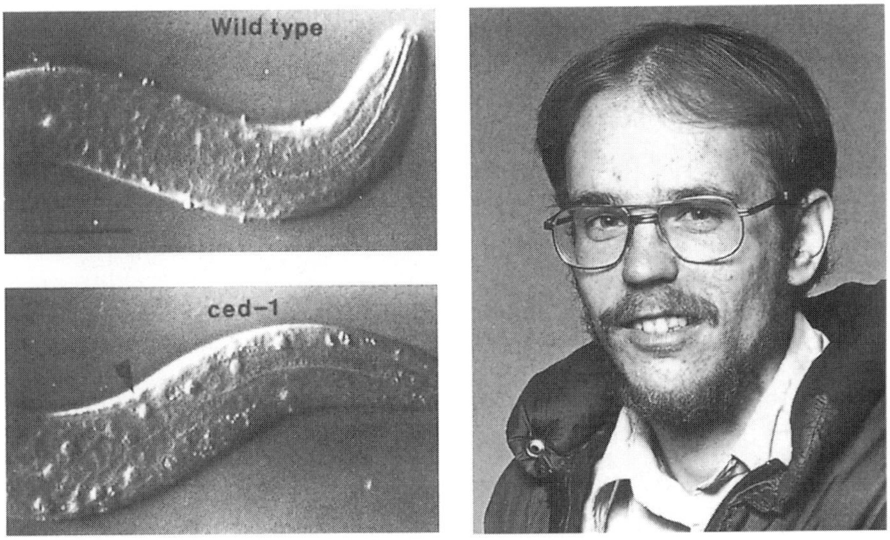

Science 220, 1277-1279 (1983)

Figure 16. Ed Hedgecock, 70's, with a *ced-1* mutant. The arrow head points to one of several persistent cell corpses visible in the mutant [Hedgecock *et al.*, 1983]. Photo: MRC Laboratory of Molecular Biology.

normal fate of a dying cell is to be phagocytosed by its neighbours (Robertson and Thomson, 1982). Ed found a number of mutations in two genes, *ced-1* and *ced-2*, required for the phagocytosis; with the loss of activity of either gene the dead cells persisted as prominent refractive objects, providing an obvious phenotype (Figure 16; Hedgecock *et al.*, 1983). Those deaths that had previously been identified as murders failed to occur in these mutants, confirming their special character. Subsequently Hilary Ellis used the same procedure of DIC screening to find revertants of CED-1 mutants, and so discovered ced-3, the first gene known to actually control cell death (Horvitz *et al.*, 1982). This was the first of many (Hengartner, 1997), and that story is taken up in Bob Horvitz's lecture.

One further type of cell death was discovered by Marty Chalfie in the course of his programme on *mec's* (Figure 17). These abnormal deaths are caused by mutations in the *deg* genes, which result in overactivity of an ion channel in a certain class of cells (Driscoll and Chalfie, 1992). The phenotypes of these mutants are unaffected by loss of *ced-1* and *ced-2*.

Meanwhile, only limited progress had been made on the egg. A century previously the early lineages had been worked out by observations of fixed and DNA stained specimens (e.g. Boveri, 1892; reviewed Chitwood and Chitwood, 1974), but now, to go further, living specimens would have to be used. Roger Freedman and Simon Pickvance had tried recording with DIC optics on 16mm film, as had T. Kaminuma in Tokyo, and G. von Ehrenstein and E. Schierenberg in Göttingen (Deppe *et al.*, 1978). The cells were visible, but retrieval of the information about cell divisions, especially when the division axis was vertical, proved impossible after the first few rounds (Figure 18).

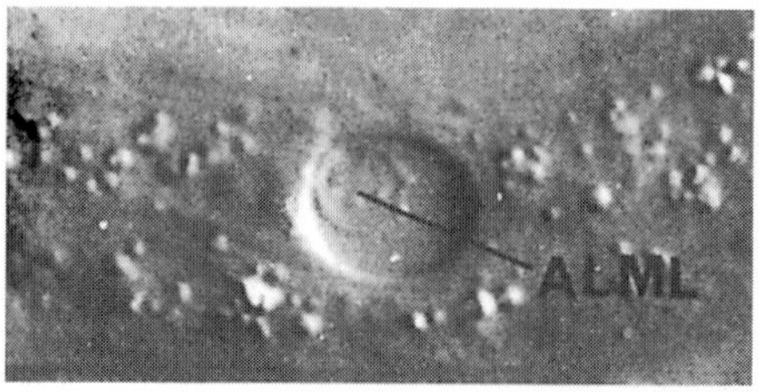

Develop Biol 82, 358-370 (1981)

Figure 17. Marty Chalfie, 70's, with a cell dying in a dominant mec-4 mutant [Chalfie and Sulston, 1981]. Photo: MRC Laboratory of Molecular Biology.

So little by little I started looking by eye and drawing as I had for the larvae. At first it was hard, but I had the time to persist, and soon the structures became clearer in my mind. The lineage is almost invariant, so I was able to return to each stage and each group of cells and became familiar with them. One gadget that helped a lot was a traditional cross hair made from gossamer, providing a point of reference in the image without degrading it at all – much better than an engraved plate. Over the course of a year and a half it was finally done. We had the entire story of the worm's cells from fertilised egg to adult (Figure 19; Sulston *et al.*, 1983). Once again, comparison with the anatomy determined from Nichol Thomson's E/M sections was crucial, but

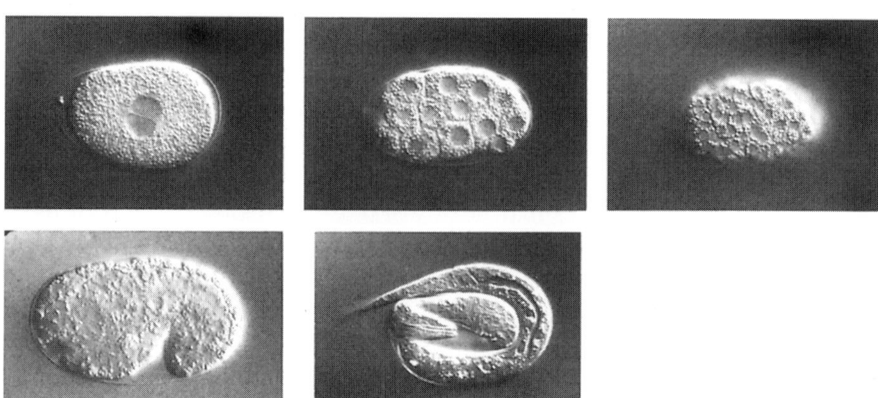

Figure 18. The egg, seen in Nomarski optics, from pronuclear fusion to hatching.

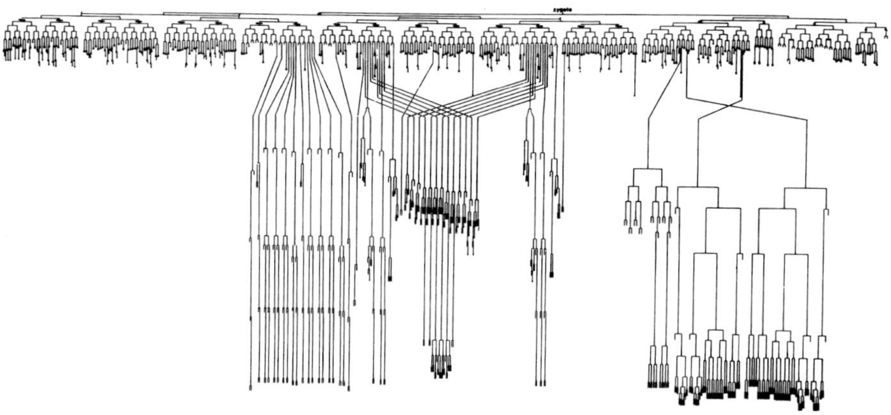

Figure 19. The complete cell lineage of the hermaphrodite, from fertilised egg to adult.

between us we were able to assign all the cells to their eventual roles (Figure 20) (White *et al.*, 1986; Albertson and Thomson, 1976).

Later John White built a device – the 4D microscope – that records on optical discs and allows much better retrieval, and there are now hopes of fully automating the observation of the lineage by means of histones linked to GFP (R. Waterston, R. Durbin, personal communications).

At this point, we discussed our findings mainly in observational terms, because there was little else we could do. For example, we were intrigued by cells that changed their function during development. Thus the DD motor neurons innervate ventral muscle in the youngest (L1) larva, but then rearrange their connections to innervate dorsal muscle in the next stage (L2), by which time new ventral innervation has been provided from postembryonically generated neurons (White *et al.*, 1978). In mutants that have no postembryonic cell division, the DD's rearrange regardless, and the mutant becomes abruptly uncoordinated at L2. Clearly this was telling us something about development, but what?

Again, there was always a lot of discussion around the lab about autonomous versus non-autonomous determination of cell fate – rather reminiscent of the politically charged debated about nature-versus-nurture for hu-

PARTS LIST FOR THE NEMATODE *C. ELEGANS*

AB	P_0 a		Embryonic founder cell
ADAL ADAR	AB plapaaaapp AB prapaaaapp		Ring interneurons
ADEL ADER	AB plapaaaapa AB prapaaaapa		Anterior deirids, sensory receptors in lateral alae, contain dopamine
ADEshL ADEshR	AB arppaaaa AB arpppaaa		Anterior deirid sheath cells "
ADEso	H2 aa	L&R	Anterior deirid socket
ADFL ADFR	AB alpppppaa AB praaappaa		Amphid neurons, dual ciliated sensory endings, probably chemosensory, enter ring via commissure from ventral ganglion, take up FITC
ADLL ADLR	AB alppppaad AB praaapaad		Amphid neurons, dual ciliated sensory endings, probably chemosensory, project directly to ring, take up FITC
AFDL AFDR	AB alpppapav AB praaaapav		Amphid finger cells, associated with amphid sheath
AIAL AIAR	AB plppaappa AB prppaappa		Amphid interneurons "
AIBL AIBR	AB plaapappa AB praap…		Amphid interneurons "

Figure 20. The beginning of the complete list of cells, as it appeared in the first worm book [Wood *et al.*, 1988].

man development that we hear so much of now. Classical developmental biology, it seemed to me, was too often concerned with setting up an experiment to settle an issue once and for all. I felt that it was no use going on arguing about whether this or that view of development was correct, but that we had to go out and find the genes. It also seemed likely that everyone would in the end be proved correct: life doesn't follow a particular neat set of rules but uses all sorts of mechanisms.

But how to find the genes? Did we need to find all of them or just the subset that would become obvious by mutational analysis, genetics and selective cloning of those parts of the genome thus revealed? Later, as we embarked on the genome map, I was told by one enthusiastic *Drosophila* researcher in the lab:" Don't worry about doing this, in five years time it will all be over". He was inspired by the recent discovery of the homeobox (reviewed: Gehring, 1987), and indeed that was a wonderful breakthrough, but would it be the whole story? I suspected not, that life would be too complicated to solve by classical genetics alone, and that we would indeed have to find all the genes before we were done.

At this point my account diverges from the Nobel citation that the three of us have received. Bob continued along the path that brought us all to Stockholm, whilst I began a different though parallel project. My excuse for including a brief description of it here is: first that it has contributed to the scientific success of the cited enterprise, and second that it led to some philosophical conclusions on which I would like to end.

As the lineage work came to an end, I was uncertain what to do next. Many people thought that, given my knowledge of the egg, I should analyse embryonic mutants. But there were already several groups engaged in this work (reviewed: Wood, 1988; Kemphues and Strome, 1997; Schnabel and Priess, 1997), and I didn't think I could add much to their efforts. But also there was another problem: there were already too many mutants available for the rate at which the genes could be isolated by the relatively new art of molecular cloning. Finding them in the 100M base pair genome was immensely time-consuming. When Bob Horvitz was leaving for MIT I asked him what he planned to do about it. "Heavy duty molecular biology" was his pained but realistic reply. It niggled me that each person should be wasting so much time on this business with gels and filters, groping around in the immensity of the genome for the few thousand bases of interest to them, and my thoughts came to a head during a seminar by Matt Scott about his heroic mapping of the *antennapaedia* region in *Drosophila*. Surely it should be possible to do it all in parallel for much less proportionate effort? Total sequencing was at that time impossible to contemplate, but large scale mapping seemed feasible.

I came back full of enthusiasm for this concept, started to learn from Jon Karn the molecular skills that I'd missed while closeted with the microscope, and discussed approaches to characterising the clones with him and Sydney. I started making clone libraries in *lambda*. The crucial moment was the decision by Alan Coulson to join me, following the retirement of his boss Fred Sanger. We switched from *lambda* clones to cosmids, and Alan got the fingerprinting to work efficiently enough to put through thousands of clones.

Rodger Staden started us off with a simple software package for manual reading of the gels and matching, but we soon exceeded its capacity. So I learned to write Fortran programmes and developed an electronic assembly system. Frank Mallett and the workshop built a scanner for the films and we wrote a semi-automatic package to read out the positions of the bands.

It was going well, and we had a map covering the genome in a few hundred segments, but the remaining gaps were proving intractable (Coulson *et al.*, 1986). Fortunately Bob Waterston visited on sabbatical at the very time that David Burke and Georges Carle were developing the YAC (yeast artificial chromosome) cloning system in Maynard Olson's lab (Burke *et al.*, 1987). Bob speculated that YACs might clone segments that cosmids could not. He returned home and quickly constructed the first YAC library for the worm, and we saw that they filled the gaps. Yuji Kohara came on sabbatical, and that summer we all worked on the long series of hybridisations that knitted the map together at last (Coulson *et al.*, 1988). And Bob joined Alan and me in what has become a 20 year collaboration (Figure 21).

The map was by now proving its worth in gene finding, and was the most complete for any complex genome. Andy Fire's introduction of efficient transformation meant that the cosmids could be tested directly for rescue of mutants, simply by injecting them in turn into the gonad (Fire, 1986). One important thing we were learning was to ensure that the information remained open. This was standard practice in the worm community, but was especially important for the genome project. The project would only be useful if everyone cooperated in placing their genetic data on the physical map, so that the genetic and physical maps would become more and more closely aligned. Only thus would gene hunting actually be facilitated (Figure 22). So this was not a matter of open data for its own sake, but was simply the way to make the map useful, and to that end we devised simple rules so that researchers could contribute and yet still be credited for their individual accomplishments.

So when our two labs moved on to sequencing – first the worm (*C. elegans* Sequencing Consortium, 1998), and then joining in with the much larger

Bob Waterston **Alan Coulson** **Fred Sanger**

Figure 21. The architects of the genome map and sequence.

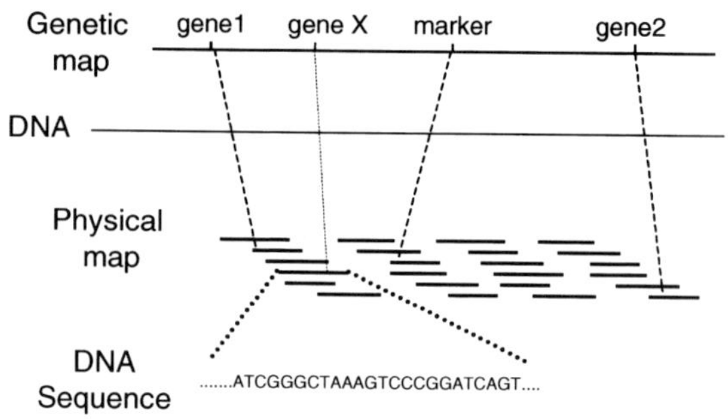

Figure 22. The principles of a genome map. The intractably long DNA molecule of the genome is converted into an overlapping series of clones, immortalised in frozen bacteria or yeast cells, and issued to researchers when needed. To maximise benefit from this physical map it is essential to accumulate publicly information from previous gene locations, so that the hunter of gene X knows where to look, using the aligned genetic map as a guide.

group for the human genome (International Human Genome Sequencing Consortium, 2001) – it was natural for us to bring the principle of free data release along. It was the pragmatic way to proceed in what was now recognised as the growing field of genomics, and in due course it became formalised in the Bermuda statement: the result of a meeting of the major human sequencing groups at Bermuda in 1996.

The human genome project was well on course for finishing in 2005, or very likely earlier, given the Wellcome Trust's commitment to it in the UK. And in 1998 the worm sequence was published, with some gaps but essentially complete, the first animal sequence to be brought to this level. But also in 1998 came a claim that a private enterprise would sequence the human faster and cheaper, outrun the public consortium, and release the data (Venter *et al.*, 2001). Apart from technical doubts, we thought it financially implausible that an industrial sponsor would release data freely. As it turned out, we were correct, but there followed a hectic three years of science politics to ensure that the human genome remained in the public domain.

That's now over, mercifully. But, in closing, I'd like to highlight the key reason for making a fuss about free release of genome data.

The fact is that proprietary databases don't work for such basic and broadly needed information as the sequence of the human genome (Figure 23). Not only do they create a class distinction between rich and poor researchers, but worse they inhibit communication between researchers in general. In order to protect the market value of a proprietary database, the owner must prohibit redistribution of the contents – otherwise the information would quickly leak out and be widely known. But often – and bioinformatics is a case in point – it's not possible to disseminate results without disseminating the underlying data at the same time. Attempts to get around this lead to fearful tangles, and it's far better to fund and distribute the information in the public domain from the start.

Additionally, this dispute was a microcosm of the debate about how scien-

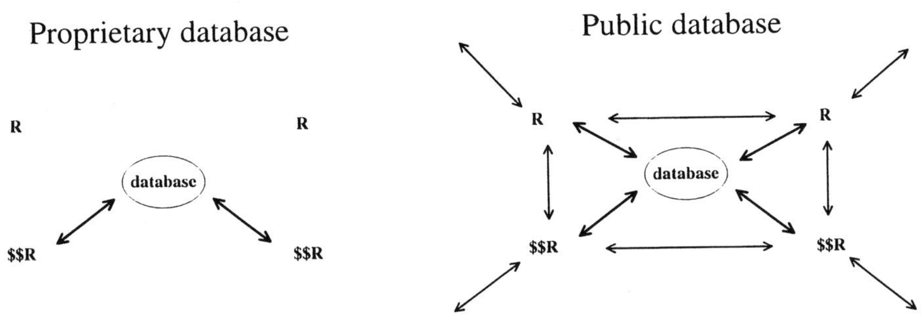

Figure 23. Contrast in communications for users of proprietary and public databases. Proprietary database can be accessed only by fee-paying researchers ($$R), who must agree not to redistribute data in order to protect its confidentiality. Public database can be accessed by all, and everyone can communicate freely with one another.

tific research should be financed. At the moment more and more funding is coming with profitable applications in mind. Whilst worthy in themselves, applications shouldn't be the only way to drive basic research. Short term goals are too limited to control exploration of the natural world, and anyway not everything we discover should be exploited. Furthermore, many of the most important potential applications – for example the neglected diseases such as tuberculosis and malaria, found mainly in the poorer parts of the world – cannot be researched through funding for profit: there is no market to repay the investment (World Health Organisation, 2002). It seems obvious that we have to revert to a greater measure of public funding, and stop imagining that somehow the free market will solve everything on its own.

This lecture has taken a meandering route and ended up far from its starting point. To loop back, I think the important purpose of science is to explore, discover and understand. I'm glad if I've been able to contribute a little to that process, and hugely grateful to all my colleagues, both here and elsewhere, for their achievements and for the fun. And I hope that we can apply our ever increasing knowledge wisely, for the good of all.

REFERENCES

Albertson, D. G. and Thomson, J. N. (1976). The pharynx of *Caenorhabditis elegans*. Phil. Trans. R. Soc. Lond. B., **275**, 299–325.

Berns, M. W. (1972) Partial cell irradiation with a tunable organic dye laser. Nature, **240**, 483–485.

Boveri, T. (1892) Über die Entstehung des Gegensatzes zwischen den Geschlechtzellen und den somatischen Zellen bei *Ascaris megalocephala*. Sitzungsber. Ges. Morphol. Physiol., **8**, 114–125.

Brenner, S. (1973). The genetics of behaviour. Brit. Med. Bull. **29**, 269–271.

Brenner, S. (1974). The genetics of *Caenorhabditis elegans*. Genetics, **77**, 71–94.

Britten, R. J. and Smith, J. (1969). In: Yrbk. Carnegie Inst. Washington, **68**, 378–386.

Burke, D. T., Carle, F. G. and Olson, M. V. (1987). Science, **236**, 806–812.

The *C. elegans* Sequencing Consortium (1998). Genome sequence of the nematode *C. elegans*: a platform for investigating biology. Science, **282**, 2012–2018.

Chalfie, M. and Au, M. (1989). Genetic control of differentiation of the *Caenorhabditis elegans* touch receptor neurons. Science, **243**, 1027–1033.

Chalfie, M., and Sulston, J. E. (1981). Developmental genetics of the mechanosensory neurons of *Caenorhabditis elegans.* Develop. Biol., **82**, 358–370.

Chitwood, B. G. and Chitwood, M. B. (1974). "Introduction to Nematology", Univ. Park Press, Baltimore.

Coulson, A., Sulston, J. E., Brenner, S. and Karn, J. (1986). Towards a physical map of the genome of the nematode *Caenorhabditis elegans.* Proc. Nat. Acad. Sci. U.S.A., **83**, 7821–7825.

Coulson, A., Waterston, R., Kiff, J., Sulston, J. E. and Kohara, Y. (1988). Genome linking with yeast artificial chromosomes. Nature, **335**, 184–186.

Driscoll, M. and Chalfie, M. (1992). Developmental and abnormal cell death in *C. elegans.* Trends Neurosci., **15**, 15–19.

Deppe, U., Schierenberg, E., Cole, T., Krieg, C., Schmitt, D., Yoder, B. and von Ehrenstein, G. (1978). Cell lineages of the embryo of the nematode *Caenorhabditis elegans.* Proc. Nat. Acad. Sci. U.S.A., **75**, 376–380.

Ferguson, E. L. and Horvitz, H. R. (1985). Identification and characterization of 22 genes that affect the vulval cell lineages of the nematode *Caenorhabditis elegans.* Genetics, **110**, 17–72.

Fire, A. (1986) Integrative transformation of *Caenorhabditis elegans.* EMBO J., **5**, 2673–2680.

Fromageot, H. P. M., Reese, C. B. and Sulston, J. E. (1968). The synthesis of oligoribonucleotides – VI: 2'-O-acyl ribonucleoside derivatives as intermediates in the synthesis of dinucleoside phosphates. Tetrahedron **24**, 3533–3540.

Fuxe, K. and Jonsson, G. (1973). The histochemical fluorescence method for the demonstration of catecholamines. J. Histochem. Cytochem., **21**, 293–311.

Gehring, W. J. (1987). Homeo boxes in the study of development. Science, **236**, 1245–1252.

Hedgecock, E. M., Sulston, J. E. and Thomson, J. N. (1983). Mutations affecting programmed cell deaths in the nematode *Caenorhabditis elegans.* Science, **220**, 1277–1279.

Hengartner, M. O. (1997). Cell death. In: *C. elegans* II (ed Riddle, D. L., Blumenthal, T., Meyer, B. J. and Priess, J. R., Cold Spring Harbor Laboratory Press), pp 383–415.

Hodgkin, J. and Brenner, S. (1977). Mutations causing transformation of sexual phenotype in the nematode *Caenorhabditis elegans.* Genetics, **86**, 275–287.

Horvitz, H. R., Ellis, H. M. and Sternberg, P. W. (1982). Programmed cell death in nematode development. Neurosci. Comment. **1**, 56–65.

Horvitz, H. R. and Sulston, J. E. (1980). Isolation and genetic characterization of cell-lineage mutants of the nematode *Caenorhabditis elegans.* Genetics, **96**, 435–454.

Horvitz, H. R. and Sulston, J. E. (1990). Joy of the Worm. Genetics, **126**, 287–292.

International Human Genome Sequencing Consortium (2001). Initial sequencing and analysis of the human genome. Nature, **409**, 860–921.

Kemphues, K. J. and Strome, S. (1997). Fertilization and establishment of polarity in the embryo. In: *C. elegans* II (ed Riddle, D. L., Blumenthal, T., Meyer, B. J. and Priess, J. R., Cold Spring Harbor Laboratory Press), pp 335–359.

Kerr, J. F. R., Wyllie, A. M. and Currie, A.R. (1972). Apoptosis: a basic biological phenomenon with wider ranging implications in tissue kinetics. Br. J. Cancer, **24**, 239–275.

Kimble, J. and Hirsh, D. (1979). The postembryonic cell lineages of the hermaphrodite and male gonads in *Caenorhabditis elegans.* Develop. Biol. **70**, 396–417.

Kohne, D. E. and Britten, R. J. (1971). Hydroxyapatite techniques for nucleic acid reassociation. In: Procedures in Nucleic Acid Research (ed Cantoni and Davies), **2**, 500–512.

Robertson, A. M. G. and Thomson, J. N. (1982). Morphology of programmed cell death in the ventral nerve cord of *Caenorhabditis elegans* larvae. J. Embryol. Exp. Morphol., **67**, 89–100.

Rubin, G. M. and Sulston, J. E. (1973). Physical linkage of the 5S cistrons to the 18S and 28S ribosomal RNA cistrons in *Saccharomyces cerevisiae.* J. Mol. Biol., **79**, 521–530.

Saunders, J. W. (1966). Death in embryonic systems. Science, **154**, 604–612.

Schnabel, R. and Priess, J. R. (1997). Specification of cell fates in the early embryo. In: C. elegans II (ed Riddle, D. L., Blumenthal, T., Meyer, B. J. and Priess, J. R., Cold Spring Harbor Laboratory Press), pp 361–382.

Sulston, J. E. (1976). Post-embryonic development in the ventral cord of *Caenorhabditis elegans*. Phil. Trans. R. Soc. Lond. B., **275**, 287–297.

Sulston, J., Albertson, D. G. and Thomson, J. N. (1980). The *Caenorhabditis elegans* male: postembryonic development of non-gonadal structures. Develop. Biol., **78**, 542–576.

Sulston, J. E. and Brenner, S. (1974). The DNA of *Caenorhabditis elegans*. Genetics, **77**, 95–104.

Sulston, J., Dew, M. and Brenner, S. (1975). Dopaminergic neurons in the nematode *Caenorhabditis elegans*. J. Comp. Neur., **163**, 215–226.

Sulston, J. E. and Horvitz, H. R. (1977). Post-embryonic cell lineages of the nematode *Caenorhabditis elegans*. Develop. Biol., **56**, 110–156.

Sulston, J. E. and Horvitz, H. R. (1981). Abnormal cell lineages in mutants of the nematode *Caenorhabditis elegans*. Develop. Biol., **82**, 41–55.

Sulston, J., Lohrmann, R., Orgel, L. E. and Miles, H. T. (1968). Specificity of oligonucleotide synthesis directed by polyuridylic acid. Proc. Nat. Acad. Sci. U.S.A., **60**, 409–415.

Sulston, J. E., Schierenberg, E., White, J. G. and Thomson, J. N. (1983). The embryonic cell lineage of the nematode *Caenorhabditis elegans*. Develop. Biol., **100**, 64–119.

Sulston, J. E. and White, J. G. (1980). Regulation and cell autonomy during postembryonic development of *Caenorhabditis elegans*. Develop. Biol., **78**, 577–597.

Venter, J. C. *et al.*, (2001). The sequence of the human genome. Science, **291,** 1304–1351.

Wessing, A. (1953). Histologische Studien zu den Problemen der Zellkonstanz: Untersuchungen an *Rhabditis anomala* P. Hertwig. Zool. Jahrb., Abt. Arat. Ontog. Tiere, **73**, 69–102.

White, J. (2000) Worm tales. Int. J. Dev. Biol. **44**, 39–42.

White, J. G. and Horvitz, H. R. (1979). Laser microbeam techniques in biological research. Electro-Optical Systems Design, **11**, 23–24.

White. J. G., Southgate, E., Thomson, J. N. and Brenner, S. (1976). The structure of the ventral nerve cord of *Caenorhabditis elegans*. Phil. Trans. R. Soc. Lond. B., **275**, 327–348.

White. J. G., Southgate, E., Thomson, J. N. and Brenner, S. (1986). The structure of the nervous system of the nematode *C. elegans*. Phil. Trans. R. Soc. Lond. B., **314**, 1–340.

White, J., Albertson, D. and Anness, M. (1978). Connectivity changes in a class of motor neurones during the development of a nematode. Nature, **271**, 764–766.

Wood, W. B. (1988). Embryology. In: The nematode *Caenorhabditis elegans* (ed Wood, W. B. and the community of *C. elegans* researchers, Cold Spring Harbor Laboratory Press), pp 215–241.

World Health Organisation (2002). Genomics and World Health, pp 123–146 (Report of the Advisory Committee on Health Research, Geneva).

Physiology or Medicine 2003

PAUL C. LAUTERBUR and Sir PETER MANSFIELD

"for their discoveries concerning magnetic resonance imaging"

THE NOBEL PRIZE IN PHYSIOLOGY OR MEDICINE

Speech by Professor Hans Ringertz of the Nobel Assembly at Karolinska Institutet.

Your Majesties, Your Royal Highnesses, Honoured Nobel Laureates, Ladies and Gentlemen,

To be able to visualize the inner organs of humans without invasive techniques is of paramount importance to modern medicine. Medical imaging has undergone a dramatic development during the last 30 years. A whole series of imaging modalities have been discovered and developed, among which the development of computer assisted tomography was awarded a Nobel Prize in Physiology or Medicine in 1979. Magnetic Resonance Imaging, MRI, represents a new modality for obtaining diagnostic medical images. The technique still has extensive potential for further development but MRI is important to monitor many diseases of most organs in the human body. Imaging with magnetic resonance is an invaluable aid in the whole healthcare chain from screening and detection, diagnosis and treatment, to follow up of diseases.

Felix Block and Edward Mills Purcell first demonstrated the physical phenomenon of nuclear magnetic resonance in 1946. These discoveries were awarded a Nobel Prize in Physics in 1952. Magnetic resonance occurs in magnetic fields between atomic nuclei and electromagnetic waves of radio frequencies. Atomic nuclei have a magnetic moment and in the magnetic field, their spin depends on the strength of the field. The direction of magnetization resulting from the magnetic moments can change. This happens when the nuclei are in resonance with radio waves of the same frequency as the frequency of their own rotation. In the same way the nuclei can send back radio waves, when there is a change in the direction of the magnetic moment.

Initially, magnetic resonance was mostly used for spectroscopy, to study structures of chemical compounds. In the early 1970s Paul Lauterbur discovered the possibility to create a two-dimensional image by introducing gradients in the magnetic field. By analysis of the characteristics of the emitted radio waves, he was able to determine their origin. This made it possible to build up images of structures that could not be visualized with other methods.

Peter Mansfield discovered further possibilities to utilize gradients in the magnetic field. He showed how the radio signals could be mathematically analysed, which made it possible to develop a useful imaging technique. Mansfield also showed how images could be achievable extremely fast using magnetic resonance. This became technically possible in clinical medicine about a decade later.

Using a metaphor, magnetic resonance spectroscopy is like listening to a radio broadcast of a symphony in the 1940s. Imaging would then be like sitting in a concert hall listening to the symphony, and not only hearing but also seeing the instruments, how they play and where they are located, like organs in the human body. And when you hear the violins, you can even recognise, as in a magnetic resonance image, a false note like a disease process in that organ.

Professor Lauterbur and Professor Mansfield,

Your discoveries of imaging with magnetic resonance have played a seminal role in the development of one of the most useful imaging modality in medicine today. All indications are that it will be even more important in the future of both medical practice and research, and, above all for the patient.

On behalf of the Nobel Assembly at Karolinska Institutet, I wish to convey to you our warmest congratulations, and I now ask you to step forward to receive the Nobel Prize from the hands of His Majesty the King.

PAUL C. LAUTERBUR

My ancestors apparently emigrated from Europe in the middle of the 19th century; the Lauterburs probably from Luxembourg, and my mother's people, Wagners and Weingartners, from Baden-Baden or nearby. They settled in northern Ohio, where my mother's father, Hans Christian Wagner, married Margaret (Maggie) Weingartner. They lived in Tiffin, Ohio when I was a child, where they had raised my mother, Gertrude Frieda Wagner, her twin brother Joseph, and their youngest child, who became a nun with the name Mary Monica. Nearby lived my grandfather Paul Lauterbur who married a woman of Irish descent, Margaret Hillan. They eventually moved south to Sidney, Ohio and had a number of children, of whom my father, Edward Joseph Lauterbur, was the youngest. He later married Gertrude Wagner (the families seem always to have been acquainted) and they had four children, Thomas who died shortly after birth, me, my younger brother Edward Joseph Lauterbur II (Joe) and my sister Margaret.

We grew up in a house in Sidney complete with a series of dogs, and as the years went by, birds, turtles, newts, fish, snakes, and other animals, and with interesting yards full of trees, bushes and flowers, as well as a nearby park, open spaces and neighbors, some of whom did not resent children trespassing on their property. It was, in memory, an idyllic time. My father worked in the town, as an engineer and part-owner of the Peerless Bread Machinery Company, and my mother kept house with help of a young woman who did some domestic chores and sometimes cared for the children. Although I attended a parochial school, Holy Angels School, I recall little of it except that the nuns who taught there seemed to value order and discipline over all else, which made it especially desirable to evade their control. More influential in my later interests was, I believe, my aunt Anna Lauterbur, who taught in the demonstration school at Ball State Teachers College (now Ball State University) in Muncie, Indiana, just west of the Ohio-Indiana border. She was fascinated by natural history, always kept a terrarium in her elementary school classroom, and gave me a subscription to Natural History magazine. A very gentle person, always willing to listen to a child, she was my favorite aunt.

Because of my parents' hobby of horseback riding, they had bought a farm just outside of town, and we moved there just as I was transferring to the public high school. The farm, with an old but remodeled house, a barn, various outbuildings, fields, woods, and a little creek, was a small paradise to a teenage boy, even though I acquired many duties, such as caring for the horses, mowing the lawn, cultivating the garden, and helping with harvesting. There was also time, of course, for hunting and fishing, collecting snakes, turtles and caterpillars to raise to butterflies or moths, and for general exploration. School was now more interesting also. Not only did I take up the game of chess as a freshman, but I beat the local champions at it, to their great disgust because they were seniors, and then moved on to play a local adult expert, one of the teachers. Classes were a mixture of pleasure and boredom. One of my teachers, who taught biology and chemistry, had the foresight to excuse me and some of my classmates, who were members of the local science club, from his lectures, so that we were free to use the time to do experiments, both standard and wild, in the school lab. He also had the courage to intervene when some of the dangerous ones came to the attention of the school authorities and we could have been expelled.

I met him again recently, and his son recalled that when told of my Nobel Prize, he said, "I always knew he would do something like that." After graduating from high school, I went on to Case Institute of Technology, an engineering school now part of Case Western Reserve University, in Cleveland, Ohio, about 200 miles north-east of Sidney. My father had recommended it, because, as he observed, he didn't know what scientists did for a living, but engineers could always get a job. But, given a choice of majors, I chose chemistry.

I had had so-called "chemistry sets" of simple chemicals and apparatus since my earliest years (I particularly liked the pungent smell of burning sulfur), and my own home laboratories even before high school. The curriculum at Case was quite general, including all forms of science (except biology) and engineering, including civil, electrical, mechanical, and chemical, and all of the related technologies such as surveying, mechanical drawing, as well as seemingly endless labs of all kinds, for which I have always been grateful. In addition to the excitement and drudgery of academics, there were also the pleasures and stresses of fraternity life, girls, and culture, as well as new friends and foods. Continuing my habit of doing things a little differently than expected, I wrote a Senior Thesis on my attempt to make an organosilicon free radical, but the advisor for it was an organic chemist who specialized in natural products.

When I graduated (with a B.S. in chemistry, because I did not qualify for an engineering degree as I had replaced a Unit Operations laboratory course with a graduate course in Quantum Chemistry), I was tired of lectures and professors, and determined to get back to lab work. I knew little about graduate study and the structure of a scientific career, so I accepted an offer to work for the Dow Corning Corporation in their Mellon Institute laboratories, where the emphasis was more scientific than technical. I was also told that I could take graduate courses at the University of Pittsburgh free as an Institute employee. There was much interesting work, I found, in our group at the Institute. Organosilicon synthesis, theories of rubber elasticity, techniques of vacuum distillation, elastomer testing, all were new to me and endlessly stimulating. I was particularly fascinated by the puzzle of how small particles strengthened rubber. I even managed to overcome my distaste for academics and take a few courses.

It had long been known that "carbon black" dramatically improved the properties of natural or synthetic organic rubbers, and it had been found that the same was true for silicone elastomers if small particles of silica were used instead of carbon, but not whether surface chemistry was involved or simply physical properties. I addressed one aspect of the problem by substituting phthalocyanine dyes for silica, and they worked perfectly, with their effectiveness decreasing as predicted when the particle size was increased by recrystalization from liquid hydrogen fluoride. Unfortunately, I never achieved a theoretical understanding of the effect, despite intense study of elastomer theory, but I had bright blue rubber and skin.

During that period, I also began to learn about nuclear magnetic resonance (NMR) from various visitors and speakers, and to read a little about that new form of spectroscopy as well. It seemed ideally suited, even at that early date, for investigating the structures and electron distributions in molecules, and various physical properties of materials. Therefore, as part of my graduate education at the University of Pittsburgh, in addition to a "literature seminar" on interstellar molecules, I gave one on a paper describing NMR properties of rubber. Before I could begin a planned collaboration on the hydrogen NMR spectroscopy of silicon compounds, however, my deferments came to an end and I was drafted into the Army and my eventual assignment was proposed to be in the SPP (Scientific and Professional Personnel) program, which my B.S. and two years of work experience qualified me for.

First, however, I was assigned by mistake to a tank battalion at Fort Knox, Kentucky. After hastily correcting that error, I was given eight weeks of minimal basic training and assigned to the SPP program, as planned, at the Army Chemical Center in Edgewood, Maryland. My specific assignment there was in the Medical Laboratories, where I learned to operate an electron microscope to measure the properties of small aerosol particles meant to carry chemical warfare agents deep into the lungs, and I also proposed, and began to set up, a light scattering apparatus to quantitate vapor absorption on aerosol particles. Another aspect of my duties was to capture and weigh experimental animals meant for chemical weapons testing, so that I became skilled, for example, at catching goats in an open field, for which my farm experience was useful. In time, I learned, from a fellow draftee in my barracks, a Columbia Ph.D., that his unit had purchased an NMR machine, but didn't know how to use it. I said, "Hey, I know all about that!", and managed a transfer to help set it up, and arranged for one of my science club buddies, Marlon Shepard, from high school, who had also just been drafted, to join me in the lab, where, among others, we had a drafted Harvard Ph.D. in physical chemistry, Norbert Muller, later a professor at Purdue for many years. We got to work enthusiastically, and I eventually published four papers from our work there, which had turned into a rather unusual opportunity for a young soldier. Perhaps, even more important for my future, I received at least second-hand scraps of a Harvard education, especially the attitudes, from Nobby Muller.

When I was mustered out of the Army, I had to decide where to go next. I even considered regular full-time graduate school, but the appeal of Mellon

Institute as a familiar supportive working environment won out, especially after my group agreed to buy me my own NMR machine. When I returned to the Institute I arranged that requisition, tested the machine on a standard organosilicon compound (polydimethylsiloxane) at the manufacturer's laboratory and factory, and impatiently did the initial installation itself when it was delivered. The first critical experiments I did, however, were on 13-C NMR by retuning the instrument, as I had calculated that, if 29-Si resonances could be seen, so could those of 13-C, and a much larger variety of stable carbon compounds existed than of silicon compounds.

My first work in that area, a broad survey of carbon compounds, led to many other publications on various classes of organic chemicals, work that absorbed much of my attention for several years and eventually provided the basis for my Ph.D. dissertation. Finally completing those requirements was stimulated in part by my learning of an academic job offer to me that was planned but never made, because the department learned that I did not yet have the degree, and I had begun to be dissatisfied with Mellon Institute because of some restrictions they had placed on my activities. After I obtained that degree, I looked at several opportunities and selected one in academia, because, as I remarked, "I wanted to be free to try any silly thing I decided to do." One unexpected feature of the job offer, at the State University of New York at Stony Brook, was that it was for the rank of associate professor, so that I went directly to that level, and almost automatic tenure soon after, without even a post-doctorate appointment. I set up another new NMR lab there, and also began to learn the duties and problems of university life while helping to build the department and the institution, and especially, learning to work with students, by that time having gotten over my own distaste for professors by becoming one myself.

During the academic year 1969–1970, I took my first sabbatical leave, spending it in Palo Alto, California, in the group of John Baldeschwieler in the Chemistry Department at Stanford. In addition to the scientific opportunities and satisfactions, there were personal activities as well. I had married Rose Mary Caputo in 1962, and although she was not in good health, we sometimes visited San Francisco and we had two children, Dan and Sharon (who later renamed herself Sharyn) who enjoyed the nearly perpetual summer there. I had an undergraduate student doing work back in Stony Brook who began a new project in my lab there, calculating hypothetical 13-C spectra of denatured proteins from data for amino acid spectra. Two graduate students, José Ramirez and Skip Hutton, also remained to continue their research, mostly of isotope effects on NMR spectra, and I flew back to Stony Brook almost once a month to stay in touch with these activities.

Back in Stanford, I was trying some new NMR-related things. I went up to the Syntex research labs nearby and began research on 3-H NMR of tritium-labeled pharmaceuticals. Only one paper on tritium NMR of organic compounds, by George Tiers, had appeared, so our discovery that one of the "standards" provided to us by Syntex was apparently not labeled in the position they thought it was interfered with our publishing those observations in the limited time we had available, but led to my later setting up a lab, with a chemistry colleague at Stony Brook, to do more such work. I also began collaborative studies at Varian Associates, in that manufacturer's service labs, of natural abundance 13C NMR spectroscopy of the protein lysozyme in their experimental new superconducting spectrometer, and published the first paper on that subject. I was also working in a lab in the Stanford Medical Center to learn to label a protein, ribonuclease A, with 13C at each of its four methionine residues for eventual NMR study. And, I suppose just to keep busy, I was working with my host, John Baldeschiwieler, and a previous visitor, Barry Shapiro, to his group and friend of mine from Mellon Institute days, to commercialize 13C isotope-enrichment technology developed at Los Alamos National Laboratories. We even started a company, "Kivatec," to use Los Alamos underground distillation methods for that purpose.

It is clear that I was actively beginning to consider biomedical NMR as a new area for application of my skills and knowledge of NMR, partly stimulated by the activities of Oleg Jardetzky, a new member of the Stanford faculty. My intense and detailed involvement in biomedical applications of NMR came, however, from an entirely unexpected direction.

After I returned to Stony Brook, by a long, leisurely automobile drive from California with my family, and settled in again to my department (where I found the same arguments continuing that had been going on when I left) another unexpected event occurred. It had its beginning several years earlier, when a field service engineer for Varian, the leading NMR company, saw an opportunity and asked for my opinion on his idea of starting his own company to make or distribute specialized NMR equipment and supplies. His business plan seemed reasonable, and I encouraged him to go ahead. For a time the company thrived, and I was a member of the Board of Directors.

In May of 1971, however, some other members of the board compared notes with the company's banker and found that the company had engaged in some very dubious business practices and was, in fact, bankrupt. At a hastily-called Board meeting, appropriate actions were weighed, and the banker, there as a guest, threatened to close the company that day unless someone he trusted could be persuaded to take over as President, Chairman of the Board, and Chief Executive Officer. I was the only academic on the Board, the semester had just ended, and the others believed that I was free for the summer, so that I was asked to take the job. I agreed, flew to the company headquarters in New Kensington, PA, near Pittsburgh, at the beginning of each week and back to Stony Brook and my family and students for the weekend.

The developments at the company could supply the plot for a novel, but the incident that is important for my purpose here is that a post-doc arrived with tumor-bearing rats to check the proton NMR relaxation times of their tumors and normal tissues and organs. I was there to observe the experiments, and noted that large and consistent differences were observed for specimens from all parts of the sacrificed animals and that the experiments seemed well-done. Some individuals were speculating that similar measurements might supplement or replace the observations of cell structure in tissues by pathologists, but the invasive nature of the animal procedure was distasteful to me, the data too complex, and the sources of differences too obscure, to be relied upon for medical decisions. As I pondered the problem that evening, I realized that there might be a way to locate the precise origins of NMR signals in complex objects, and hence to form an image of their distributions in two or even three dimensions. That story, and its consequences, is told more fully elsewhere.

Shortly afterwards, I returned to my university for the fall semester, and a colleague took over my company responsibilities. The beginning of the new academic year was a very busy time, and I found some quiet moments to test my ideas about a mathematical approach to such imaging during attendance at seminars and then to consider other practical aspects of the idea as the semester proceeded. In the meantime, I began dropping in on the new medical library of the university, which I passed each morning on the way to work, to spend a few minutes reading, in journals and books, about new developments, problems, and questions in medicine that a new imaging method might address. As I became more confident that these techniques could be both practical and useful, I gradually reoriented most of my research in that direction, then spent almost 30 years on developing its techniques and applications, while chemistry as such became mostly a subject to be taught to students.

An exception, later to be significant, was my general interest in evolution and the origin of life, a topic that I addressed in guest lectures in my university and in selected teaching experiments for undergraduate laboratories. During this period, of course, my children were growing up, as they do, but my marriage was disintegrating. I began to be recognized for my imaging work, and my earlier scientific accomplishments began to be overshadowed by this new direction. At the same time, my efforts to expand the imaging studies, now named MRI by medical doctors, began to be seriously inhibited by administrative and political problems at Stony Brook. My marriage ended in divorce, and I formed a new personal attachment with Joan Dawson, an American physiologist, working at University College, London, whose field was muscle biophysics and physiology, as studied mostly by NMR. If we were to be together, either she needed a new position at Stony Brook or we both needed new jobs elsewhere. After looking at several possibilities, and getting married in 1984, we accepted offers at the University of Illinois.

We moved to Urbana in 1985, with a new baby and high hopes for our professional lives, which were immediately dashed. A plan to share our time between the Urbana and Chicago campuses was foreclosed by technical and political problems in Chicago, and my intended equipment in Urbana, a new whole-body MRI machine associated with a hospital there, was unavailable because of a legal dispute. That problem was never resolved. The hospital eventually sold the machine, but I obtained a small animal-scale machine from the university and began new experiments. My laboratory was organized as the Biomedical Magnetic Resonance Laboratory, initially located in a rented building near campus. When the landlord, a hospital, decided to demolish the building to further its own plans, a small new building was built for my laboratory.

In the late nineteen nineties, that building, including my office, my laboratories, my staff, and all of my equipment, including that provided from university funds in 1985 and those items purchased from external grants over the years, were transferred to another university operation. My wife and I considered looking for new positions, but, in addition to having spent a great deal of time and money building a house, our daughter was in a very good high school, so we stayed. I had a joint appointment in the Department of Chemistry, and moved there, because I had already begun to think about a new approach to the origin of biology from chemistry and wanted to pursue that line of research. Thus, by the time the long-awaited Nobel Prize for MRI was awarded, I had left that field for another (and my daughter had entered college). I am now not only actively pursuing my new research interests, but leaning the new skills in time management required of a Nobel Laureate.

SELECTED REFERENCES (OUT OF 319):

"Filler Phenomena in Silicone Rubber," E.L. Warrick and P.C. Lauterbur, Ind. Eng. Chem. *47*, 485–491 (1955).

"Nuclear Magnetic Resonance Field Shifts of Si29 in Various Materials," G.R. Holzman, P.C. Lauterbur, J.H. Anderson, W. Koth, J.Chem. Phys. *25*, 172–173 (1956).

"Nuclear Magnetic Resonance Spectra of Phosphorus Compounds," N. Muller, P.C. Lauterbur and J. Goldenson, J. Am. Chem. Soc. *78*, 3557–3561 (1956).

"C^{13} Nuclear Magnetic Resonance Spectra," P.C. Lauterbur, J. Chem. Phys. *26*, 217–218 (1957).

"Some Applications of C^{13} Nuclear Magnetic Resonance Spectra to Organic Chemistry," P.C. Lauterbur, Ann. N. Y. Acad. Sci. *70* (4), 841–857 (1958).

"Anisotropy of the C^{13} Chemical Shift in Calcite," P.C. Lauterbur, Phys. Rev. Letters *1*, 343 (1958).

"Sn^{119} Nuclear Magnetic Resonance Spectra," J.J. Burke and P.C. Lauterbur, J. Am. Chem. Soc. *83*, 326–331 (1961).

"Magnetic Shielding and the Electronic Structures of Aromatic Molecules," P.C. Lauterbur, Tetrahedron Letters, 274–279 (1961).

"Nuclear Magnetic Resonance Spectra of Elements Other than Hydrogen and Fluorine," P.C. Lauterbur, Chapter 7 in *Determination of Organic Structures by Physical Methods*, Vol. 2, edited by F.C. Nachod and W.D. Phillips, Academic Press, New York, NY, 1962, pp. 465–533.

"^{13}C Nuclear Magnetic Resonance Spectroscopy. VI. Azines and Methyl Azines," P.C. Lauterbur, J. Chem. Phys. *43*, 360–363 (1965).

"Solvent Isotope Effects on Chemical Shifts of Ions in Aqueous Solutions," A. Loewenstein, J. Shporer, P.C. Lauterbur and J.E. Ramirez, Chem. Commun., 214–215 (1968).

"Pseudorotation in Trigonal-Bipyramidal Molecules," P.C. Lauterbur and F. Ramirez, J. Am. Chem. Soc. *90*, 6722–6726 (1968).

"^{13}C NMR Spectroscopy of Biopolymers," P.C. Lauterbur, E.J. Runde and B.L. Blitzer, in *Magnetic Resonances in Biological Research*, C. Franconi, editor., Gordon and Breach, London, England, 1971, pp. 355–364.

"Image Formation by Induced Local Interactions: Examples Employing Nuclear Magnetic Resonance," P.C. Lauterbur, Nature *242*, 190–191 (1973).

"Zeugmatographic High Resolution Nuclear Magnetic Resonance Spectroscopy. Images of

Chemical Inhomogeneity within Microscopic Objects," P.C. Lauterbur, D.M. Kramer, W.V. House, Jr. and C.-N. Chen, J. Am. Chem. Soc. *97*, 6866–6868 (1975).

"*In Vivo* Zeugmatographic Imaging of Tumors," P.C. Lauterbur, C.-M. Lai, J.A. Frank and C.S. Dulcey, Jr., Physics in Canada *32*, Special July Issue: Digest of the Fourth International Conference on Medical Physics, Abstract 33.11 (1976).

"NMR Studies of the Protein-Solvent Interface," P.C. Lauterbur, B.V. Kaufman and M.K. Crawford, in *Biomolecular Structure and Function*, P.F. Agris, editor, Academic Press, New York, NY, 1978, pp. 329–351.

"Augmentation of Tissue Water Proton Spin-Lattice Relaxation Rates by *In Vivo* Addition of Paramagnetic Ions," P.C. Lauterbur, M.H. Mendonca Dias, and A.M. Rudin, in *Frontiers of Biological Energetics*, P.O. Dutton, J. Leigh and A. Scarpa, editors, Academic Press, New York, NY, 1978, pp. 752–759.

"The Sensitivity of the Zeugmatographic Experiment Involving Human Samples," D.I. Hoult and P.C. Lauterbur, J. Magn. Reson. *34*, 425–433 (1979).

"On Two Approaches to 3D Reconstruction in NMR Zeugmatography," R.B. Marr, C.-N. Chen and P.C. Lauterbur, in *Mathematical Aspects of Computed Tomography*, Vol. 8, G.T. Herman and F. Natterer, editors, Springer-Verlag, 1981, pp. 225–240.

"The Use of Paramagnetic Contrast Agents in NMR Imaging. II. *In Vivo* Studies," M.H. Mendonca Dias, P.C. Lauterbur and E.J. Brown, Jr., *Abstracts*, First Annual Meeting of the Society of Magnetic Resonance in Medicine, Boston, MA, 1982, pp. 105–106.

"Aspects of Cardiac Diagnosis Using Synchronized NMR Imaging," E. Heidelberger, S.B. Petersen and P.C. Lauterbur, Europ. J. Radiol. *3*, 281–285 (1983).

"NMR Technology for Medical Studies," T.F. Budinger and P.C. Lauterbur, Science 226, 288–298 (1984).

"Ferromagnetic Particles as Contrast Agents for Magnetic Resonance Imaging," M.H. Mendonca Dias, M.L. Bernardo, Jr., R.N. Muller, V. Acuff and P.C. Lauterbur, *Abstracts*, Fourth Annual Meeting of the Society of Magnetic Resonance in Medicine, London, England, 1985, p. 887.

"Cancer Detection by Nuclear Magnetic Resonance Zeugmatographic Imaging," P.C. Lauterbur; Accomplishments in Cancer Research, 1985 Prize Year, General Motors Cancer Research Foundation, J.B. Lippincott Co., Philadelphia, (1986); also in Cancer *57*, pp. 1899–1904 (May 1986).

"Microscopic NMR Imaging," P.C. Lauterbur and L. Kyle Hedges, *Abstracts*, XXIII Congress Ampere on Magnetic Resonance, Rome, Italy, 1986, pp. 24–27.

"SLIM: Spectral Localization By Imaging," X. Hu, D.N. Levin, P.C. Lauterbur and T. Spraggins, Magn. Reson. Med. *8*, 314–322 (1988).

"Three Dimensional Electron Spin Resonance Imaging," R.K. Woods, G. Bacic, P.C. Lauterbur and H.M. Swartz, J. Magn. Reson. *84*, 247–254 (1989).

"Relaxivity and Stabilities of Metal Complexes of Starburst Dendrimers: A New Class of MRI Contrast Agents," E. Wiener and P.C. Lauterbur, *Works-in-Progress Abstracts*, Ninth Annual Meeting of the Society of Magnetic Resonance in Medicine, New York, NY, 1990, p. 1106.

"NEUROVISION: A Software Tool for Functional MRI Neuroimaging Analysis," C.S. Potter, M. Banich, N. Cohen, A. Kramer, P.C. Lauterbur and H.D. Morris, *Abstracts*, SMRM/SMRI Functional MRI of the Brain Workshop, Arlington, VA, 1993, p. 243.

"ChickScope: An Interactive MRI Classroom Curriculum Innovation for K-12," B.C. Bruce, B.O. Carragher, B.M. Damon, M.J. Dawson, J.A. Eurell, C.D. Gregory, P.C. Lauterbur, M.M. Marjanovic, B. Mason-Fossum, H.D. Morris, C.S. Potter and U. Thakkar, Computers & Education Journal *29*, pp. 73–87 (1997).

Principles of Magnetic Resonance Imaging: A Signal Processing Perspective, Z.-P. Liang and P. C. Lauterbur, IEEE Press, Piscataway, NJ (1999).

"The Structure of Chemical Matter and the Germs of Life," Second Astrobiology Symposium, NASA Ames Research Laboratory, April 7–11, 2002, poster.

"The Chemical Origins of Biologies: Bootstrapping toward Life," P.C. Lauterbur, (in preparation).

ALL SCIENCE IS INTERDISCIPLINARY – FROM MAGNETIC MOMENTS TO MOLECULES TO MEN

Nobel Lecture, December 8, 2003

by

PAUL C. LAUTERBUR

Biomedical Magnetic Resonance Laboratory, University of Illinois, Urbana, IL 61801, USA.

The title is not a tribute to some trendy hybrid field, but an introduction to a lecture on a Nobel Prize for Physiology or Medicine that is shared by a chemist and a physicist.

Few events could illustrate more clearly the blending both at the boundaries, and sometimes through the bodies, of our disciplines. For that is what they are, disciplines, not natural categories with rigid boundaries to be defended against intrusions, but guides to instruction and efficient administration.

Historically, the record is clear. Chemistry, for example, was cobbled together from mystical alchemy, metallurgy, physics, mineralogy, medicine, and cookery, eliminating incompatibilities as it evolved and consolidated into a more-or-less unified discipline. Physics has been formed, and enriched, by contributions from astronomy, mechanics, mathematics, chemistry, and other sciences. We have recently observed the rationalization of much of biology by chemistry, with the help of physics.

Nuclear magnetic resonance began within physics, at a confluence among particle physics, condensed matter physics, spectroscopy, and electromagnetics. Discovery of ways to observe the subtle properties of atomic nuclei in solids, liquids, and eventually gases, earned Felix Bloch and Edward Purcell a Nobel Prize in Physics in 1952. Applications to studies of molecular motions and structures began almost immediately. The discoverers themselves, it is told, even used their own bodies as samples. In an early predecessor to MRI, Jay Singer measured blood flow in a human arm, and actual medical measurements were started when Erich Odeblad, a Swedish M.D., constructed apparatus and devised methods to study very small quantities of human secretions for medical purposes. Other biological studies followed, in other labs, using animal tissues, including hearts, and entire small animals.

In 1971, Raymond Damadian observed that some malignant tissue, obtained from implanted tumors removed from rats, had longer NMR relaxation times than many normal tissues. This observation caught the attention of several people, and Hollis decided to attempt to confirm and extend it by a study of a related system, Morris hepatomas in rats, readily available to him at Johns Hopkins University. At one point, a post-doctoral fellow in his laboratory, Leon Saryan, brought some of the animals to the small company in western Pennsylvania which had actually carried out the earlier Damadian work. There, the rats were sacrificed and dissected and the tissue samples studied by NMR. I happened to be present to observe the entire process, for reasons described elsewhere, and, as a chemist not ordinarily involved with animal experiments, found them rather distasteful. All such NMR experiments were subject to error from non-uniformities in sample composition, the static magnetic field, and the radiofrequency magnetic field. However, the differences in the NMR signals from one tissue to another, normal as well as diseased, seemed robust in the experiments I observed. I thought they might actually be reproducible and useful, especially if the signal intensities, relaxation times, etc., could be measured from outside the living body with sufficient spatial resolution.

That evening, over dinner, it occurred to me that, as the frequencies of NMR signals depended on the local magnetic field, there might be a general way to locate them in a non-uniform magnetic field. I knew, however, that a static field could not have a unique value in each location in three dimensions, but that a complex shape could be represented by an expansion in a set of functions such as those provided by the correction, or "shim" fields, available on NMR machines to cancel unwanted magnetic field non-uniformities, term by term, with linear gradients, quadratic ones, etc. Could this be the answer? A little reflection made me doubt it. I recalled that single-center expansions of molecular wave-functions had been tried in quantum chemistry, but converged on useful solutions slowly and poorly. An alternative occurred to me. What if one used a large set of simple linear gradients, oriented in many directions in turn in three dimensions? I knew of no examples in any field. This was early September 1971, and X-ray CT was not yet widely known, and neither had I encountered the similar ideas that were being tested in radioastronomy by Braceewell and in electron microscopy by Herman and Gordon, and by others in different fields. Nor did I know of any mathematics to solve such problems, but I recalled another idea from quantum chemistry, that when equations were not solvable in analytic form, an iterative method, in which approximate solutions were compared to known properties and systematically adjusted to a closer and closer fit, could be used.

To test this idea, I wrote down small arrays of numbers such as 1s and 0s, in small arrays 4×4 or even 8×8 square, and added them along the vertical and horizontal directions, representing the 1-dimensional data that would be generated by linear magnetic field gradients perpendicular to those directions, as well as data, at 45° and 135°, that could be generated similarly. The "data" could then be "back-projected" across the image space as a series of bands and summed where they crossed, from which the trial image of summed intensities could be projected in each of the original directions for comparison with the actual "data", and modified by added or multiplicative terms to agree with it. The procedure might then be repeated, in hopes that the next computed trial image would be a closer approximation on each cycle to the synthetic original one. I asked local mathematicians whether such a procedure was known and would work. All said they knew of no examples, and some said it was obviously valid, while others said it was clear that it would not converge, so I just tried it myself, with pencil and paper calculations. The result, with such simple mathematical "phantoms" (test objects) at least, was that the calculations converged very rapidly indeed. Later, a computer scientist with whom I had consulted came across a paper in a subsequent issue of a journal that used exactly my algorithm. This simultaneously validated the method and eliminated my claims to priority. I later learned that much work on the so-called "reconstruction from projections" problem had been published, by many people in many contexts, in recent years, almost all developed independently in different fields. My real interest, however, was in the magnetic resonance imaging problem, and that remained unique.

I then turned my attention to the question of whether there would be enough NMR signal-to-noise ratio with large enough radiofrequency coils to surround a human body and the low magnetic fields I thought might be practical in resistive magnets over such large volumes. The standard reference, "Nuclear Magnetism", by Abragam provided equations that suggested that the answer was favorable. At about the same time, my review of the magnet literature revealed that resistive magnets with fields of the order of 1000 gauss (0.1 tesla) and diameters of about 1 meter could be constructed and operated economically with enough field uniformity to support the NMR experiments I had in mind.

It then appeared that all the requirements could be met if the right research and development could be done, so that a new and useful medical diagnostic tool would be available. But first, there was another matter to resolve. A patent attorney at the company had advised me to do no experiments at my university, as they would compromise my patent position. He and I had been actively working on preparing patent application documents, in exchange for his fee of a percentage of any financial returns that might result. Unfortunately, a business dispute developed between us in connection with the company, and he declined to continue with our agreement. When that happened, I made a patent disclosure to my university, which in turn sent it to the organization they used to evaluate such documents and prepare patent applications.

In the meantime, I began experiments, preparing test objects by attaching 1 mm diameter glass melting point capillaries to the inside of 5 mm glass NMR sample tubes, the capillaries filled with ordinary water (H_2O) and the outer tubes with heavy water (D_2O). The reason for the D_2O was to roughly match the magnetic susceptibility across the sample so that the capillary signals would be less distorted than they would have been with air in that space. I first tried three capillaries, but the signals were too complex for easy interpretation, so I tried just two. I also tried using the linear gradients in the magnetic field supplied by the appropriate "shim" controls on a small analytical NMR spectrometer in the Chemistry Department, with 5 mm sample tubes filled with ordinary water. As expected, their projections were half-ellipses, or semi-circles for the proper adjustment of the strength of the gradient. For an actual test of the image mathematics on real data, I attached a paper disc, marked at intervals of 45°, to the outer tube containing two capillaries and rotated it to orientations of 0°, 45°, 90° and 135° relative to the gradient direction while recording the NMR signal on a pen and ink recorder. I then digitized the recordings by measuring the height of the curves at intervals with a ruler and recording the numbers on a piece of paper, with the intervals corresponding to the projections of a square grid at each angle. The numbers were then transferred manually to punch cards that could be fed to a reader attached to the departmental instrument control computer, originally intended only to operate an X-ray diffractometer for single crystal structure determinations. Its memory (ferrite cores) was so limited that all calculations had to be made in integer ("fixed point") mode, and intermediate results had to be punched out on a deck of cards to be reentered later for the next step, and each subprogram had to be kept on a separate deck of punched cards. The final result was then printed by a typewriter as a 20x20 array of numbers, and the "image" produced by hand-drawn contours on that array. This seems tedious on later description but was exciting then because the whole process and its results were being encountered for the first time, especially as I recognized that the "pictures" were a new kind of image, based on principles completely different from those behind other imaging methods. To emphasize this point, I coined the new word "zeugmatography" as a description, checking with a classical scholar for its fidelity to ancient roots and with a speaker of contemporary Greek to ensure that the meaning of "zeugma" had not shifted during later centuries.

Reassured, I used it in a manuscript I wrote for the journal *Nature*, which was summarily rejected. I felt this was a mistake, not because I foresaw all of the medical applications that would follow, but because of the physical uniqueness of the concept. I was also trying to think of another example that would work in practice, but it was to be over a quarter of a century later that an example, involving the differential shift of the spectra of two closely-spaced atoms by an inhommogeous electric field, was published, but the authors did not notice the similarities. My appeal to *Nature* was followed by submission of a revised version of my manuscript containing references to cancer and other more obviously relevant topics, and this time it was accepted. Almost thirty years later, *Nature* publicly celebrated its appearance there. Slightly earlier, I had presented my results in a short contributed paper at an American Physical Society meeting, which then had a policy of accepting any meeting talk by a member, but it was attended by only a handful of listeners, one of whom was a graduate student who told me that his professor had done the same thing, but I never found any evidence that he had. A similar pattern repeated itself several times in later years, with people telling me that they had the same ideas but had not followed them up with experiments and publication.

This work, and its subsequent elaborations, became the subject of my lectures afterward at most meetings I attended, including seminars. Before I began describing it in detail everywhere, however, the University's agent rejected the patent application because they felt that it could not generate enough funds to pay for the application process. I then asked my university for permission to pursue the application independently but never received a reply. I was not in a financial position to quit my job and defy the university, and the grace period for applying for a U.S. patent after publication had nearly expired, so I abandoned that idea and decided instead to encourage others to pursue this new technology, inviting everyone interested to visit my laboratory to observe our efforts and learn from us. People did come, from industry, academia, and government laboratories, foreign and domestic, and I began supplying a bibliography of such work to all and helping to organize meetings on the subject to compare our methods and results. Among these were Professor Raymond Andrew and members of his group at Nottingham University, and Drs. Mansfield, Moore, and others there, as well as representatives of medical instrument companies and medical doctors and medical physicists themselves. As I hoped, interest began building as many other groups were involved.

We continued our work, which shortly involved graduate students and postdoctoral fellows as well as undergraduate research students, and, as I had hoped, more contributions were published by other laboratories, with some remarkable early images from Waldo Hinshaw in Andrew's group at Nottingham. As the depth and breadth of application grew, both large and small companies began to see opportunities, and within less than ten years commercial instruments began to come to market, large enough to hold a human being and to support true clinical research. Competitive pressures among physicians, industrial interest, and multiplying applications and techniques began to generate the explosive growth that was to characterize the past twenty years, leading, among other things, to the recognition of this phenomenon by the Royal Swedish Academy of Sciences. I and my group continued to make contributions through this period as well, some of them significant, but the most gratifying experiences emotionally were those when a stranger would volunteer "you saved my daughter's life", or "your machine saved me from an unnecessary operation". By the end of the millennium, despite the continuing excitement of the field, almost thirty years of a detour from chemistry to medical imaging began to pall, and I changed my focus to a field of chemical research, just in time for my past to catch up with me in the form of a Nobel Prize. All detours should be so productive!

SELECTED REFERENCES (OUT OF 319):

"On The Signs of CH and HH Coupling Constants," P.C. Lauterbur and R.J. Kurland, J. Am. Chem. Soc. *84*, 3405 (1962).

"Pseudorotation in Trigonal-Bipyramidal Molecules," P.C. Lauterbur and F. Ramirez, J. Am. Chem. Soc. *90*, 6722–6726 (1968).

"Image Formation by Induced Local Interactions: Examples Employing Nuclear Magnetic Resonance," P.C. Lauterbur, Nature *242*, 190–191 (1973).

"Stable Isotope Distributions by NMR Zeugmatography," P.C. Lauterbur, *Proceedings* of the First International Conference on Stable Isotopes in Chem., Biol. and Med., U.S.A.E.C. CONF-730525, 1973, p. 255–260.

"Magnetic Resonance Zeugmatography," P.C. Lauterbur, Pure Appl. Chem. *40*, 149–157 (1974).

"Reconstruction from Selectively-Excited Signals in Nuclear Magnetic Resonance Zeugmatography," P.C. Lauterbur, W.V. House, Jr., D.M. Kramer, C.-N. Chen, F.W. Porretto and C.S. Dulcey, Jr., in *Image Processing for 2-D and 3-D Reconstruction from Projections: Theory and Practice in Medicine and the Physical Sciences*, Opt. Soc. Am., 1975, p. MA10-1-2.

"Zeugmatographic High Resolution Nuclear Magnetic Resonance Spectroscopy. Images of Chemical Inhomogeneity within Microscopic Objects," P.C. Lauterbur, D.M. Kramer, W.V. House, Jr. and C.-N. Chen, J. Am. Chem. Soc. *97*, 6866–6868 (1975).

"Water Proton Spin-Lattice Relaxation Times in Normal and Edematous Dog Lungs," P.C. Lauterbur, J.A. Frank and M.J. Jacobson, Physics in Canada *32*, Special July Issue: Digest of the Fourth International Conference on Medical Physics, Abstract 33.9 (1976).

"Nuclear Magnetic Resonance Zeugmatography for Medical Imaging," C.-M. Lai, W.V. House, Jr., and P.C. Lauterbur, *Proceedings* of IEEE Electro/78 Conference, Session 30, "Technology for Non-Invasive Monitoring of Physiological Phenomena," May 25, 1978, paper 2.

"Augmentation of Tissue Water Proton Spin-Lattice Relaxation Rates by *In Vivo* Addition of Paramagnetic Ions," P.C. Lauterbur, M.H. Mendonca Dias, and A.M. Rudin, in *Frontiers of Biological Energetics*, P.O. Dutton, J. Leigh and A. Scarpa, editors, Academic Press, New York, NY, 1978, pp. 752–759.

"The Sensitivity of the Zeugmatographic Experiment Involving Human Samples," D.I. Hoult and P.C. Lauterbur, J. Magn. Reson. *34*, 425–433 (1979).

"31P Spectroscopic Zeugmatography of Phosphorus Metabolites," P. Bendel, C.-M. Lai and P.C. Lauterbur, J. Magn. Reson. *38*, 343–356 (1980).

"Zeugmatography by Reconstruction from Projections," P.C. Lauterbur and C.-M. Lai, IEEE Trans. Nucl. Sci., Vol. NS-27, No. 3, 1227–1231 (June 1980).

"On Two Approaches to 3D Reconstruction in NMR Zeugmatography," R.B. Marr, C.-N. Chen and P.C. Lauterbur, in *Mathematical Aspects of Computed Tomography*, Vol. 8, G.T. Herman and F. Natterer, editors, Springer-Verlag, 1981, pp. 225–240.

"True Three Dimensional Nuclear Magnetic Resonance Zeugmatographic Images of a Human Brain," D.M. Kramer, J.S. Schneider, A.M. Rudin and P.C. Lauterbur, Neuroradiology *21*, 239–244 (1981).

"Nuclear Magnetic Resonance Zeugmatographic Imaging of the Heart: Application to the Study of Ventricular Septal Defect," M.A. Heneghan, T.M. Biancaniello, E. Heidelberger, S.B. Petersen, M.J. Marsh and P.C. Lauterbur, Radiology *143*, 183–186 (1982).

"Three-dimensional Nuclear Magnetic Resonance Zeugmatographic Imaging of Surgical Specimens," M.H. Mendonca Dias, W.J. Mann, J. Chumas, M.L. Bernardo, Jr. and P.C. Lauterbur, Bioscience Reports *2*, 713–717 (1982).

"3D Synchronized Proton NMR Imaging of the Beating Heart," E. Heidelberger, S.B. Petersen and P.C. Lauterbur, *Abstracts*, First Annual Meeting of the Society of Magnetic Resonance in Medicine, Boston, MA, 1982, pp. 72–73.

"Gas Phase 19F NMR Zeugmatography: A New Approach to Lung Ventilation Imaging," E. Heidelberger and P.C. Lauterbur, *Abstracts*, First Annual Meeting of the Society of Magnetic Resonance in Medicine, Boston, MA, 1982, pp. 70–71.

"The Use of Paramagnetic Contrast Agents in NMR Imaging. I. Preliminary *in vitro* Studies," M.H. Mendonca Dias, P.C. Lauterbur and A.M. Rudin, *Abstracts*, First Annual Meeting of the Society of Magnetic Resonance in Medicine, Boston, MA, 1982, pp. 103–104.

"Rapid Medium-Resolution 3-D NMR Zeugmatographic Imaging of the Head," M.L. Bernardo and P.C. Lauterbur, Europ. J. Radiol. *3*, 257–263 (1983).

"True 3-D Imaging of Limbs by NMR Zeugmatography with Off-Resonance Irradiation," R.N. Muller, M.J. Marsh, M.L. Bernardo and P.C. Lauterbur, Europ. J. Radiol. *3*, 286–290 (1983).

"NMR Technology for Medical Studies," T.F. Budinger and P.C. Lauterbur, Science *226*, 288–298 (1984).

"Theory and Simulation of NMR Spectroscopic Imaging and Field Plotting by Projection Reconstruction Involving an Intrinsic Frequency Dimension," P.C. Lauterbur, D.N. Levin and R.B. Marr, J. Magn. Reson. *59*, 536–541 (1984).

"Experimental Example of NMR Spectroscopic Imaging by Projection Reconstruction Involving an Intrinsic Frequency Dimension," M.L. Bernardo, Jr., P.C. Lauterbur and L.K. Hedges, J. Magn. Reson. *61*, 168–174 (1985).

"Microscopic NMR Imaging of the Magnetic Fields Around Magnetite Particles," P.C. Lauterbur, M.L. Bernardo, Jr., M.H. Mendonca Dias and L.K. Hedges, *Works in Progress Abstract*, Fifth Annual Meeting of the Society of Magnetic Resonance in Medicine, Montreal, Canada, 1986, pp. 229–230.

"Image Reconstruction," R.K. Woods, W.B. Hyslop, R.B. Marr and P.C. Lauterbur, in *EPR Imaging and In Vivo EPR*, G.R. Eaton, S.S. Eaton and K. Ohno, editors, CRC Press, Inc., Boca Raton, FL, 1991, pp. 91–117.

"Simultaneous 3D-Localized, 31P Spectroscopy of Brain and Muscle in the Intact Head of the Living Rat Using the SLIM Technique," E.S. Fletcher, C.D. Gregory, H. Lee, W.T. Greenough, P.C. Lauterbur and M.J. Dawson, *Works-in-Progress Abstracts*, Ninth Annual Meeting of the Society of Magnetic Resonance in Medicine, New York, NY, 1990, p. 1332.

"Effects of Restricted Diffusion on Microscopic NMR Imaging," W.B. Hyslop and P.C. Lauterbur, J. Magn. Reson. *94*, 501–510 (1991).

"Constrained Reconstruction Methods in MR Imaging," Z.-P. Liang, F.E. Boada, R.T. Constable, E.M. Haacke, P.C. Lauterbur and M.R. Smith, Rev. of Mag. Res. in Med. *4*, 67–185 (1992).

"RF Microcoils with Micron-scale Feature Sizes for NMR Microscopy," T.L. Peck, R.L. Magin, L. LaValle, I. Adesida and P.C. Lauterbur, *Abstracts*, 12th Annual Meeting of the Society of Magnetic Resonance in Medicine, New York, NY. 1993, p. 296.

"Dendrimer-Based Metal Chelates: A New Class of MRI Contrast Agents," E.C. Wiener, M.W. Brechbiel, H. Brothers, R.L. Magin, O.A. Gansow, D.A. Tomalia and P.C. Lauterbur, Mag. Res. in Med. *31*, 1–8 (1994).

"Is There a Role for Interactive Brain Mapping Over the World Wide Web?" P.C. Lauterbur, C.S. Potter, C.D. Gregory and H.D. Morris, *Abstracts*, International Symposium on Brain Mapping, Oiso, Japan, p. 42 (1996).

"ChickScope: An Interactive MRI Classroom Curriculum Innovation for K-12," B.C. Bruce, B.O. Carragher, B.M. Damon, M.J. Dawson, J.A. Eurell, C.D. Gregory, P.C. Lauterbur, M.M. Marjanovic, B. Mason-Fossum, H.D. Morris, C.S. Potter and U. Thakkar, Computers & Education Journal *29*, pp. 73–87 (1997).

"Apparent Water Diffusion Measurements in Electrically Stimulated Neural Tissue," V. Gulani, G.A. Iwamoto and P.C. Lauterbur, J. Magn. Reson. in Med., *41*, pp. 241–246 (1999).

"Lactate Quantitation in a Gerbil Brain Stroke Model by GSLIM of Multiple-Quantum-Filtered Signals," J. A. Kmiecik, C.D. Gregory, Z.-P. Liang, P.C. Lauterbur and M.J. Dawson, J. Magn. Reson. Img. , *9*, pp. 539–543 (1999).

Principles of Magnetic Resonance Imaging: A Signal Processing Perspective, Z.-P. Liang and P. C. Lauterbur, IEEE Press, Piscataway, NJ (1999).

"Apparent Diffusion Tensor Measurements in Myelin Deficient Rat Sprinal Cords," V. Gulani, A.G. Webb, I.D. Duncan and P.C. Lauterbur, Magn. Reson. In Med., *45*:2, pp. 191–195 (2001).

"The Structure of Chemical Matter and the Germs of Life," Second Astrobiology Symposium, NASA Ames Research Laboratory, April 7-11, 2002, poster.

"The Chemical Origins of Biologies: Bootstrapping toward Life," P.C. Lauterbur, (in preparation).

PETER MANSFIELD

INTRODUCTION

I was born on 9th October 1933 in Lambeth, London the youngest of three brothers. I grew up in Camberwell, ten minutes walk from Camberwell Green the epicentre of the Borough. My father, Sidney George Mansfield was the eldest son in a family of five sons and four daughters.

My father worked for the South Metropolitan Gas Company, as a gas fitter. My mother, Rose Lillian, was the youngest of three daughters. At the time of my birth she was not working, but looked after me and my two brothers, Conrad William, the eldest, and Sidney Albert.

OUTBREAK OF WORLD WAR II

In September 1939 when war broke out I was nearly 6 years old and my memory of this period is now hazy. However, I do recall very clearly the first time the air raid siren sounded on the day war was declared. I was playing in the street nearby and ran home asking what the strange wailing sound was. In the months that followed plans were instigated to evacuate all children from the London area.

I was evacuated from London on three occasions during the war years. The first was a relatively short period spent in Seven Oaks. Later I was sent to Torquay, Devon and stayed with the Rowland family. During a lull in the air raids I came back to London for approximately one year then returned to Torquay shortly after the commencement of the V2 bombardment of London and was fortunate to be able to stay again with the Rowlands.

SECONDARY EDUCATION

Back in London after the war I was hurriedly told by my school master that I should take the 11+ examination, something I had never heard of before. There was no preparation because time was too short. I took the exam and failed, but not completely. The mark I received was not quite high enough to get me into the local Grammar school, but it was sufficient for me to go to a Central School in Peckham. However, this situation lasted for approximately one year. Then all Secondary schools in London at least, were completely reorganized so that Central schools were dispensed with altogether and pupils from Secondary schools and Central schools were merged into new schooling arrangements which were called Secondary Modern schools, (these were later renamed Comprehensive schools, but after I had left).

ROCKET SCIENCE

I left school at age 15 and started work as a printer's assistant. This took me almost to the age of 18. Because I had developed an interest in rocketry I applied for and obtained a job in the then Ministry of Supply at the Rocket Propulsion Department in Westcott, Buckinghamshire. I remained in this position for around 18 months and was called up for National Service to serve in the army for two years, returning to Westcott after National Service in 1954. I studied for Advanced Levels part time for approximately two years and gained University entrance.

UNIVERSITY EDUCATION

In 1956 I entered Queen Mary College, University of London. My subject was Physics and in my third undergraduate year we were given individual projects to pursue instead of the normal practical laboratories which we had taken during the first two years of the course. My individual topic was set by Dr. Jack G Powles. It was to build a portable NMR spectrometer to measure the earth's magnetic field. This had been done previously using valve technology, but my task was to build a transistorized version. I knew very little about transistors at this time since it had not formed any part of the electronics course that we had undergone in physics. My experience was very much in valve technology. Nevertheless, I enjoyed the project very much and of course learned a great deal about transistors. Towards the end of my undergraduate studies I was approached by Jack Powles, who offered me a position in his research group working on NMR.

At that time Jack Powles was a Reader in Physics and I felt greatly honoured to be asked to join his research group. Jack Powles' main interest then was in studying molecular motion in a range of materials, mainly liquids. My task was to build a pulsed NMR spectrometer to study solid polymer systems. It was during this period that I discovered what we later called 'solid echoes'. This was towards the end of my three year doctoral studies and we were only able, therefore, to produce a short paper on solid echoes observed in a single crystal of Gypsum.

THE AMERICAN ADVENTURE

Towards the end of my studies I was asked whether I would be interested in going to the United States for a short post-doctoral period. Jack offered to arrange a position working with Professor Charlie Slichter in the University of Illinois at Urbana, Illinois.

I married my wife, Jean Margaret Kibble, on 1st September 1962 and we left to join the University of Illinois in October 1962. Charlie and his first wife Nini met us at the airport at Urbana-Champaign and installed us in what turned out to be temporary post-graduate accommodation at Orchard Downs. Orchard Downs was about two miles from the campus centre and was therefore within walking distance, so I thought. Of course I had not taken in to account the weather. In late October early November the typical air temperature in early morning could reach minus 20 degrees Centigrade even though the skies above were clear blue. I remember attempting to walk in one morning and I got roughly half way to the campus when the frost began to effect me. I continued to walk briskly and by the time I got to the Physics Department my limbs were beginning to sieze up with the cold. My eyelids crunched when I blinked and I was extremely thankful that I managed to get to the Department before anything serious happened.

The research work that I was involved in during my stay at Urbana was the NMR study of doped metals. A scientific paper by Kohn and Vosko had predicted that conduction electron scattering around zinc centres would behave in a periodic manner so that the resonance shifts of copper atoms would vary in shells around the zinc scattering centres and these shifts would be measurable by looking for low amplitude side bands in the copper resonances of the two copper isotopes. My task was first to build a double resonance spectrometer capable of looking at the copper resonances in a pulsed mode and secondly, to produce single crystals of doped copper as suitable samples to be studied. Building the apparatus was relatively straight forward since I had already built a pulsed spectrometer back in London, but the production of a single crystal of copper with sufficient surface area to give a measurable free induction decay signal was a novel experience for me. It involved growing a single crystal and slicing it into thin plates to give sufficient surface area. The slicing process was done with an electro-spark cutting process which left the surfaces severely disturbed so that it was necessary to etch the damaged surfaces away to reveal a clean undamaged surface.

I worked as a post-doc for approximately two years and made several samples and did many experiments on these samples but could not see the predicted effect. But the experience that I gained on this project and the knowledge and background that I learnt during my stay in Illinois were invaluable.

In Charlie Slichter's laboratory, I was not able to pursue the solid echo work started in London, but a colleague from Jack Powles' laboratory, Doug Cutler, who came over at roughly the same time as we did was working with Professor Ted Rowland in the Metallurgy Department at Urbana. The apparatus that Doug Cutler was using was very similar to the equipment that I had left behind in London. Towards the end of my stay in Urbana I persuaded Doug Cutler to perform a couple of experiments for me on pure aluminium powder. He was able to see solid echo effects and I produced a short mainly theoretical paper extending the theory of solid echoes to systems with spin greater than a half. In the same paper I also included solid echo effects on the two spin species system, NaF. My colleague, Dr John Strange (now

Professor), while working as a post-doc at Cornell University, also allowed me to include some of his preliminary experimental data.

While I was busy in the Physics Department my wife took a post working in the University Health Centre as a Secretary. She quickly made friends with a number of other secretaries and introduced me to people with whom I would not otherwise have come in contact. We saw her friends on several occasions and had an enjoyable interaction with them. Of course we also interacted with the Slichters and the other members of the group.

THE WANDERERS RETURN

After the two year period we were sad to leave Urbana and the various friends that we had made there during our stay. But we were also pleased to return to London to see our respective parents after such a long period away. I had been approached by Professor Raymond Andrew while we were still in the States. He had recently moved from Bangor to Nottingham and in a letter to me in July or August of 1964 he wrote to offer me a Lectureship at the University of Nottingham. I had had other offers including one from Jack Powles who had recently taken the founding Professorship in Physics at the University of Kent at Canterbury, however I decided that I liked the offer from Nottingham and accepted. After several weeks in London my wife and I made our way to Nottingham to meet Professor Andrew and to take up his offer. I was still quite interested in pursuing my earlier ideas developed during my PhD studies in London and so I was given a room where I was able to set up equipment to pursue my studies in multiple-pulse NMR.

LIFE IN NOTTINGHAM

I was extremely fortunate to be offered a Canadian research student shortly after my arrival at Nottingham. His name was Don Ware and he came from UBC in Canada. He had an MSc in NMR from the Chemistry Department at UBC. His MSc supervisor was Basil Dunnell. Don had little experience in pulsed NMR but was keen to learn so I proposed that he build a pulsed NMR spectrometer capable of performing multiple-pulse experiments. This was in 1964. A year or so later Jack Powles came to Nottingham to give a colloquium talk in the Physics Department. Afterwards he was shown around the various research groups. By that time Don and I had managed to perform our first multiple-pulse experiments and we were pleased to demonstrate these results to Jack. During the course of his visit he said he thought he had seen something similar from a pre-print that he had been sent by John Waugh from MIT. But he wasn't sure whether it was the same or not, and that he would contact me when he got back to Canterbury. A week or so later a copy of John Waugh's pre-print to Jack arrived on my desk. By then I had already drafted a letter of our own which was duly sent off for publication to Physics Letters. Our paper was published shortly after John Waugh's paper and its appearance triggered a period of inter-group squabbling that continued on up to the early 70's.

In 1972 Dr. Alan Garroway, an American post-doc, joined my group to study multi-pulse techniques. He had carried out his PhD studies at Cornell University. His work had been concerned with using NMR to study fluid flow. He and his first wife, Mary (now deceased), arrived in February 1972, and on the day they arrived we invited them to dinner, but the country was in the grip of a series of electricity power cuts so that Alan and Mary's first experience of Britain was a candle lit dinner. Earlier in 1971 I had received an SRC research grant to pursue multi-pulse NMR and included in the equipment support was money for a mini-computer. I had, therefore, purchased a Honeywell computer with 4 k bytes of magnetic memory and had spent quite a bit of the summer of 1971 adapting the machine so that we could connect it to the experimental apparatus in a computer controlled mode. One of the first things that Alan was engaged in was the implementation of the Cooley-Tukey fast Fourier transform algorithm so that transient signals could be rapidly captured and transformed into a spectrum. This work was very successful and later in 1972 we were regularly using the computer controlled spectrometer to study the response of a number of suitable compounds including Calcium Fluoride. Computerization of the NMR spectrometer meant that we were able to pop specimens in for extremely rapid analysis and as a result we quickly ran out of suitable compounds to study with interesting chemical shift anisotropies. It was at this point in early summer of 1972 that I started to think seriously about applications of the line narrowing process.

THE BIRTH OF MRI

In those days coffee and tea breaks were taken in the tea room of the Physics Department. It was an opportunity for members of staff and students, sometimes in different research groups, to interact and exchange ideas. On this occasion all other people had left the tea room except for me, Alan Garroway and my student Peter Grannell. We had exhausted all the readily available materials for studying chemical shift anisotropy. Among the results that we had obtained was an extreme elongation of the free induction decay of Fluorine in a single crystal of Calcium Fluoride. It suddenly occurred to me that by removing the dipole-dipole interaction in Calcium Fluoride it ought to be possible, at least theoretically, to look at the atomic structure of Fluorine by imposing an external magnetic field gradient. Alan, who had worked with gradients during his PhD studies, was skeptical, but Peter Grannell was more amenable to the idea. I rushed away from this meeting and wrote up several pages of calculations showing the theoretical possibilities and handed them to Peter Grannell. Peter was in the last year of his PhD studies which finished formally on the 1st October 1972. We had already discussed the possibility of him staying on as a post-doc and I suggested that one of the topics he should work on was NMR diffraction studies along the lines of the calculations I had already produced.

ALT HEIDELBERG DU FEINE

I had also made arrangements for sabbatical leave to work in Professor Karl Hausser's group in Heidelberg starting in October 1972. My family and I left for Heidelberg in late September 1972 with the plan that Peter and I would continue to interact via post while he prepared to do some experiments on NMR diffraction. I had already decided that there were extremely difficult problems to overcome using real single crystals of Calcium Fluoride and that a simulated approach was called for initially. In this approach a model lattice was to be constructed with several plates of camphor separated by thin sheets of plastic. Although a solid, camphor contains a large number of highly mobile protons which rotate to give a relatively narrow absorption line. The sheets separating the camphor plates contained far fewer mobile protons which gave virtually no signal component to the free induction decay. Our first experimental results were obtained as early as November 1972, and they showed the diffraction effects when the magnetic field gradient was switched on. When the gradient was turned off a single absorption line shape was observed with no splitting. After repeating these results many times during the early part of 1973, the results were written up for presentation at the First Specialized Colloque Ampère in Krakow, Poland in September 1973. At the same time a more formal publication was submitted to the Journal of Physics, C which appeared in November 1973.

JOURNEY TO KRAKOW

My wife and I made the long journey to Krakow by coach stopping overnight in Leibzig. My talk at the Conference was scheduled as the first invited lecture entitled "Multi-pulse Line Narrowing Experiments: NMR Diffraction in Solids?". My recollection is that the paper was well received and created some discussion afterwards. One person in the resulting discussion asked if I was aware of similar work that had been carried out by Professor Paul Lauterbur. In fact I was completely unaware and asked where I might find this work. I also asked whether it was concerned with imaging in solids and was told that it was not, that it was concerned with imaging of water in test tubes. The person raising the matter was none other than my old sparing partner, Professor John Waugh. When I returned to England I looked up the relevant publication in the library and decided that the approaches though completely different were not entirely unrelated. The paper by Paul Lauterbur had been published several months earlier in the year. But there were several concerns that I had concerning Paul's approach to imaging liquids and indeed our own approach to imaging in solids. The first which applied to both techniques, was the question of defining an active slice of material. In the case of Paul's work he had used the method of projection reconstruction to obtain an image and in that technique a large number of experiments needed to be conducted in order to extricate the image with reasonable resolution. This took valuable time. In my own approach my first concern was the complication of producing images from multi-pulse experiments. There was also the question of defining a slice. We very soon decided that the pursuit of imaging in solids was perhaps ahead of its time and could be deferred for future work. It would be so much easier to look at biological specimens where the relaxation times were shorter and where the line widths were generally speaking narrower.

PASSAGE TO INDIA

Early in 1974 a number of colleagues from the Physics Department travelled to India for an International Conference on NMR. Among these were Dr. Bill Moore and Raymond Andrew's post-doc, Dr. Waldo Hinshaw. At this conference Paul Lauterbur gave a talk on his work on imaging which created some excitement and provoked quite a bit of discussion among the Nottingham group during their return flight to England. Shortly afterwards they conceived a different approach to imaging which was called the "sensitive point" method of imaging. By applying time dependent magnetic field gradients along two orthogonal axes they were able to define a point volume of the specimen which could be swept thereby interrogating the whole plane, point by point, to produce an image. Inevitably this was an extremely slow process.

THE SLICE OF LIFE

I was still very much concerned with imaging speed and also the question of sliced definition. After a lot of thought and discussion with Peter Grannell we came up with a method of slice selection which looked as though it might work reasonably well. Alan Garroway also came up with a different method of slice selection using a string of short pulses to define the slice and between us we thought that the sensible approach would be to combine our efforts and publish a short note on the general technique of slice selection. This was sent to the Journal of Physics and was published in the form of a letter. The question of imaging times was still concerning me and during the course of 1974 I spent a great deal of time turning over my thoughts on how this may be achieved. One way forward was what I called line scan imaging. In this method a line of magnetization in a specimen was selectively excited and read out. This process was repeated until the object had been scanned. The technique was much faster than the sensitive point scan method of Hinshaw and also turned out to be faster than the projection reconstruction method of Paul Lauterbur, but I was still not satisfied. Nevertheless, line scanning was used to produce a number of images and in particular it was used to scan the finger of one of my early research students, Dr Andrew Maudsley. The scan times for these finger images were 15–23 minutes. These were the first images of a live human subject and were presented at a special meeting of the Medical Research Council which was convened in 1976 to review the work of the several imaging groups that had sprung up both at Nottingham and also in Aberdeen. All groups were vying for MRC support and this meeting was called specially to review the topic and to decide how best to support the work. The images demonstrating live human anatomy were annotated by Professor Rex Coupland, then head of the Department of Human Morphology. They produced a startling response at this meeting and convinced the MRC that our work should be supported. We produced a grant application requesting a substantial sum of money to produce a whole body MRI machine.

The application was handed to Professor Andrew for his comments before sending it off to MRC. In fact the application was delayed by a month or so with no comment or explanation. However I learnt subsequently that Raymond Andrew himself had already send in a research grant application to MRC and he had decided that he should wait until he had received a decision from MRC before allowing my application to go forward. His application, which was granted, was concerned with building an intermediate size imaging machine with a sample access diameter of about 10 cm. The intention was to establish an intermediate step between the small scale approaches that we had already demonstrated and the whole body machine which I was keen to press on with. My grant application was sent in subsequently and considered at the following round of grants. The application was successful and the result was announced in 1977.

The delay in the submission of my MRC grant created some acrimony which continued in one form or another for several years until Raymond took early retirement and left for the University of Florida in 1984. During this period strife within the Andrew group occurred when Bill Moore decided that he wanted to split his imaging activities away from Raymond's group. This occurred in the late 70's when a whole body magnet was obtained with a grant from the Wolfson Foundation and Bill Moore together with Neil Holland decided that the results of this new work should be independently managed. Thus a period of internecine squabbling broke out between Raymond Andrew's remaining group, which included Dr Paul Bottomley and a member of staff, Dr Peter Allen, who was loosely associated, and the third group headed by Bill Moore. After a while, under this new regime, matters began to settle down again in a quasi steady state.

BLACKSBURG

Dr Peter Morris, who in 1977 had just finished his PhD studies on multi-pulse NMR in solids and liquid crystals, stayed on at Nottingham as a post-doc for two years or so. This occurred at approximately the time when the MRC funding was granted. He together with Dr Ian Pykett were heavily involved in the installation of our first 0.1 T electro-magnet and subsequent RF and gradient coil designs. We had all been working flat out to get a large scale image in time for the ENC Meeting in Blacksburg, Virginia in April 1978. In the event, the night before we were due to fly out, I volunteered to climb into the machine for an attempt at imaging my abdomen.

Just a day or so before, I had received a typewritten draft of a piece from

Professor Tom Budinger of the University of California at San Francisco suggesting that the gradient levels and switching rate that we were proposing to use to produce my abdominal image, were potentially dangerous. I personally did not believe the calculations and had performed alternative calculations which suggested that the expected induced transient currents flowing in response to the gradient switching were much lower than those predicted by Tom.

I climbed into the machine and signaled to Peter and Ian to push the button for a single pulse. There was an audible crack but I felt nothing. I then signaled to start the scan. The magnet was enclosed in aluminium sheeting forming an RF screen. Due to lack of time there was no light inside. I was therefore clamped in the magnet vertically and in pitch darkness for 50 minutes until the procedure was completed. Our wives and fiancées were present ready to haul me out of the magnet in an emergency, but the whole experiment went well and images were recorded. Photographs of the raw images were taken, but the film was processed in a local store in the USA a day or so before the presentation.

TROUBLE AHEAD

During the early 80's there were growing difficulties within the University of Nottingham concerned mainly with the promotional prospects in the various departments on campus. This meant that aspiring members of staff were being artificially held back because of arbitrary constraints introduced by the government of the day and forced upon the Vice-Chancellor, Professor Basil Weedon. These constraints induced a feeling of great uncertainly throughout the University which was heightened when calls were made to introduce an early retirement scheme in order to reduce the over staffing levels that were considered extant in many Departments. These considerations produced a rash of requests for early retirement which in some cases carried a substantial early retirement premium especially for younger members of staff. In the exodus that followed the Department of Physics lost four members of staff to early retirement. These were Professor Andrew, Dr Moore, Dr Peter Allen and Dr Bill Derbyshire. Dr Moore went to the United States to take up a Professorship at the Brigham and Women's Hospital in Boston, Massachusettes. Dr Allen went to the University of British Colombia in Canada to set up a medical imaging group there and Dr Derbyshire took a Senior Lectureship at Sunderland Polytechnic. Sadly Dr Moore collapsed and died while playing tennis approximately six months after taking up his new position. Raymond Andrew died of cancer in 2001 at the age of 78. Dr Peter Allen successfully created his own MRI group and continues to research his particular interests in medical imaging. Dr Derbyshire later took a position working with Rank-Hovis and is now retired.

With all these staff changes occurring over a 1–2 year period, my group was the only one in MRI to survive in Physics at Nottingham. In fact in 1984, I myself was courted to return to the States with no fewer than three offers, from the University of Alabama, University of Illinois at Urbana and the University of Maddison, Wisconsin. I decided to stay at Nottingham despite the attractive offers.

A NEW START

Shortly after my return to Nottingham, I received a call from the Department of Health telling me that there was some funding available to purchase a 0.5 T superconductive magnet – was I interested? I accepted immediately and later in 1984/5 the magnet arrived and was installed in a small extension of the Physics Department. These events helped to settle me firmly in Nottingham and removed all doubts I may have had about moving elsewhere.

I had been greatly concerned about our proposals to use the 0.5 T magnet for EPI. My major worry was whether rapid magnetic gradient switching within the close confines of the magnet would induce the static magnetic field to quench. I had been thinking seriously about the problem for several months prior to the arrival of the magnet, but with no clear solution in mind when it arrived. Shortly after installation of the magnet, I recall rushing down to the laboratory in great excitement to announce to my post-doc, Dr Barry Chapman, that I believed I had solved the magnetic problem by introducing a magnetic screen between the gradient coil and the inner bore of the static magnet. The idea was that the screen, itself, would be another coil designed to make the total magnetic flux from the screen and the primary gradient coil zero beyond the screen, thereby removing any interaction between the gradient coil and the magnet.

We hurriedly put together some calculations and filed the idea as a Patent. Meanwhile, Dr Robert Turner, who had joined my research group to learn about MRI in general and EPI in particular, had been thinking about the problem from a different view point, using a so-called annealing algorithm, but without success. I mentioned my approach to him and within three months or so, he together with Dr Roger Bowley, a theoretician in Physics, had come up with a much more rigorous mathematical analysis of my idea. This material was added to the Patent within the one year allowed before final filing. So active magnetic screening was born and quickly applied to the development of EPI at Nottingham. Once published active magnetic screening was rapidly taken up by the MRI industry and forms the basis of all commercial MRI machines today.

THE GOLDEN ERA

The 10 year period from 1980–1990 was exceptionally fruitful in terms of research output, development and medical applications of high speed imaging. Considerable effort was put into the implementation of EPI, initially by two research students, Roger Ordidge and Richard Rzedzian (now deceased), then later by several other students including David Guilfoyle and Mark Doyle. While this was going on, other work continued in non-medical applications of MRI principally by Stephen Blackband and Richard Bowtell. Other types of whole body imaging were tried out by Volker Bangert, a German student from Berlin who had a grant from the Deutsche Akademische Austauschdienst. The major effort made in EPI produced increasingly better images as a result of further work carried out by research students Martin Cawley and Alistaire Howseman. Whole body EPI work with a series of patients became possible with the help of Dr Michael Stehling, a medical graduate from Germany who came to Nottingham, initially as a visitor with a grant from the Deutsche Forschungsgemeinschaft, but later decided to stay on to study for a PhD in MRI.

By the end of the 80's the quality of EPI data had improved dramatically so that most regions of the human body could be studied rapidly, bringing relief to many patients. It was at about this time that Dr Penny Gowland joined the group and made substantial contributions to our work on foetal imaging.

This was the end of a Golden Era in the development of MRI at Nottingham. There were, of course, further developments well into the 1990's and beyond including the trial experiments using EVI at 0.5 T, carried out by Paul Harvey, and at 3.0 T, carried out by Paul Glover, Ron Coxon and Jonathan Hyking.

Following my retirement in 1994 from teaching in the University, I continued my research activities both at the University and with the help of my small company, General Magnetic. Our main interest has been concerned with MRI safety matters. In particular we have tried to reduce the acoustic noise levels generated by the gradient coils when pulsed in EPI. This work, started in the University by Joanna Beaumont and Ron Coxon, has subsequently been further developed by Dr Barry Chapman, and my son in law, Brett Haywood.

Another safety issue also related to gradient coils is the unwelcome level of electric field that naturally accompanies high levels of changing magnetic field. Electric fields are responsible for neural stimulation induced in patients when being imaged. Currents flowing within the patient may induce peripheral muscular twitch for relatively mild stimulation levels. However, for larger and faster gradients applied to the thorax, induced cardiac fibrillation is a real and serious danger. This aspect of our work is a joint venture between the University of Nottingham and General Magnetic and involves Professor Roger Bowley and Brett Haywood.

None of the work in MRI could have been achieved without the enthusiasm and dedicated support of a highly motivated team of technical and academic staff, research students and post-docs sustained over the period from 1972 – to the present day. I wish also to thank my three secretaries, Mary Newsum-Smith, Lesley Key and Pamela Davies for their unstinting help and the patience shown to me over the course of a long and exciting career. Last, but no means least, I would like to thank my wife, Jean, and my family for their unfailing and unswerving support during the good as well as the not so good times.

CURRICULUM VITAE

1. University Education
   1956–1959 Full time honours course in physics at Queen Mary College, London. Awarded First Class Honours.
   1959–1962 Full time course for PhD at Queen Mary College, London. Awarded PhD in September 1962.

2. Appointments held
   1962–1964 Appointed Research Associate at the Physics Department, University of Illinois, Urbana, Illinois, U.S.A.
   1964 Appointed Lecturer in Physics, University of Nottingham.
   1968 Promoted to Senior Lecturer in Physics.
   1970 Promoted to Reader in Physics.
   1972–1973 Senior Visitor at the Max-Planck-Institute for Medical Research, Heidelberg.
   1979 Appointed Professor of Physics at Nottingham.
   1983–1988 Medical Research Council Professorial Research Fellow.
   1988–1994 Resumed Professorship in Physics at Nottingham.
   1994 Emeritus Professor of Physics, Nottingham.

3. Awards and Prizes
   1983 Gold Medal of the Society of Magnetic Resonance in Medicine.
   1984 Royal Society Wellcome Foundation Gold Medal and Prize.
   1988 Duddell Prize and Medal, Institute of Physics.
   1988 British Institute of Radiology, Silvanus Thompson Medal.
   1989 International Society of Radiology and the Antoine Béclère Institute in Paris, Antoine Béclère Medal for 1989 in Radiology.
   1990 Joint award of the Royal Society Mullard Medal and Prize.
   1992 International Society of Magnetic Resonance (ISMAR) prize, awarded jointly with P Lauterbur.
   1993 Barclay Medal, British Journal of Radiology.
   1993 First Silver Plaque of the European Society of Magnetic Resonance in Medicine and Biology.
   1995 Garmisch-Partenkirchen Prize for MRI.
   1995 Gold Medal of the European Congress of Radiology and the European Association of Radiology.
   1995 Gold Medal of the Journal of Clinical MRI.
   1996 Joint award of First Annual Editor's Excellence Award, J CAT.
   1997 Joint award of the Rank Prize.
   2003 Joint Award of the Nobel Prize in Physiology or Medicine with Professor Paul Lauterbur.

4. Honours
   Elected Fellow of Queen Mary College (now Queen Mary and Westfield College), London University in June 1986.

Elected Fellow of the Royal Society, London in February 1987.
Elected President of the Society of Magnetic Resonance in Medicine, 1987–88.
Knighted in the New Year's Honours for 1993.
Elected Honorary Fellow of the Royal College of Radiology in 1993.
Elected to Honorary Membership of the British Institute of Radiology in 1993.
Elected Honorary Member of the Society of Magnetic Resonance Imaging in 1994.
Elected Fellow of the Society of Magnetic Resonance in 1994.
Conferred Doctor *Honoris Causa*, University of Strasbourg, 1995.
Conferred Honorary Doctorate, University of Kent at Canterbury 1996.
Honorary Fellow of the Institute of Physics, Jan 1997.
Conferred Doctor of Science *Honoris Causa*, Jagellonian University, Department of Physics, Krakow, Poland, June 2000.
Conferred Doctor of Science *Honoris Causa*, University of Nottingham, June 2004.
Elected Fellow (hon) Royal College of Physicians, July 2004.

SNAP-SHOT MRI

Nobel Lecture, December 8, 2003

by

PETER MANSFIELD

Sir Peter Mansfield Magnetic Resonance Centre, Department of Physics and Astronomy, University of Nottingham, Nottingham, NG7 2RD, U.K.

INTRODUCTION

The topic of magnetic resonance imaging (MRI) started for us at Nottingham in the early summer of 1972. During a discussion with one of my graduate students, Peter Grannell and my post-doc Dr Allan Garroway, concerning multiple-pulse line narrowing experiments in solids, the idea occurred to me to use the line narrowing technique as a means of effectively removing dipolar interactions in a material like CaF_2 and at the same time impose an external linear gradient on the sample thus broadening the line shape to reveal the atomic or molecular structure within the sample.

It soon became apparent, however, even with the achievable narrowed line-widths of around 1 Hz for CaF_2, corresponding to a line width reduction of 3×10^4, that the residual line width was still too broad using practical external gradients to resolve the atomic structure in a single crystal of CaF_2. Despite this setback, work continued with artificial one-dimensional lattices made up of several thin plates of camphor.

Peter Grannell and I continued this work during the course of 1972 and it resulted finally in a paper presented at the First Specialized Colloque Ampère, Krakow in 1973 (1). Formal publication appeared shortly after (2). These papers emphasized the Fourier transform approach used, even though the images of the camphor stacks were one-dimensional. It was clear that we had made our task much more difficult by choosing to work with solids. Thoughts rapidly turned to liquid-like spin systems where the line narrowing approach would be unnecessary.

The imaging approach considered so far was essentially one or two-dimensional. The next step was to define a thin slice of material so that this would be imaged without spill over to adjacent planes. This was achieved using a technique called selective irradiation (3).

One of the major practical difficulties encountered with MRI to this point was the time it took to acquire the data. Line-scanning, for example, took typically 10–20 min to acquire an image comprising 64×64 pixels (4).

The breakthrough came in 1977 with the introduction of echo-planar imaging (EPI) (5). This snap-shot technique meant that in principle complete two-dimensional images could be achieved in extremely short times ranging from 20–50 ms. However, to achieve these acquisition times, yet another inventive step was required. This was the introduction of active magnetic screening (6,7). Ordinary magnetic gradient coils, necessary to define the slice thickness and image axes, were found to interact strongly with the metal cryostat structure of the super-conductive magnet. Time dependent currents induced in these structures would decay away with their own independent time constant, adding an undesirable and unpredictable time dependence to the otherwise static magnetic field.

By actively shielding the gradient coils, all extraneous time dependence is obviated, together with all undesirable reflected static magnetic fields. Of course, the magnetic screening process itself introduces magnetic fields which change the character of the gradient field but in a well defined and calculable manner.

Magnetically screened gradient coils now form an integral part of virtually all commercial MRI scanners.

EXPERIMENTAL APPARATUS

Figure 1 shows a home built 0.5 T whole body MRI scanner together with a patient bed support. Figure 2 shows a doubly screened gradient coil assembly used in the experimental scanner of Figure 1. With a doubly screened gradient coil in which the primary coil comprises a single current loop, it can be shown that the unshielded loop carrying current I produces magnetic flux

Figure 1. Photograph of home built magnetic resonance imager based on a 0.5 T super-conductive magnet.

Figure 2. Photograph of a doubly screened active magnetic shielded gradient coil set for insertion in the super-conductive magnet of Figure 1.

lines which form a series of elliptic like magnetic field loops with displaced centres (7). However if the current loop or the primary coil is now magnetically screened with the double screen arrangement, the magnetic field lines within the inner screen are exactly the same as produced by the primary coil in free space. However, between the screens the magnetic flux is confined and the field inverts. The magnetic field at the center of the coil assembly takes the same form and magnitude as the unscreened coil. That is to say, the magnetic field B within a current loop is as expected for an unscreened loop. Outside the outer screen the field B = 0. If the inner magnetic screen is removed leaving the outer screen, flux confinement between the screens can no longer occur and the magnetic field now leaks beyond the outer screen.

ECHO-PLANAR IMAGING (EPI)

Imaging Sequence

The mathematical relationship between real-space and k-space, first described in 1750 by the French mathematician Joseph Fourier, is known as the Fourier transform (FT). It takes the general form

$$S(\mathbf{k}) = \int d\mathbf{r}\, \rho(\mathbf{r}) \exp(i\mathbf{k}.\mathbf{r}) \qquad \text{Eq. [1]}$$

where

$$\mathbf{k} = \int_0^t \gamma \partial \mathbf{G}(t')/\partial t'\, dt' \qquad \text{Eq. [2]}$$

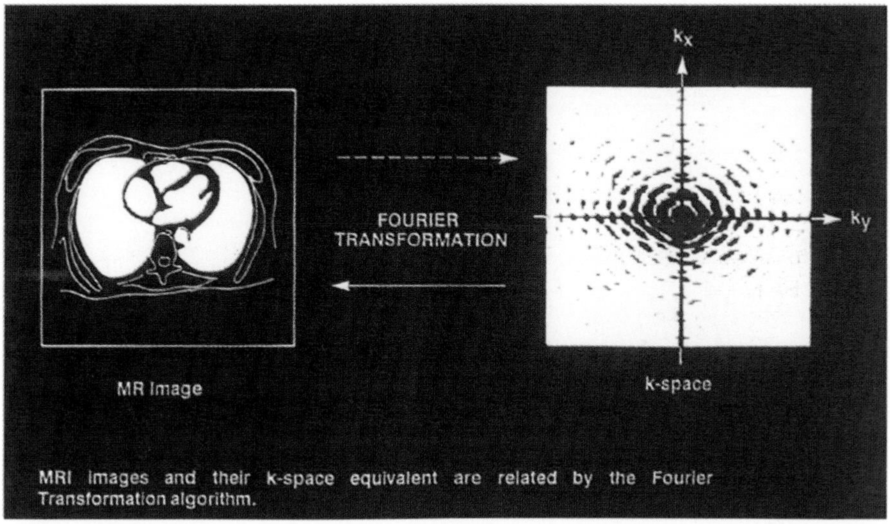

Figure 3. Diagram of a slice through the mediastinum showing the two lung fields and heart mass, also shown is the Fourier transform of this real-space image to the k-space map. (Reproduced with permission from M K Stehling, R Turner and P Mansfield, SCIENCE 253, 43–50 (1991).)

and where **k** and **r** may both be one-, two- or three-dimensional. The density $\rho(\mathbf{r})$ describes the image in real space and $S(\mathbf{k})$ describes the image in k-space. In our case $\mathbf{G}(t)$ is an externally applied time dependent magnetic gradient, γ is the magneto-gyric ratio which is constant for a particular nuclear spin species and t is the evolution time.

Using the Fourier transform expression in Eq. [1] allows a reversible transformation from k-space to real or r-space. This is exemplified in the forward and inverse Fourier transforms shown in Figure 3.

In fact the two dimensional k-space image is built up from a series of one dimensional free induction decays, (FID's), suitably stacked to give an image which is in effect the diffraction pattern of the object. The inverse FT of the k-space image produces the r-space image. In this case a transverse cross-sectional image through the mediastinum showing a diagram of the heart mass and lung fields.

In order to obtain the k-space map in a single experiment a specially designed pulse sequence is applied as shown in Figure 4. In general terms this comprises an initial spin preparation phase followed by a transverse slice selection pulse. This creates an active magnetic signal or FID which is allowed to decay away in the presence of the spatial encoding gradients G_x and G_y. For EPI these are applied in the form of a square wave or a trapezoidal waveform for G_y and either a long low level pulse or a train of short blipped pulses for G_x. In either case the areas under the long low pulse or the string of blips must be equal. The effect of these gradient waveforms causes the FID follow-

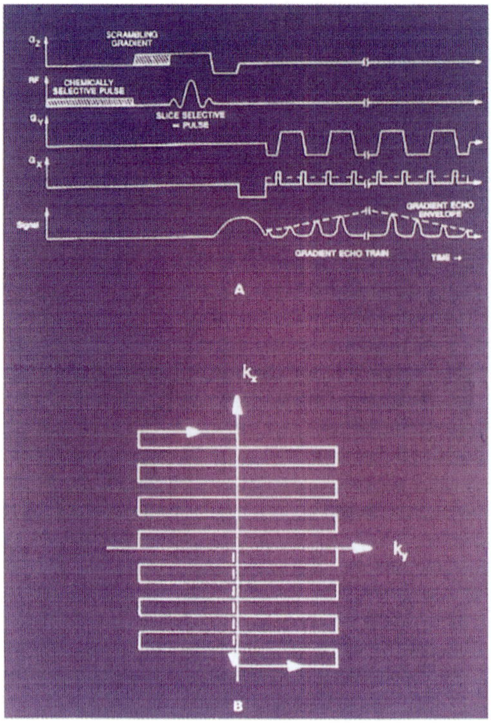

Figure 4. (Upper diagram) MBEST pulse sequence diagram and (lower diagram) k-space trajectory (Reproduced with permission from M K Stehling *et al.*, Brit J Radiol 63, 430–437 (1990).)

ing slice selection to dephase and rephase in a series of spin echoes. The amplitude of these spin echoes is initially low but grows to a maximum and then decays provided an initial dephasing gradient pulse is arranged immediately before the blipped or low level G_x sequence starts.

Also included in Figure 4 is the k-space trajectory. This starts at $k_x = k_y = 0$. The pre-pulse described above displaces the trajectory from 0 to $-k_{xmax}$ at which point the locus starts from $k_y = 0$ moving right to k_{ymax}. At this point the trajectory is displaced upwards with the first positive blip. The transverse scan then proceeds from right to left traversing the full k_y scan from k_{ymax} to $-k_{ymax}$. In this way the whole of the **k** plane is scanned. During this journey the signal is regularly sampled to produce the first step in obtaining the k-space scan as sketched in Figure 3. The remaining operation is an editing function which reverses the order of all data in alternate lines of the k-space image. At this point there are two options to choose from in order to produce the r-space image. The first is to perform a two-dimensional FT on the k-space map thereby producing the r-space image. However, to do this in general requires an overhead in time for data manipulation in addition to the two dimensional FT. The alternative approach is to take the first and last points in the reordered k-space scan and, metaphorically speaking, pull the whole array out, like a string of beads, to form a one dimensional array. The whole string is Fourier transformed using a one dimensional FT with less computing overhead.

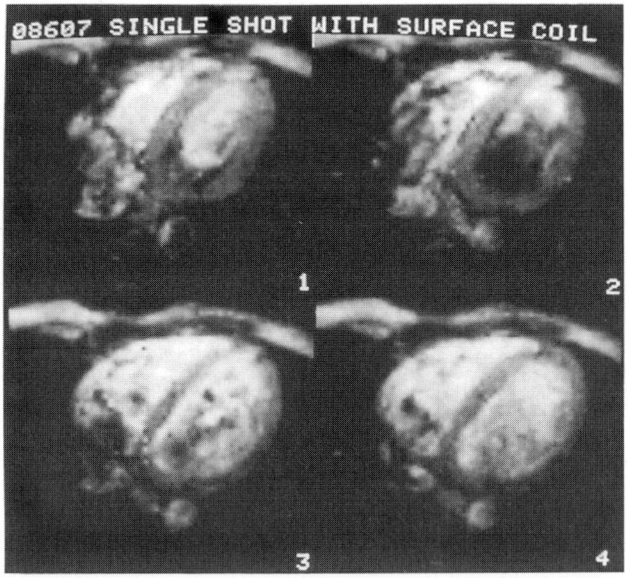

Figure 5. Snap-shot EPI images through the heart obtained with use of a surface coil. (1) Transection during systole shows left ventricular myocardial wall thickening. (2) Rapid ventricular filling in late systole. (3, 4) Transections obtained during diastole show thinner myocardial walls. The spatial resolution of these images is less then 2 mm. (Reproduced with permission from M J Stehling *et al.*, RADIOLOGY, 170: 257–263, (1989).)

General Two Dimensional Imaging Results at 0.5 T

The above described EPI sequence was used to image a series of patients and volunteers during the early eighties through to the nineties. Whole body scans may be performed comprising 64 transections commencing in the upper thorax and moving down the torso in 5 mm steps through the mediastinum, liver, kidney into the lower abdomen and finishing just below the bladder. They are an example of a quick sighting scan, the whole imaging process taking approximately one minute.

Figure 5 shows four EPI snap-shots taken through the heart using a surface coil on the chest wall (8). Each image is acquired at a different phase of the cardiac cycle. Figures 5.1 and 5.2, for example, correspond to different phases during systole, when the heart is in contraction and pumping. Figure 5.2 shows loss of signal (black) due to turbulence within the left ventricle. Figures 5.3 and 5.4 were acquired during the relaxed phase or diastole. In all images the myocardium gives a less intense signal allowing differentiation of the muscle tissue from blood.

Figure 6 shows four transections through the liver of a patient with a series of hydatid cystic lesions (9). These show as bright regions within the darker liver. An imaging variation on the same patient using the inversion-recovery or IR-EPI sequence uses an initial spin inversion pulse as the preparation phase. After a short delay or inversion time TI, the EPI sequence follows. When TI = 0 we have the normal sections as in Figure 6. By varying TI, short

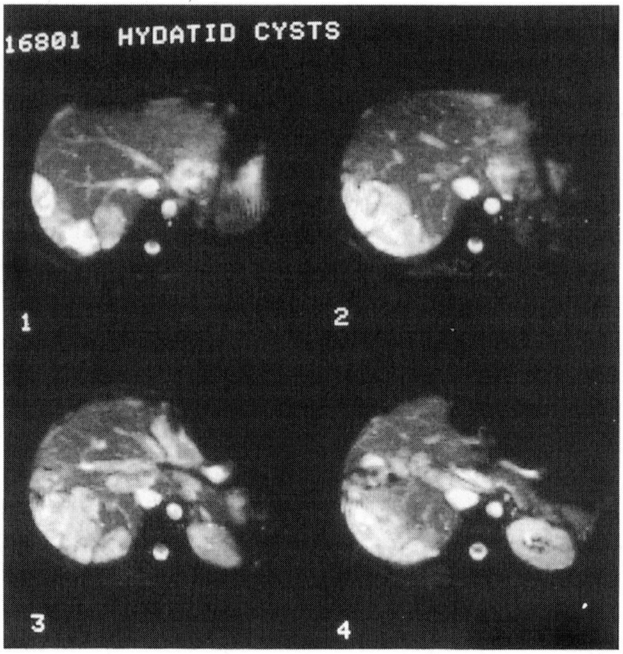

Figure 6. Four 128 ms acquisition time snap-shot MBEST images of a liver with hydatid cysts. These images are based on 128 × 128 pixel arrays with approximately 2.3 mm in-plane resolution. (Reproduced with permission from M J Stehling *et al.*, Brit J Radiol 63: 430–437 (1990).)

T_1 signal components within the image may be effectively removed thereby delineating the normal tissue. This procedure can be used to indicate clear margins between the lesions and the normal tissue. By increasing TI long relaxation time components may be eliminated thereby revealing the relatively fast relaxing hydatid cystic lesions.

Foetal Imaging Results at 0.5 T

An important area of application of EPI has been foetal imaging during the 3rd trimester. This has been valuable in assessing cases of foetal growth retardation (10–12).

Figure 7 shows a maternal coronal view of a foetal saggital section at 37 weeks gestation. The foetal section, taken from a set of scans, clearly shows the head, brain and spinal column and also the right leg. The darker signal from within the amniotic sac is the placenta. The brighter signal regions surrounding the foetus come from the amniotic fluid.

In other transectional images at 37 weeks gestation, highlighting of the lung field can be used to measure foetal lung volume. The same technique can be used to measure the subcutaneous fatty tissues surrounding the foetus.

Foetal birthweight versus the EPI estimated foetal volume has also been measured for 12 babies with birthweight spanning the range 1.5–3.5 kg

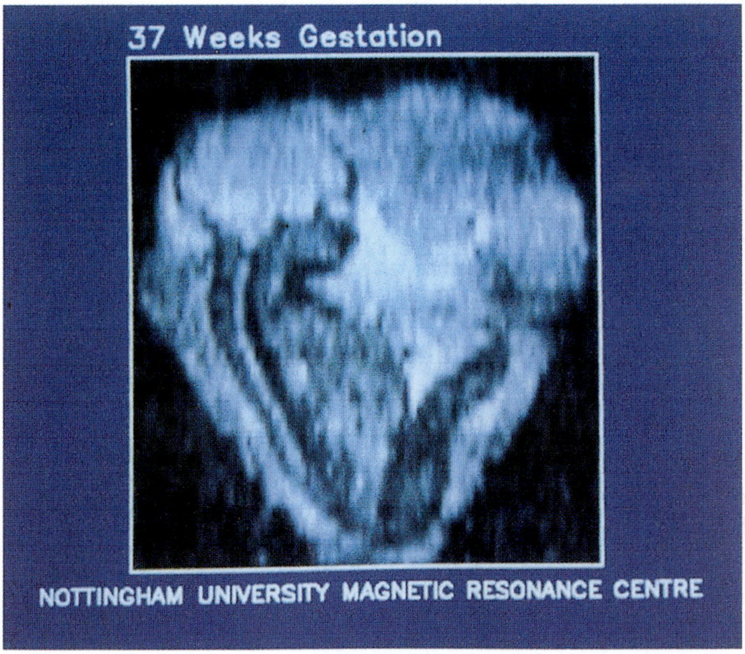

Figure 7. Sagittal section taken from a three-dimensional data set of a foetus *in utero* at 37 weeks gestation. The foetal brain and spinal canal can be clearly seen. Also shown to the right and slightly darker is the placenta. The bright region between the placenta and foetus corresponds to amniotic fluid.

(11,12). The foetal volume is assessed from a series of contigious EPI slices spanning the foetus. Using the same technique as described above, the foetal cross-sectional image is traced out and highlighted. The elemental volume is the highlighted area multiplied by the slice thickness. The total foetal volume is then the sum of all cross-sectional slices. The results indicate a good linear correlation between volume and weight.

Paediatric Imaging at 0.5 T

In the course of imaging young children suffering from cyanotic heart disease, (13–16), snap-shot transectional images have been taken through the mediastinum for a child with a normal heart. They all show the lung fields and heart mass at the base of the heart, moving down through the left and right ventricles towards the apex. Also shown is the classification system for a normal heart. This is contrasted with an example taken from a set of images for a child with a truncus arteriosus. Towards the apex the left and right ventricles can be clearly seen as dark regions together with the intact ventricular septum. The left and right atria show as bright regions. The classification diagram indicates the coalescence of the pulmonary artery and the aorta into a common truncus.

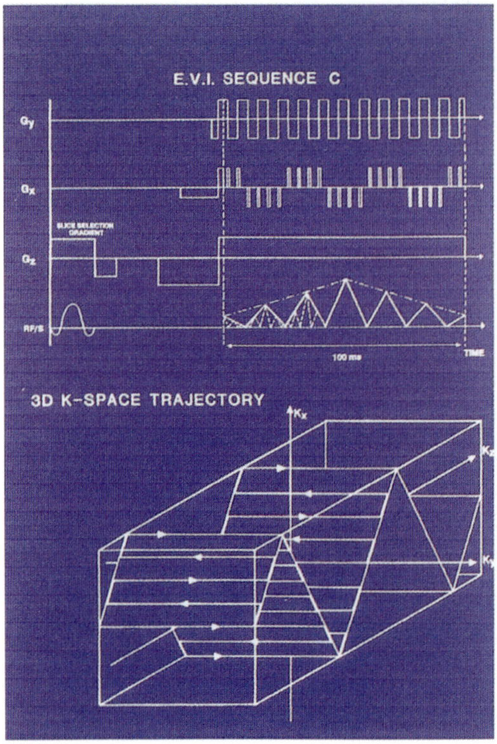

Figure 8. (A) Gradient wave form sequence for (B) (below). T_A = total data acquisition time; T_X = single echo acquisition time; and T_y = a single pulse acquisition time. Dotted portions of G_x and G_y indicate the alternative starting phases. (B) One pass of a general four pass 3D k-space trajectory for EVI. (Reproduced with permission from P R Harvey and P Mansfield, MRM 35:80–88 (1996).)

EPI Results at 3.0 T

A number of snap-shot images of various objects have been obtained at 3.0 T (17), including a snap-shot image of a phantom comprising 256 × 256 pixels. The slice thickness was 0.5 cm. The image acquisition time was approximately 90 ms. Other examples include images through the brain of a normal volunteer. The posterior horns of the ventricles together with part of the anterior horns are just visible in these images.

ECHO-VOLUMAR IMAGING (EVI)

Imaging Sequence

Details of the pulse-timing sequence for EVI (18) are shown in the upper part of Figure 8. The top trace shows the G_y gradient modulation waveform. The next modulated waveform is the gradient G_x. The third waveform G_z includes the slice selection gradient G_z together with a negative pre-positioning pulse immediately followed by a long low level gradient. In the bottom trace the slice selection pulse is indicated and slightly later the modulated spin signals arising from the sequence are sketched.

In the lower half of Figure 8 the k-space trajectory, corresponding to the above described EVI sequence, is sketched. Because of the pre-positioning

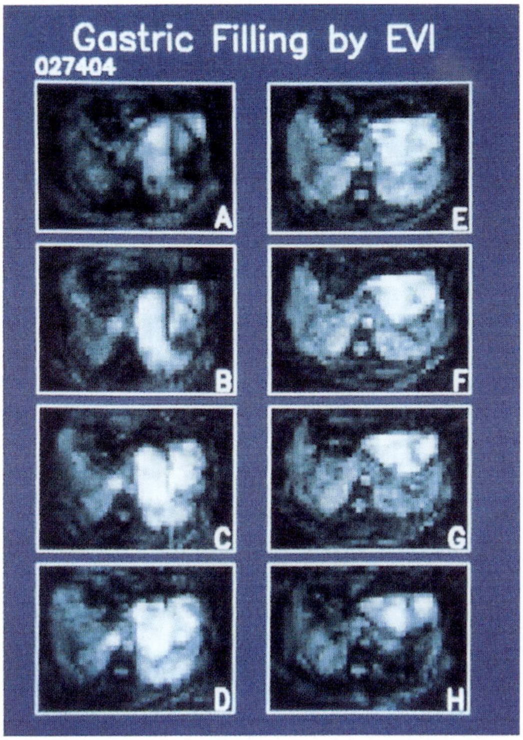

Figure 9. An EVI snap-shot taken from a movie loop showing gastric filling in a volunteer imbibing 1 litre of water. The data set comprises 8 planes spanning approximately an eight cm thick slice through the stomach. The water surface and gaseous void within the stomach are indicated in images (e,f,g). These images correspond approximately to the antrum. Other morphological land marks in these images show the kidneys, liver and spinal canal. (Reproduced with permission from P R Harvey and P Mansfield, MRM 35:80–88 (1996).)

pulse the k-space trajectory starts at $k_x = k_y = 0$ and $k_z = - k_{zmax}$. As shown in the diagram the trajectory moves right to k_{ymax} then to and fro uphill alternately to $\pm k_{ymax}$ until the trace reaches the first peak and k_{xmax}. The trajectory then continues its oscillation as it drops down the negative slope to reach $-k_{xmax}$. The whole process continues uphill and down dale as k_z evolves along the z axis to k_{zmax}.

During the whole process data are sampled (in this case at a constant rate). Because the evolution arrows of the k-trajectory reverse periodically it is necessary to re-order the sampled data so that the arrows are all pointing in the same direction before performing the Fourier transformation to obtain the image. In fact one re-ordering of the k-space data is required producing two slightly different images of the data. However, these maybe combined into a single image with no loss of clarity. Other versions of EVI can also be implemented in which the blips used for G_x are replaced by a square wave modulation of G_x. Even G_z may be replaced by a sequence of blips. However, reasonable quality multi-pass images may be obtained by using square wave modulation of G_x with G_z unmodulated as shown, provided the number of lines in an individual image is large. But these complications can be completely avoided by obtaining one pass snap-shot images using blipped G_x and G_z gradients rather than the square wave or trapezoidally modulated waveform described here.

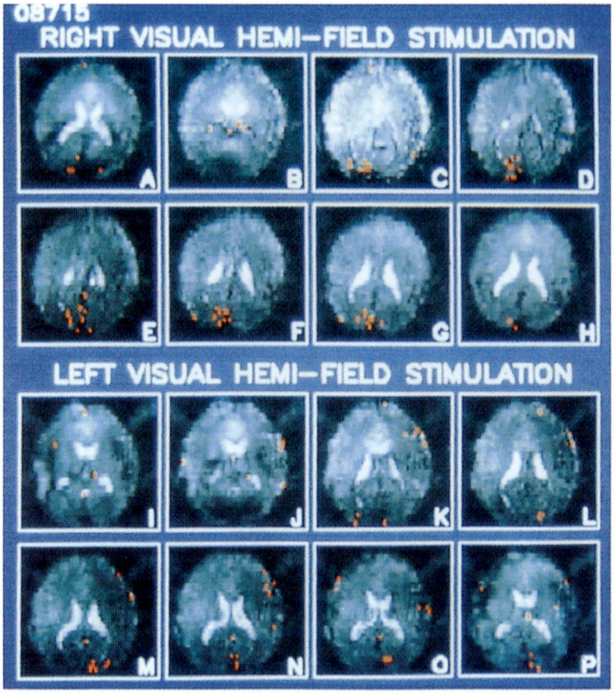

Figure 10. Two complete echo-volumar imaging (EVI) snap shots, each comprising eight contiguous slices. (A–H) One volume set with the functionally active regions highlighted in red corresponding to right visual hemi-field stimulation. (I–P) EVI set 2 shows one volume set with the functionally active regions highlighted in red corresponding to left visual hemi-field stimulation. Transections H and P correspond to the most caudal in each set, respectively. The experimental EVI parameters and resolution details are as follows. Each image comprises 64×64 pixels and the slice thickness corresponds to 2.5 mm. (Reproduced with permission from P Mansfield, R Coxon and J Hykin, J CAT 19:847–852 (1995).)

The result of this data manipulation is, in our case, the production of eight simultaneous images. Each image comprises 64×64 pixels and each pixel has an in-plane resolution of 6×4 mm^2 and the voxel thickness is 10 mm. The total acquisition time for these eight images is 128 ms. The first EVI images were obtained at 0.5 T with a volunteer subject. The EVI image set shown in Figure 9 shows gastric filling with approximately 1 litre of water (18). The water surface is clearly seen in frames E–H. An EVI snap-shot was also obtained through the bladder of the same volunteer subject. In this case, the images indicate a very full bladder characterized by the very rectangular shape observed (18).

EVI has also been tried at 3.0 T. The first results at this field strength, corresponding to the brain of a normal patient, show the ventricles at slightly different levels (19). The imaging time for all eight cross-sections was 115 ms. In Figure 10 are shown the results of the first EVI functional imaging experiments conducted with a normal volunteer. The upper half of Figure 10 shows eight slices through the brain of a volunteer subjected to right visual hemi-field stimu-

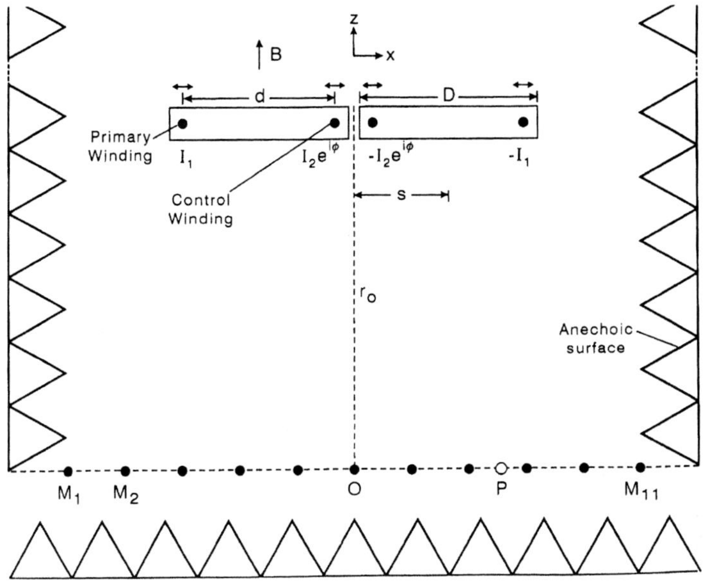

Figure 11. Diagram showing a plan sectional view of a split plate arrangement suspended in an anechoic chamber perpendicular to the main field B (not to scale). Each half of the plate-pair has a width D = 0.126 m. Embedded in each block are two conductors with separation d = 0.1 m. The plate centres are displaced from the assembly origin by ± s = ± 0.063 m. The outer wire pair comprises the primary winding carrying current I_1. The inner wire pair comprises the control winding carrying current $I_2e^{I\varphi}$ where ϕ is the relative phase difference of the two currents. The microphone array lies in a line parallel to the plate through point O. Point P represents a single microphone. The distance r_o = 1.25 m. The plate length I = 0.58 m. The material is GRE, and the plate thickness is 0.012 m. The plate-pair has an overall width of 0.255 m, including an air gap of 0.003 m separating the plates. The compressional wave velocity of sound is taken as 2500 m/s. (Reproduced with permission from B L W Chapman, B Haywood and P Mansfield, MRM 50:931–935 (2003).)

lation. Highlighted on these images are the red dots marking changes in signal arising predominately from the left side of the visual cortex. In the lower eight brain slices the volunteer is subjected to left visual hemi-field stimulation. Also highlighted on these images are red dots marking signal changes in this case arising predominately from the right side of the visual cortex.

ACTIVE ACOUSTIC CONTROL

One of the problems in the implementation of EPI and EVI is the high levels of acoustic noise generated by the gradient coils when pulsed rapidly. The noise originates from movement of the gradient coil windings resulting from the Lorentz forces created by the high currents flowing in the gradient coil assembly. Even if the wires are firmly cemented into the coil former, the former itself deforms slightly and the tiny deformation in the former surface is enough to couple with the surrounding air thereby launching undesirable acoustic waves into and around the MRI apparatus.

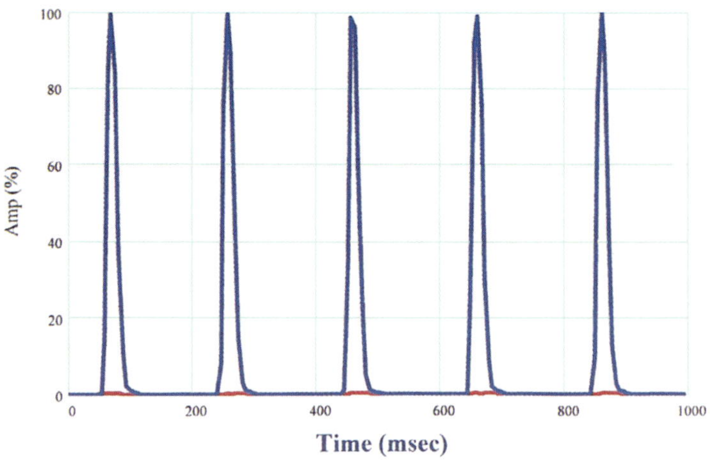

Figure 12. Sound recording before (blue) and during active acoustic control (red) for an optimized gradient pulse sequence running at 5 pulses per second.

The typical noise levels for EPI can reach 140 dB or more. However, the safety levels for noise are limited in the work place to a maximum of 85 dB (20,21). Prolonged exposure to very high level noise can lead to irreversible hearing damage. Of course, some local protection may be introduced to mitigate the effects of high level noise. This often takes the form of ear defenders worn by the patient. These give typically a 30 dB protection. But even these are unreliable for small children because of poor fitting or looseness. For pregnant women there is virtually no reliable protection for the foetus.

It is for these reasons that we have been considering other ways of dealing with the problem. Our current approach is effectively to try to prevent the coil assembly from radiating noise.

Rectangular Gradient Coil

Our initial approach to the acoustic noise problem has been to consider a rather basic coil arrangement comprising four rectangular plates or sectors. Each sector carries a current I_1. In addition there is a re-entrant loop, which constitues the control winding. The control winding current is $I_2e^{i\phi}$ where ϕ is the phase of I_2 relative to I_1. Only one sector so far has been built and tested (22–25).

The test arrangement is shown in Figure 11. The single sector is supported within an anechoic chamber which itself fits into a 3.0 T magnet. A linear array of microphones receives acoustic signals across one end of the magnet.

Theoretical calculations for the acoustic output from a signal sector plate assembly have been performed. These show that when $\phi = 180^\circ$ the plate produces the full acoustic output. This reduces considerably when the phase is adjusted to $\phi = {\sim}0^\circ$.

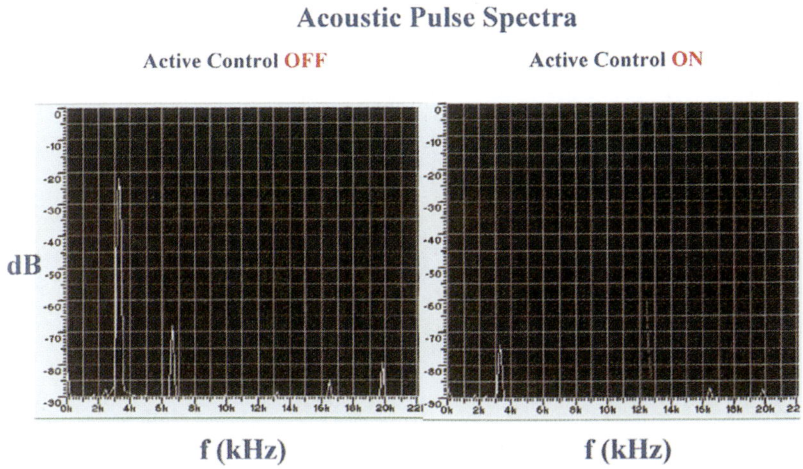

Figure 13. Logarithmic spectra from a 22 m/s Kaiser 7 filtered single pulse with (left) active control off and (right) active control on, showing a resulting reduction in noise of 50 dB (99.7%) in the fundamental noise component at 3.2 kHz. (Reproduced with permission from B L W Chapman, B Haywood and P Mansfield, MRM 50:931–935 (2003).)

Results

A specially designed optimized pulse corresponding to an EPI sequence was used to test the acoustic efficiency of the gradient sector assembly (25). The EPI pulse length was approximately 10 ms. A series of EPI pulses were applied to the gradient sector at a rate of 5 pulses per second in the acoustically optimized mode (red) and non-optimized mode (blue). These results are shown in Figure 12 for five pulses lasting 1 second. The acoustic signal in the optimized mode (red) has been reduced to almost zero level on this linear trace. However, one pulse from the same data has been Fourier analyzed to produce a frequency spectrum of the EPI pulse with and without acoustic control switched on. These results are shown in Figure 13. With active control switched off, acoustic noise of approximately –20 dB occurs primarily at 3.2 kHz, the fundamental carrier frequency of the EPI pulse. Some second harmonic content is present at 6.4 kHz together with higher harmonics. In the optimized mode, the fundamental component of noise is reduced to approximately –70 dB with no second harmonic and reduced higher harmonics. The total noise reduction of the fundamental frequency at 3.2 kHz is 50 dB. Further noise reduction in the harmonic frequencies is also indicated.

CONCLUSIONS

The basis of ultra-high speed snap-shot MRI is reviewed and illustrated with early examples of actual EPI results obtained at 0.5 T and 3.0 T. Also described are some original three dimensional results obtained using snap-shot EVI. Both techniques show promise as useful medical imaging modalities. However it is fair to say that so far only EPI has found its way into regular

Acknowledgements

Physics

Dr. P. K. Grannell; Prof. A. A. Maudsley; Prof. P. G. Morris; Dr. I. L. Pykett; Prof. R. J. Ordidge; Dr. R. Rzedzian*; Dr. M. Doyle; Dr. D. N. Guilfoyle; Dr. S. J. Blackband; Dr. M. G. Cawley; Prof. R. W. Bowtell; Dr. A. M. Howseman; Dr. R. J. Coxon; Dr. M. K. Stehling; Dr. A. M.Blamire; Dr. P. Gibbs; Dr. P. R. Harvey; Dr. M. Symms; Dr. M. McJury; Dr. M. Clemence; Dr. B. Issa; Dr. P. Glover; Dr. A. Freeman; J. Hykin; Dr. A. Rodriguez; Dr. A. Peters; Dr. P. Boulby; Dr. M. Al-Mugheiry; Dr. J. Beaumont; Dr. M. Bencsik; B. Haywood✻; Prof. R. Bowley.

Postdoctoral Visitors

Dr. B. Chapman✻; Dr. A.N. Garroway; Dr. P. Gowland; Dr. J.C. Sharp; Prof. R. Turner; Dr. P. Tokarczuk; Prof. A. Jasinski; Dr. G. Planinsic; Dr. A. Snaar

✻ Special thanks for help in preparation of this presentation. * Deceased.

Medicine

Dr. K. Morris; Prof. I. Johnson; Prof. R. Coupland; Prof. B. Worthington; Dr. P. Small; Dr. A. Chrispin*; Dr. C. O'Callaghan; Mr. J. Firth; Prof. M. Symonds

University of Nottingham

Figure 14. List of past students, post-doctoral assistants and visitors and medical collaborators.

medical use. Quite often this use is limited to a role of rapid sighting of the region of interest within a patient before slower techniques are used to obtain the final image.

The reason for doing things this way round is partly historical. The slower MRI techniques were developed and honed by the manufacturing companies while ignoring the advances in ultra-high speed imaging methods. However, given the same intensive development, EPI and possibly EVI, will move up from a rapid sighting technique to a fully diagnostic technique in its own right.

Such a move would remove the dread felt by some claustrophobic patients at the prospect of being confined in the MRI machine for between a half and one hour. Of course, ultra-high speed imaging brings with it its own problems which need addressing. One of these is the high level of acoustic noise generated by the large and rapidly switched magnetic gradients required for EPI and EVI. Some progress has been made in ameliorating this situation in principle. But it remains to be demonstrated whether the active acoustic control principle can be fully implemented on an operational gradient coil set. For further reading see Refs (26–31).

ACKNOWLEDGEMENTS

I am grateful to my early research supervisors, Professor J G Powles, originally at Queen Mary College, University of London and Professor C P Slichter, at the University of Illinois, Urbana, Illinois, USA for teaching me the art and practice of NMR. I am also greatly indebted to Professor R Bowley and my other colleagues at Nottingham and to my own research students and post-doctoral assistants listed in Figure 14 all of whom contributed extensively to the development of MRI at Nottingham. Also listed are the key medical col-

Funding and Equipment Support

Medical Research Council

Department of Health and Social Security
G. Hickson, Dr. N. Slarke, J. Williams

British Technology Group,
I. Harvey, N. Davis, G. Blunt, R. Sutherland

University of Nottingham
Sir Colin Campbell

British Heart Foundation

Oxford Magnet Technology

General Magnetic

University of Nottingham

Figure 15. List of funding and equipment support agencies.

leagues whose collaboration and support helped to bring about acceptance of MRI as a diagnostic tool both nationally and internationally.

I am especially grateful to the technical and workshop staff of both the Magnetic Resonance Centre and the Department of Physics and Astronomy for their untiring and unstinting support in the construction and testing of several prototype MRI systems over the years.

I am also most grateful for the financial support received from the Institutions listed in Figure 15. Without their generous and sustained support over many years there would be little to celebrate today.

Finally, I wish to express my sincere appreciation and thanks to my personal secretaries over the years, Mary Newsum-Smith, Lesley Key and most recently, Pamela Davies, for their unflagging efforts in typing and re-typing the many manuscripts, reports and general correspondence.

REFERENCES

1. Multi-pulse line narrowing experiments. P Mansfield, P K Grannell, A N Garroway and D C Stalker, Proc First Specialised Colloque Ampere, Poland (1973).
2. NMR `Diffraction' in solids? P Mansfield and P K Grannell, J Phys C 6, L422 (1973).
3. Image formation in NMR by a selective irradiative process. A N Garroway, P K Grannell and P Mansfield, J Phys C 7, L457, (1974).
4. Medical imaging by NMR. P Mansfield and A A Maudsley, Brit J Radiol 50, 188 (1977).
5. Multi-planar imaging formation using NMR spin echoes. P Mansfield, J Physics C Solid State Phys 10, L55–L58 (1977).
6. Active magnetic screening of coils for static and time-dependent magnetic field generation in NMR imaging. P Mansfield and B Chapman, J Phys E: Sci Instrum 19, 541–546 (1986).
7. Double active magnetic screening of coils in NMR. B Chapman and P Mansfield, J Phys D: Appl Phys 19, L129–L131 (1986).

8. Whole Body Echo-Planar MR Imaging at 0.5 T. M J Stehling, A M Howseman, R J Ordidge, B Chapman, R Turner, R Coxon, P Glover, P Mansfield and R E Coupland, Radiol 170, 257–263 (1989).
9. Ultra-fast magnetic resonance scanning of the liver with echo-planar imaging. M K Stehling, R M Charnley, A M Blamire, R J Ordidge, R Coxon, P Gibbs, J D Hardcastle and P Mansfield, Brit J of Radiol, 63, 430–437 (1990).
10. Estimation of lung volume in infants by echo planar imaging and total body plethysmography. B Chapman, C O'Callaghan, R J Coxon, P M Glover, G Jaroszkiewicz, A M Howseman, P Mansfield, P Small, A D Milner and R E Coupland, Arch Dis in Childhood 65, 168–170 (1990).
11. Fetal weight estimation by echo-planar magnetic resonance imaging. P N Baker, I R Johnson, P A Gowland, J Hykin, P R Harvey, A Freeman, V Adams, B S Worthington, P Mansfield, Lancet 343, 644–45 (1994).
12. Obstetrics. Echo-planar magnetic resonance imaging in the estimation of fetal weight. J Hykin, P Gowland and P Mansfield, Contemp Rev Obstet Gynaecol 6, 173–177 (1994).
13. Transectional echo planar imaging of the heart in cyanotic congenital heart disease. A Chrispin, P Small, N Rutter, R E Coupland, M Doyle, B Chapman, R Coxon, D Guilfoyle, M Cawley and P Mansfield, Pediatr Radiol 16, 293–297 (1986).
14. Echo-planar imaging of normal and abnormal connections of the heart and great arteries. A Chrispin, P Small, N Rutter, R E Coupland, M Doyle, B Chapman, R Coxon, D Guilfoyle, M Cawley and P Mansfield, Pediatr Radiol 16, 289–292 (1986).
15. EPI of the infant heart (1). Demonstration of connection between ventricle and artery by EPI construction. P Mansfield with P Small, A Chrispin, N Rutter, R E Coupland, M Doyle, B Chapman, D N Guilfoyle and M G Cawley, Paediatric Radiol 15, 280 (1985).
16. EPI of the infant heart (2). Demonstration of ventriculo/bulbar septation by EPI transection. P Mansfield with A Chrispin, P Small, N Rutter, R E Coupland, M Doyle, B Chapman, D N Guilfoyle and M G Cawley, Paediatric Radiol 15, 280 (1985).
17. Echo-Planar Imaging of the Brain at 3.0 T: First Normal Volunteer Results. P Mansfield, R Coxon and P Glover, J CAT, 18(3): 339–343 (1994).
18. Echo-Volumar Imaging (EVI) at 0.5 T: First Whole-Body Volunteer Studies. P R Harvey and P Mansfield, Mag Res Med 35, 80–88 (1996).
19. Echo-Volumar Imaging (EVI) of the Brain at 3.0 T: First Normal Volunteer and Functional Imaging Results. P Mansfield, R Coxon and J Hykin, J CAT 19(6): 847–852 (1995).
20. NRPB Board Statement on Clinical MR Diagnostic Procedures, 2:1 (1991).
21. Price RR The AAPM/RSNA physics tutorial for residents: MR imaging safety considerations. Radiographics 19, 1641-1651 (1999).
22. Sound Generation in Gradient Coil Structures for MRI. P Mansfield, P M Glover and J Beaumont, Mag Res Med 39, 539–550 (1998).
23. Principles of Active Acoustic Control in Gradient Coil Design. P Mansfield and B Haywood, MAGMA 10, 147–151 (2000).
24. Active Acoustic Control in Gradient Coils for MRI. P Mansfield, B Haywood and R Coxon, Mag Red Med 46, 807–818 (2001).
25. Optimized Gradient Pulse for Use with EPI Employing Active Acoustic Control. B L W Chapman, B Haywood and P Mansfield, Mag Res Med 50, 931–935 (2003).
26. Snapshot echo-planar imaging methods: current trends and future perspectives. P Mansfield, A M Blamire, R Coxon, P Gibbs, D N Guilfoyle, P Harvey and M Symms, Phil Trans. R. Soc. Lond. A 333, 495–506 (1990).
27. Snapshot Imaging at 0.5 T Using Echo-Planar Techniques. R Ordidge, A Howseman, R Coxon, R Turner, B Chapman, P Glover, M Stehling and P Mansfield, Mag Res Med 10, 227–240 (1989).
28. Echo-Planar Imaging: Magnetic Resonance Imaging in a Fraction of a Second. M K Stehling, R Turner and P Mansfield, Science 254, 43–50 (1991).
29. 'N.M.R. Imaging in Biomedicine', P Mansfield and P G Morris, Academic Press, New York, (1982).

30. 'Echo-Planar Imaging: Theory, Technique and Application'. F Schmitt, M K Stehling and R Turner, Springer-Verlag, Heidelberg, (1998).
31. 'NMR Imaging in Medicine and Biology'. P G Morris, Clarendon Press, Oxford (1986).

Physiology or Medicine 2004

RICHARD AXEL and LINDA B. BUCK

"for their discoveries of odorant receptors and the organization of the olfactory system"

THE NOBEL PRIZE IN PHYSIOLOGY OR MEDICINE

Speech by Professor Sten Grillner of the Nobel Assembly at Karolinska Institutet. Translation of the Swedish text.

Your Majesties, Your Royal Highnesses, Honoured Nobel Laureates, Ladies and Gentlemen,

We perceive the world through our senses – we enjoy the brilliant colours around us, we listen to well-known sounds in the environment, but we also experience the world through a symphony of scents. We can perceive no less than 10 000 different smells, at this very moment not only the flowers from Nobel's San Remo, but also perfumes and other fragrance. Soon we shall enjoy this evening´s dinner, in particular the sense of smell will help us identify the gastronomic subtleties, and evaluate the qualities of the wines. A scent from our childhood can make us remember long forgotten memories, the tar of the fishing-boats from long ago, or the scent of honeysuckle rose. The sense of smell is in some way more primitive than other senses – less verbal – often we do not have words to describe the qualities of a new smell. If our sense of smell is damaged, however, we lose much of our quality of life - life becomes greyer – the food does not taste – but we will survive. For most animals, the sense of smell is a matter of life and death. A newborn pup finds its mother with the help of olfactory cues – without the sense of smell, no survival.

For a long time, it has been known that the olfactory receptor cells are located far up in the nose, and that they send their thin neural processes through small canals in the bone directly to the part of the brain called the olfactory bulb. Many important details of how the sense of smell is designed were also known, but up to the elegant studies of Richard Axel and Linda Buck that we award here today, the basic mode of operation of the sense of smell was not understood. In their studies on mouse, they showed that no less than 3% of the genes were coding for olfactory receptors – docking stations for different odorants. They are large molecules located in the cell membrane of the olfactory cells that react to different odorants that pass through the nose. This large gene family "manufactures" around 1000 different types of docking stations, each of which responds to only a few odorants. It was subsequently shown that each type of olfactory cell only expresses a single type of docking station in its cell membrane. In the mouse there are thus 1000 different types of olfactory receptor cells. For each type, there are a large number of copies spread in the mucosa of the nose.

I will now shift over to the lingua franca of science.*

The mouse thus has no less than 1000 different types of receptor cells or sensors that react to different odorants. In humans some of the genes have degenerated and we may have as little as 350 types of sensors. The olfactory world of a mouse or a dog is most likely, infinitely richer than our own. Nevertheless our sense of smell adds importantly to our quality of life. Each population of receptor cells sends their processes to the same specific microregions in the olfactory bulb, which in turn relays the information to the next set of nerve cells. They project to the cortical regions concerned with the perception of smell. In technical terms there is a "labelled line" from each receptor subtype to the cortex, which thus will be continuously informed about the degree of activation of each of the many subtypes of receptors.

Each characteristic scent is, as a rule, a mix of odorant molecules that activates a combination of different sensors – types of receptor cells that send their signal via the olfactory bulb to the cortex. The specific combinatorial code that reaches the cortex makes us perceive a given odorant. Richard Axel and Linda Buck have unravelled the enigma of our ability to discern so many different scents.

Professors Axel and Buck,

On behalf of the Nobel Assembly at Karolinska Institutet, I wish to convey to both of you our warmest congratulations for having unravelled the function of one of our senses, the olfactory system. You will now receive the Nobel Prize from the hands of His Majesty the King.

*In the first two paragraphs, the speech was given in Swedish.

RICHARD AXEL

New York City is my world. I was born in Brooklyn, the first child of immigrant parents whose education was disrupted by the Nazi invasion of Poland. Although not themselves learned, my parents shared a deep respect for learning. I grew up in a home rich in warmth, but empty of books, art or music. My early life and education were centered on the streets of Brooklyn. Stickball, baseball with a pink ball and broom handle, and schoolyard basketball were my culture. In stickball, a ball hit the distance to one manhole cover was a single, and four manhole covers, a home run, a Nobel Prize. My father was a tailor. My mother, although quick and incisive, did not direct her mind to intellectual pursuits and I had not even the remotest thought of a career in academia. I was happy on the courts. In those days, we worked at a relatively young age. At eleven, I was a messenger, delivering false teeth to dentists. At twelve, I was laying carpets, and at thirteen, I was serving corned beef and pastrami in a local delicatessen. Vladimir, the Russian chef, was the first to expose me to Shakespeare which he recited as we sliced cabbage heads for coleslaw.

My local high school had the best basketball team in Brooklyn but the Principal of my grade school had a vision different from my own and insisted that I attend Stuyvesant High School, far away in Manhattan. Stuyvesant High advertised itself as a school for intellectually gifted boys but had the worst basketball team in the city. I was unhappy about the prospect of attending for it seemed antithetical to my self–image. Shortly after I entered, however, my world changed. I embraced the culture and aesthetics of Manhattan. The world of art, books and music opened before me and I devoured it. In school, I heard bits of an opera for the first time. I remember it distinctly, the Letter Duet from Mozart's *Marriage of Figaro.* The next night I attended *Tannhauser* at the Metropolitan Opera and thus began a love affair, bordering on an obsession, that has had no end. Twice a week, I stood on line for standing room tickets at the Metropolitan Opera where I was exposed to a cult of similarly obsessed but far more knowledgeable afficionados who taught me the intricate nuances of this rich genre. The great Italian tenor, Franco Corelli, would serve us coffee as we waited and the diva, Joan Sutherland, would invite us backstage.

On other days, I would read in a most beautifully appointed place, the Reading Room of the Central New York Public Library on 42nd Street. One passes the pair of sculpted lions, ascends a flight of stairs into a huge high-ceilinged room of impressive silence where I read incessantly without direction but with a newfound fascination that made up for years of illiteracy. I met a coterie of library dwellers, men and women of New York, who spent all of their days in the Reading Room. I did not know who they were or how they came to be there, but they had an insight and understanding of literature that amazed and still perplexes me and they were my teachers. This was New York for me, a city of the culturally obsessed that opened up before me and framed my new world.

To support a seemingly extravagant life for a young high school student, I worked. I used my skills as a waiter in a delicatessen in Brooklyn, to wait tables in the cafes and nightspots of Greenwich Village. In the sixties, the Village was the home of the beat generation that through music and poetry and ultimately protest translated discord into meaningful changes in both America and the world. Stuyvesant High School was on the fringe of Greenwich Village and some of its teachers were artists, writers, performers who fueled the politically-fired student body, many the sons of Marxist immigrants. With this array of artistic faculty Stuyvesant nourished my new and voracious appetite.

But old worlds die hard. I continued to play basketball in high school and this led to a most memorable and humbling experience. I came onto the court as the starting center, and the center on the opposing team from Power Memorial High School lumbered out on the court, a lanky 7 foot 2 inch sixteen year old. When I was first passed the ball, he put his hands in front of my face, looked at me and asked, "What are you going to do, Einstein?" I did rather little. He scored 54 points and I scored two. He was the young Lew Alcindor, later known as Karim Abdul Jabar, who went on to be among the greatest basketball legends and I became a neurobiologist.

My decision to remain in New York and attend Columbia College revealed the provincial but endearing quality of my family. When I chose to accept a gracious scholarship offered by Columbia, my father was disappointed. It was a fact well known that the brightest children of Brooklyn immigrants attended City College. My freshman year at Columbia, I lived with abandon. The opera, the arts, the freedom, the protest left little time for study. In the first semester, I met a student from Tennessee, Kevin Brownlee, who remains a dear friend and is now a Professor of Medieval French at the University of Pennsylvania. Brownlee urged me to redirect this intensity to learning. The world of the arts will remain, but my time at Columbia University was limited. Once again, a new world opened before me. With Kevin as my guide, I became a dedicated, even obsessed, student. My life was spent in a small room lined with volumes of Keats' poetry at the Columbia Library and I immersed myself in my studies. The study of literature at Columbia in the sixties was exciting in the presence of the poet, Kenneth Koch, the critics, Lionel Trilling, Moses Hadas, and Jacques Barzun. It was largely chance, however, that led me to biology.

To support myself in college, I obtained a job washing glassware in the laboratory of Bernard Weinstein, a Professor of Medicine at Columbia University. Bernie was working on the universality of the genetic code. The early sixties was a time shortly after the elucidation of the structure of DNA and the realization that DNA is the repository of all information and from which all information flows. The genetic code had just been deciphered and the central dogma was complete. I was fascinated by the new molecular biology with its enormous explanatory power. I was a terrible glassware washer because I was far more interested in experiments than dirty flasks. I was fired and was rehired as a Research Assistant and Bernie spent endless hours patiently teaching this scientifically naïve, but intensely interested young student. I was torn between literature and science. Dubious about my literary ambitions and fascinated by molecular biology, I decided to attend graduate school in genetics.

My plans were thwarted by an unfortunate war and to assure deferment from the military, I found myself a misplaced medical student at Johns Hopkins University School of Medicine. I entered medical school by default. I was a terrible medical student, pained by constant exposure to the suffering of the ill and thwarted in my desire to do experiments. My clinical incompetence was immediately recognized by the faculty and deans. I could rarely, if ever, hear a heart murmur, never saw the retina, my glasses fell into an abdominal incision and finally, I sewed a surgeon's finger to a patient upon suturing an incision. It was during this period of incompetence and disinterest that I met another extremely close friend, Frederick Kass, now a Professor of Psychiatry at Columbia University. Fred was an unusual medical student, a Texan with a degree in art history from Harvard, who remains a kindred spirit.

It was a difficult time, but I was both nurtured and protected by Howard Dintzis, Victor McCusick, and Julie Krevins, three professors at Johns Hopkins who somehow saw and respected my conflict. Without them, there is little question that I would not have been tolerated but they urged the deans to come up with a solution. I was allowed to graduate medical school early with an M.D. if I promised never to practice medicine on live patients. I returned to Columbia as an intern in Pathology where I kept this promise by performing autopsies. After a year in Pathology, I was asked by Don King, the Chairman of Pathology, never to practice on dead patients.

Finally, I was afforded the opportunity to pursue molecular biology in earnest. I joined the laboratory of Sol Spiegelman in the Department of Gentics at Columbia University. Spiegelman was a short, incisive, witty man with a tongue as sharp as his mind. Spiegelman was the first to synthesize infectious RNA *in vitro* and this led to a series of extremely interesting and clever experiments revealing Darwinian selection at the level of molecules in a test tube. Sol recognized the importance of the early RNA world in the evolution of life and had recently turned his laboratory to a study of RNA tumor viruses. An immediate bond formed between us and Sol taught me how to think about science, to identify important problems, and how to effect their solution.

Although I felt a growing confidence in my abilities in molecular biology, I was naïve in other areas of biology, notably biophysics. Importantly, I had a sense early in my career that my interest in biology was eclectic and that I would need a concomitantly broad background to embrace the different areas of biology without trepidation. I left to begin a second postdoctoral fellowship at the National Institutes of Health, working with Gary Felsenfeld on DNA and chromatin structure. Since I entered medical school to avoid the draft, I had a military obligation that was fulfilled by my years at the NIH and was endearingly termed a "yellow beret." Gary was great, but the NIH was alien, a government reservation with a fixed workday. As a night person, I found it strange and at some level difficult since I arrived at noon after all the parking spaces were occupied, left at midnight and accumulated an increasing number of parking tickets. In the midst a molecular hybridization reaction, I was arrested by two FBI agents (the NIH is a federal reservation) for 100 summonses for parking violations.

As a fellow in Felsenfeld's lab studying how chromatin serves to regulate gene expression, I formed close friendships that continue to the present. On the beach at Cold Spring Harbor, I sat with Tom Maniatis and Harold Weintraub and talked about chromosome replication and gene expression and within a few hours a bond formed, a respect for one another and for one another's thinking, that has lasted for thirty years. Hal, unfortunately, died ten years ago of a brain tumor, but his warmth, his creativity persist.

Sol Spiegelman invited me to return to Columbia as an Assistant Professor in 1974 in the Institute of Cancer Research. I was ecstatic to occupy a lab and office adjacent to his. Sol had many visitors in those years, and when he felt bored in a meeting he would excuse himself and hide in my office where we talked science until his visitors finally gave up and left. I was studying the structure of genes in chromatin and had the good fortune of participating in a revolution made possible by recombinant DNA technology. I spend a great deal of time with Tom Maniatis, who pioneered many of the techniques in recombinant DNA. Tom left Harvard for Cal Tech, because he was restricted from performing recombinant DNA experiments in Cambridge, Massachusetts. We learned how to cut and paste DNA, to isolate genes and to analyze their anatomy down to the last detail. We recognized that to understand gene control and gene function, however, required a functional assay. Within months of establishing my own laboratory in 1974, Michael Wigler, my first graduate student along with Sol Silverstein, a Professor at Columbia, developed novel procedures that allowed DNA-mediated transformation of mammalian cells. Michael, even at this very early stage in his career, was conceptually and technically masterful and within a few years he devised procedures that permitted the introduction of virtually any gene into any cell in culture. He developed a system that not only allowed for the isolation of genes, but also for detailed analysis of how they worked. We now had a facile assay to study the sequences regulating gene expression as well as gene function.

Michael went off to the Cold Spring Harbor Laboratories and simultaneous with Bob Weinberg at MIT identified the mutant ras gene as the gene responsible for malignant transformation in many cancer cells. My laboratory went off in many directions, first identifying the regulatory sequences responsible for control of specific gene expression. At the same time, a fellow, Dan Littman, now a Professor at NYU, joined the lab interested in two molecules that characterize the major classes of T cells. Dan, along with a student, Paul Maddon, succeeded in exploiting the gene transfer to isolate these two molecules. As often in science, serendipity heightened the interest in these molecules: we demonstrated that one of these receptors, CD4, was the high affinity receptor for HIV, allowing attachment and infection of immune cells.

This early work on recombinant DNA was a period of enormous excitement, for it led to a revolution in both thinking and technology in biology. It provided a new tool for the study of fundamental problems and spurred a new and valuable industry, biotechnology. We, who were involved at its inception, were perhaps a bit haughty, aggressive and proud, and were accused by many of playing "God." As evidence, the press noted that "I baptized my first child, Adam."

Recombinant DNA aroused a good deal of passion and hostility. The notion of tinkering with life was thought to endanger life and this cry became one of the major indictments of modern biology. These experiments raised endless debate because the idea that genes can be taken out of one organism and introduced into the chromosome of another is by itself upsetting. The very notion of the performance of recombinant DNA was linked with the mysterious and supernatural. This conjured up myths that elicited intense anxiety. Recombinant DNA, it was feared, would permit biologists to alter individual species as well as the evolution of species. This controversy emphasized the fact that advances in science may indeed bring harm as well as benefit. In the case of recombinant DNA, as François Jacob said, "Apocalypse was predicted but nothing happened." In fact, with recombinant DNA, only good things happened. At a practical level, the ability to construct bacteria replicating eucaryotic genes has allowed for the production of an increasingly large number of clinically important proteins. At a conceptual level, gene cloning has permitted a detailed look at the molecular anatomy of individual genes and from a precise analysis of these genes we have deduced the informational potential of the gene and the way in which it dictates the properties of an organism.

At a personal level, the emergence of a new discipline, biotechnology, introduced me to a world outside of academia. This important excursion showed me that brilliance is not limited to universities. I met and remain very close to two dynamic leaders of technology development, Fred Adler and Joe Pagano. Despite disparate histories, we remain very close and they continue to fascinate me with lives quite different from that of a university professor.

In 1982, I began to think about the potential impact of the new molecular biology and recombinant DNA technology on problems in neuroscience. Molecular biology was invented to solve fundamental problems in genetics at a molecular level. With the demystification of the brain, with the realization that the mind emerges from the brain and that the cells of the brain often use the very same principles of organization and function as a humble bacterium or a liver cell, perhaps molecular biology and genetics could now interface with neuroscience to approach the tenuous relationship between genes and behavior, cognition, memory, emotion, and perception. This thinking was the result of a faculty meeting at which Eric Kandel and I overcame our boredom with administration by talking science. Eric was characteristically exuberant about his recent data that revealed a correlation between a simple form of memory in the marine snail, *Aplysia* and cellular memory at the level of a specific synapse. Molecular biologists had encountered cellular memory before in the self-perpetuating control of gene expression. This led to the realization that this was the moment to begin to apply the techniques of molecular biology to brain function and I would attempt to recruit Eric Kandel as my teacher.

A courageous new postdoctoral fellow in my laboratory, Richard Scheller, now Director of Research for Genentech, was excited about embarking on an initial effort in molecular neurobiology in a laboratory with absolutely no expertise in neuroscience. Together with Richard and Eric, we set out to isolate the genes responsible for the generation of stereotyped patterns of innate behaviors. All organisms exhibit innate behaviors that are shaped by evolution and inherited by successive generations that are largely unmodified by experience or learning. It seemed reasonable to assume that this innate behavior was dictated by genes that might be accessible to molecular cloning. It was an exciting and amusing time with myself unfamiliar with action potentials and Kandel uncomfortable with central dogma. Richard Scheller exploited the techniques of recombinant DNA to identify a family of genes encoding a set of related neuropeptides whose coordinated release is likely to govern the fixed action pattern of behaviors associated with egg laying. A single gene, the ELH gene, specifies a polyprotein that is cut into small biologically active peptides such that individual components of the behavioral array may be mediated by peptides encoded by one gene.

Watching the story unfold, observing the interface of molecular biology and neuroscience provided great pleasure. More importantly, this collaboration formed the basis of a continuing relationship with Eric Kandel, with his incisive mind, inimitable laugh and boundless energy. In 1986, neuroscience for me was made even richer when Tom Jessell came along. Tom joined the faculty at Columbia and was to occupy a lab adjacent to my own. Not surprisingly, the lab was not ready and I had the great pleasure of hosting Tom in my own laboratory and this forged a long-lasting scientific and personal relationship. Jessell, the understated British scientist with a wry wit and piercing mind, joined a fellow in my laboratory, David Julius, now at the University of California at San Francisco, and together they devised a clever assay for the isolation of genes encoding the neurotransmitter receptors. These experiments, which might have been the last performed by the hands of Jessell, led to the isolation of genes encoding the seven transmembrane domain serotonin receptor, 5HT1C, and more generally provided an expression system that permitted the identification of functional genes that encode receptors in the absence of any information on the nature of the protein sequence. With Kandel one floor above, and Jessell next door, there was no departure from neuroscience. I was surrounded and I did not want to escape. I was beginning to feel that neuroscience was indeed an appropriate occupation for a molecular biologist. To quote Woody Allen, a fellow New Yorker, "The brain is my second favorite organ."

In the late 1980's I became fascinated in the problem of perception: how the brain represents the external world. I was struck by observations from animal behavior that what an organism detects in its environment is only part of what is around it and that part can differ in different organisms. The brain functions then not by recording an exact image of the world but by creating its own selective picture. Biological reality will therefore reflect the particular representation of the external world that a brain is able to build and a brain builds with genes. If genes are indeed the arbiters of what we perceive from the outside world then it follows that an understanding of the function of these genes could provide insight into how the external world is represented in the brain. Together with Linda Buck, a creative fellow in the lab, we began to consider how the chemosensory world is represented in the brain. The problem of olfaction was a perfect intellectual target for a molecular biologist. How we recognize the vast diversity of odorous molecules posed a fascinating problem. We assumed that the solution would involve a large family of genes and Linda Buck devised a creative approach that indeed identified the genes encoding the receptors that recognize the vast array of odorants in the environment. Linda came to me with the experimental data late one night, exuberant, and I fell uncharacteristically silent. There were 1000 odorant receptor genes in the rat genome, the largest family of genes in the chromosome and this provided the solution to the problem of the diversity of odor recognition. More importantly, the identification of these 10000 genes and their expression revealed an early and unanticipated logic of olfaction. Indeed, the subsequent use of these genes to manipulate the genome of mice has afforded a view of how the olfactory world could be represented in the brain, how genes shape our perception of the sensory environment. From that late night moment to the present, it has been a joy to watch this story unfold.

It is this work for which Linda Buck and I share the profound honor and good fortune of having been awarded the Nobel Prize in Physiology or Medicine. But there are deeper, more human joys, two sons, Adam and Jonathan, my sister, Linda, a very close coterie of friends, and a new love. Watching, contributing to the growth of my children is not only moving but humbling and puts my intense life in science in perspective. Often this intensity, bordering on obsession, distracted me from fathering and this is a regret. But my sons have emerged from a frenetic teenage into very human college students, extremely unlikely to pursue a career in science. My sister remains a close and dedicated member of an increasingly small family. A new love, Cori Bargmann, a behavioral geneticist now at Rockefeller University, has entered my world. Her intensity for science hides a knowledge and passion for books, music, and art. I have learned much from her but most importantly, Cori has shown me how to combine intellectual intensity with humanity and warmth.

Finally, the Nobel Prize was awarded to me not as a man, but for my work, a work of science that derives from the efforts of many brilliant students as well as from the incisive teachings of devoted colleagues. I take equal pride in the science that has been accomplished in the laboratory as in the scientists that have trained with me and are now independently contributing to our understanding of biology. I therefore feel that I can only accept the Nobel Prize in trust, as a representative of a culture of science in my laboratory and at Columbia University. I am deeply grateful for this culture.

SCENTS AND SENSIBILITY: A MOLECULAR LOGIC OF OLFACTORY PERCEPTION

Nobel Lecture, December 8, 2004

by

RICHARD AXEL

Howard Hughes Medical Institute, Columbia University, College of Physicians and Surgeons, New York, NY 10032, USA.

INTRODUCTION

The image in the painting *La Bonne Aventure* is not a nose (Fig. 1). It is a portrayal by the surrealist René Magritte of his own brain's representation of the external world. It is a vignette that reveals a tension between image and reality, a tension that is a persistent source of creativity in art, brought to its culmination by the surrealists. The problem of how the brain represents the external world is not only a central theme in art but is at the very core of philosophy, psychology, and neuroscience. We are interested in how the chemosensory world is represented in the brain.

All organisms have evolved a mechanism to recognize sensory information in the environment and transmit this information to the brain where it then must be processed to create an internal representation of the external world. There are many ways for organisms to probe the external world. Some smell it, others listen to it, many see it. Each species therefore lives in its own unique sensory world of which other species may be partially or totally unaware. A whole series of specific devices alien to human perception have evolved: biosonar in bats, infrared detectors in snakes, electrosensitive organs in fish, and a sensitivity to magnetic fields in birds. What an organism detects in its environment is only part of what is around it and that part differs in different organisms. The brain functions, then, not by recording an exact image of the world, but by creating its own selective picture; a picture largely determined by what is important for the survival and reproduction of the species.

Sensory impressions, therefore, are apprehended through the lens of the particular perceiving brain and the brain must therefore be endowed with an *a priori* potential to recognize the sensory world (1). Our perceptions are not direct recordings of the world around us, rather they are constructed internally according to innate rules. Colors, tones, tastes, smells are active constructs created by our brains out of sensory experience. They do not exist as such outside of sensory experience (2). Biological reality, I argue, therefore

Figure 1. La Bonne Aventure. The painting *La Bonne Aventure* (Fortune Telling), by René Magritte (1937) portrays a monumental nose. I have added the inscription "Ceci n'est pas un nez" (This is not a nose) in Magritte's script to emphasize the tension between image and reality, a conflict inherent in much of his art as well as in the science of perception.

reflects the particular representation of the external world that a brain is able to build and a brain builds with genes.

If our genes are indeed the arbiters of what we perceive from the outside world, then it follows that an understanding of the function of these genes could provide insight into how the external world is represented in our brain. But what can molecular biology really tell us about so elusive a brain function as perception? Molecular biology was invented to solve fundamental problems in genetics at a molecular level. With the demystification of the brain, with the realization that the mind emerges from the brain and that the cells of the brain often use the very same principles of organization and function as a humble bacterium or a liver cell, molecular biology and genetics could now interface with neuroscience to approach the previously tenuous relationship between genes and behavior, cognition, memory, emotion, and perception.

Why would a molecular neuroscientist interested in perception choose to focus on the elusive sense of smell? In humans, smell is often viewed as an aesthetic sense, as a sense capable of eliciting enduring thoughts and memories. Smell however is the primal sense. It is the sense that affords most organisms the ability to detect food, predators, and mates. Smell is the central sensory modality by which most organisms communicate with their environment. Second, humans are capable of recognizing hundreds of thousands of different odors. For molecular neuroscientists studying the brain, the mechanism by which an organism can interact with the vast universe of molecular structures defined as odors provides a fascinating problem in molecular recognition and perceptual discrimination. Finally, the problem of perception necessarily involves an understanding of how sensory input is ultimately translated into meaningful neural output: thoughts and behavior. In olfaction, the sensory input is extremely well defined and consists of chemicals of precise molecular structure. The character of the input in olfaction is far simpler than that of a visual image, for example, which consists of contour, texture, color, movement and form of confounding complexity. Representation of an olfactory image is simpler and reduces to the problem of how precisely defined chemical structures are transformed in brain space.

As molecular neurobiologists, Linda Buck and I approached olfactory sensory perception by dividing it into two problems: First, what mechanisms have evolved to allow for the recognition of the vast array of molecular structures we define as odorants? Clearly, there must be receptors in the sensory neurons of the nose capable of associating with odor molecules. Do we have a relatively small number of "promiscuous" receptors, each capable of interacting with a large number of odorous molecules? Alternatively, olfactory recognition may involve a very large number of "chaste" receptors each capable of interacting with a limited set of odor molecules. The second problem is conceptually more difficult: how does the olfactory sensory system discriminate among the vast array of odorous molecules that are recognized by the nose? Put simply, how does the brain know what the nose is smelling? This question will ultimately require knowledge of how the different odors are represented and encoded in the brain.

A LARGE FAMILY OF ODORANT RECEPTOR GENES

We approached the problem of odor recognition directly by isolating the genes encoding the odorant receptors (3). The experimental design we employed to isolate these genes was based on three assumptions: First, the odorant receptors were likely to belong to the superfamily of receptors, the G-protein coupled receptors (GPCRs), that transduce intracellular signals by coupling to GTP binding proteins (4,5,6,7). Second, the large repertoire of structurally distinct, odorous molecules suggests that the odorant receptors themselves must exhibit significant diversity and are therefore likely to be encoded by a multigene family. Third, the expression of the odorant receptors should be restricted to the olfactory epithelium. Experimentally, we used the polymerase chain reaction (PCR) to amplify members of the GPCR gene superfamily expressed in olfactory sensory neurons. We then asked whether any of the PCR products were indeed members of a large multigene family. We observed that restriction enzyme cleavage of a single PCR band generated a set of DNA fragments whose molecular weight summed to a value significantly greater than that of the original PCR product (3). In this manner, we identified a multigene family that encodes a large number of GPCRs whose expression is restricted to the olfactory sensory neurons. The receptors were subsequently shown to interact with odors translating the energy of odor binding into alterations in membrane potential (8,9,10,11).

The completed sequence of both the murine and human genome ultimately identified 1300 odorant receptors in the mouse (12,13) and 500 in humans (14,15,16). If mice possess 20,000 genes, then as much as 5% of the genome, one in 20 genes encodes the odorant receptors. A large family of odorant receptors is observed not only in vertebrates but in the far simpler sensory systems of invertebrates. A somewhat smaller but highly diverse family of about 80 odorant receptor genes has been identified in the *Drosophila* genome (17,18,19,50,67). The invertebrate, *C. elegans*, with only 302 neurons and 16 olfactory sensory neurons expresses about 1000 odorant receptor genes (20,21). These experiments provide a solution to the first question; we recognize the vast array of molecular structures defined as odorants by maintaining in our genome a large number of genes encoding odorant receptors.

The observation that over 1000 receptors are required to accommodate the detection of odors suggests a conceptual distinction between olfaction and other sensory systems. Color vision in humans, for example, allows the discrimination of several hundred hues with only three different photoreceptors (22,23). These photoreceptors each have distinct but overlapping absorption spectra. Discrimination of color is thought to result from comparative processing of the information from these three classes of photoreceptors. Whereas three photoreceptors can absorb light across the entire visible spectrum, our data suggest that a small number of odorant receptors cannot recognize the full spectrum of distinct molecular structures perceived by the mammalian nose. Rather, olfactory perception requires a large number of receptors each capable of recognizing a small number of odorous ligands.

The large number of odorant receptor genes when compared with receptor numbers in other sensory systems, perhaps reflects the fact that in vision and hearing the character of the sensory stimulus is continuously variable. Color is distinguished by quantitative differences in a single parameter, the wavelength of light. Similarly, one important parameter of hearing, the frequency of sound, is continuously variable. The diversity of chemical structures of odors do not exhibit continuous variation of a single parameter and therefore cannot be accommodated by a small number of receptors. Rather, the full spectrum of distinct molecular structures perceived by the olfactory system requires a large number of receptors, each capable of interacting with a small number of specific odorous ligands.

A TOPOGRAPHIC MAP IN THE OLFACTORY BULB

We next turned to the question of olfactory discrimination: how does the brain know what the nose is smelling? The identification of a large family of receptor genes allowed us to pose this question in molecular terms. We could now ask how the brain knows which of the numerous receptors have been activated by a given odor. The elucidation of a mechanism by which the brain

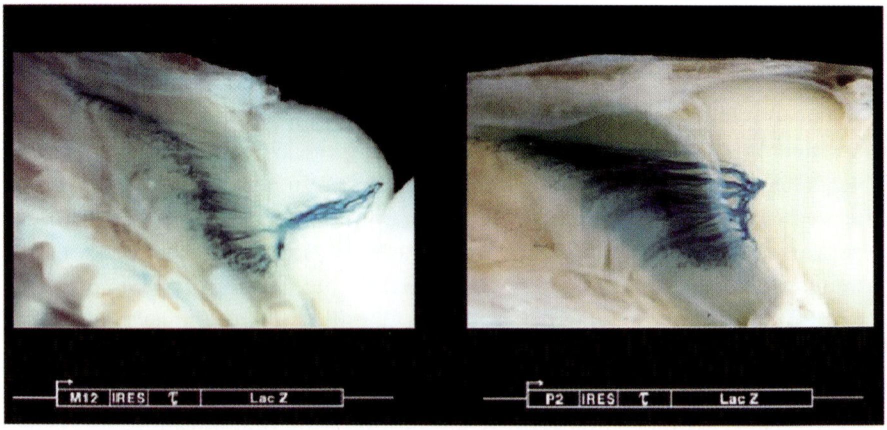

Figure 2. Convergence of Axons from Neurons Expressing a Given Receptor. Odorant receptor loci were modified by homologous recombination in ES cells to generate strains of mice in which cells expressing a given receptor also express a fusion of the microtubule associated protein, tau, with β-galactosidase. These whole mount photographs reveal neurons expressing either the M12 (left) or P2 (right) receptors along with their axons as they course through the cribriform plate to a single locus in the olfactory bulb. Neurons expressing different receptors converge on different glomeruli. The genetic modifications that assure the coordinate expression of receptor and tau-lacZ are shown beneath the whole mount view. Reprinted from *Cell*, Vol 87, 1996, pp 675-686, Mombaerts *et al.*, with permission from Elsevier.

distinguishes the different combinations of receptors activated by different odors would provide a logic of odor discrimination. This problem was further simplified by the demonstration that an individual sensory neuron expresses only one of the 1000 receptor genes (10,24). This observation emerged from single neuron cDNA cloning experiments, and allowed us to translate the problem of how the brain determines which receptor has been activated to a far simpler problem: how does the brain knows which neuron has been activated by a given odor. As in other sensory systems, an invariant spatial pattern of olfactory sensory projections could provide a topographic map of receptor activation that defines the quality of a sensory stimulus.

In other sensory systems, spatially segregated afferent input from peripheral sensory neurons generates a topographic map that defines the location of a sensory stimulus within the environment as well as the quality of the stimulus itself. Olfactory sensory processing does not extract spatial features of the odorant stimulus. Relieved of the requirement to map the position of an olfactory stimulus in space, we asked whether the olfactory system might employ spatial segregation of sensory input to encode a quality of an odorant. Robert Vassar in my lab and Kerry Ressler in Linda Buck's lab therefore analyzed the spatial patterns of receptor expression in the olfactory epithelium by *in situ* hybridization and observed that cells expressing a given receptor are restricted to one of four broad but circumscribed zones (25,26).

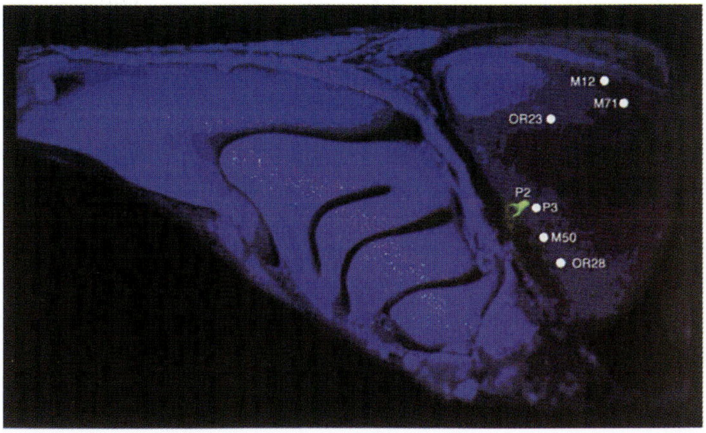

Figure 3. A Topographic Map of Olfactory Sensory Axons in the Bulb. A whole mount reveals neurons expressing two modified P2 alleles: P2-IRES-tau-lacZ (red) or P2-IRES-GFP (green). These neurons send axons that co-converge on the same glomerulus in the olfactory bulb. Neurons expressing other receptors converge on different glomerular loci that are shown schematically. All nuclei are stained blue with TOTO-3. The relative positions of the different glomeruli are maintained in different mice revealing an invariant topographic map in the olfactory bulb.

The overriding feature of this organization, however, is that within a zone neurons expressing a given receptor are not topographically segregated, rather they appear randomly dispersed. When they performed *in situ* hybridization experiments to the bulb, the first relay station for olfactory sensory neurons in the brain, they observed that topographic order was restored (27,28). Neurons expressing a given receptor, although radomly distributed in the epithelium, project to spatially invariant glomeruli in the olfactory bulb generating a topographic map.

Peter Mombaerts, then a fellow in the lab, developed a genetic approach to visualize axons from olfactory sensory neurons expressing a given odorant receptor as they project to the brain (29). We modified receptor genes by targeted mutagenesis in the germ line of mice. These genetically altered receptor genes now encode a bicistronic mRNA that allows the translation of receptor along with tau-lacZ, a fusion of the microtubule-associated protein tau with β-galactosidase. In these mice, olfactory neurons that transcribe a given receptor also express tau-lacZ in their axons, permitting the direct visualization of the pattern of projections in the brain (Fig. 2).

We observe that neurons expressing a receptor project to only two topographically-fixed loci, or glomeruli, in the bulb creating mirror image maps in each bulb. Neurons expressing different receptors project to different glomeruli. The position of the individual glomeruli is topographically defined and is similar for all individuals in a species (Fig. 3). Individual odors could activate a subset of receptors that would generate specific topographic patterns of activity within the olfactory bulb such that the quality of an olfactory stimulus could be encoded by spatial patterns of glomerular activity.

The identification of an anatomic olfactory sensory map poses four questions. The first, addresses the singularity of receptor gene choice. What mechanism assures that a sensory neuron expresses only a single receptor and then projects with precision to one of 1000 topographically fixed glomerular loci. Second, does the anatomic map translate into a functional map such that different odors elicit different patterns of activity? Third, can we relate specific spatial patterns of glomerular activity to specific behaviors? Finally how is the map read? How does the brain look down upon a spatial pattern of activity and associate this pattern of with a particular odor?

RECEPTOR CHOICE AND THE TOPOGRAPHIC MAP

The topographic map in the olfactory system differs in character from the orderly representation inherent in the retinotopic, tonotopic, or somatotopic sensory maps. In these sensory systems, the peripheral receptor sheet is represented in the central nervous system (CNS), such that neighbor relations in the periphery are preserved in the CNS (reviewed in 30,31). In this manner, peripheral receptor cells may acquire a distinct identity that is determined by their spatial position in the receptor sheet. Spatial patterning in the periphery can therefore endow individual neurons with positional information that directs their orderly representation in the brain.

The olfactory system, however, does not exhibit an orderly representation of receptor cells in the periphery. Neurons expressing a given receptor are randomly dispersed within a given zone and order is restored in the bulb where neurons expressing a given receptor converge on discrete loci to create a topographic map. Olfactory neurons differ from one another not by virtue of their position in a receptor sheet, but rather by the nature of the receptor they express. The tight linkage between the choice of an odorant receptor and the site of axon convergence suggests a model in which the odorant receptor is expressed on dendrites, where it recognizes odorants in the periphery, and also on axons, where it governs target selection in the bulb. In this manner, an olfactory neuron would be afforded a distinct identity that dictates the nature of the odorant to which it responds as well as the glomerular target to which its axon projects. If the odorant receptor also serves as a guidance molecule, this leads to two experimental predictions. First, the receptor should be expressed on axons as well as on dendrites and second, genetic modifications in the receptor sequence might alter the topographic map.

The first prediction was tested by Gilad Barnea who generated specific antibodies against two odorant receptors and examined the sites of receptor expression on sensory neurons (32). Antibodies were raised against extracellular and cytoplasmic epitopes of the mouse odorant receptors, MOR28 and MOR11-4. In the sensory epithelium, we observe intense staining in the den-

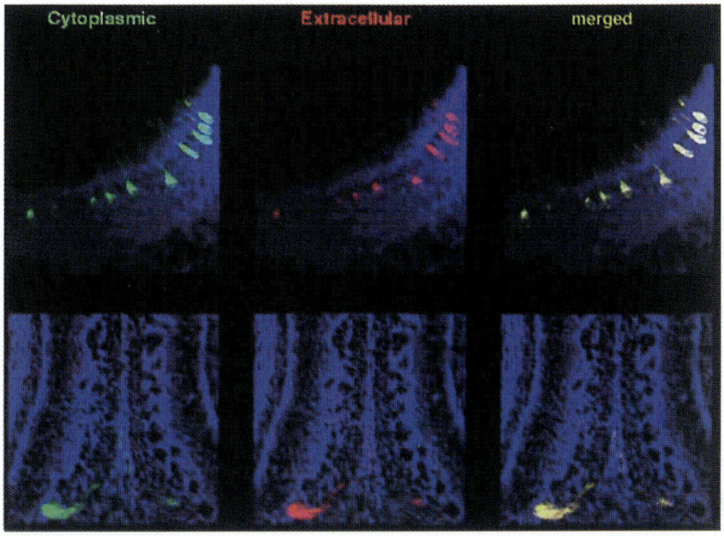

Figure 4. Odorant Receptor is Expressed on both Dendrites and Axons of Olfactory Sensory Neurons. The mouse sensory epithelium (upper panel) or olfactory bulb (lower panel) was stained with antibody to either an extracellular or cytoplasmic epitope of the MOR28 receptor. These experiments reveal the expression of odorant receptor in the cell body and dendrites in the epithelium as well as on axon termini within a defined glomerulus in the bulb. Antibody staining in the olfactory bulb coincides with the site of convergence of MOR28 axons. Adapted with permission from 32. Reprinted from *Science* 304,1468, 2004, with permission from *Science.*

dritic knobs, the site of odor binding. In the olfactory bulb, antibody stains axon termini whose arbors are restricted to two glomeruli (Fig. 4). Antibody staining of the bulb from mice bearing the MOR28-IRES-tau-lacZ allele reveals that the glomeruli stained by antibody to MOR28 also receives the tau-lacZ fibers. Thus the receptor is expressed on both dendrites and the axons of sensory neurons.

In a second series of experiments performed by a student Fan Wang, we provided genetic evidence suggesting that the receptor on axons is indeed a guidance molecule. We modified our gene targeting approach to ask whether substitutions of the P2 receptor coding sequence alter the projections of neurons that express this modified allele (33). We replaced the coding region of the P2 gene with the coding regions of several other receptors, and examined the consequences on the formation of the topographic map. Substitution of the P2 coding region with that of the P3 gene, a linked receptor gene homologous to P2 and expressed in the same epithelial zone, results in the projection of axons to a glomerulus distinct from P2 that resides immediately adjacent to the wild type P3 glomerulus. Other substitutions that replace the P2 coding sequences with receptor sequences expressed either in different zones or from different chromosomal loci also result in the convergence of fibers to glomeruli distinct from P2. These observations, along with recent experiments involving more extensive genetic modifications (34,35) provide support for the suggestion that the olfactory receptor plays an instructive role in axon targeting as one component of the guidance process.

How may the odorant receptors participate in the guidance process? In one model, the odorant receptor is expressed on the axon termini along with other guidance receptors where it recognizes positional cues elaborated by the bulb. Each of the 1000 distinct types of sensory neuron will therefore bear a unique combination of guidance receptors that define a code dictating the selection of a unique glomerular target. Such a model does not necessarily imply that there are 1000 distinct cues, each spatially localized within the bulb. Rather, a small number of graded cues may cause the differential activation of the different odorant receptors on axon termini. In this manner, the different affinities of individual receptors for one or a small number of cues, and perhaps different levels of receptor, might govern target selection. Such a model is formally equivalent to models of retinotopy in which a gradient of guidance receptor on retinal axons is matched by a positional gradient of guidance cues in the tectum (reviewed in 31).

THE SINGULAR AND STABLE CHOICE OF RECEPTOR

If the odorant receptor defines the functional identity of a sensory neuron and also determines the site of projection in the brain, then the expression of a single receptor gene in a neuron is an essential feature in models of olfactory perception. This immediately poses the question as to what mechanism has evolved to assure the expression of a single receptor gene from the family of 1000 genes in the chromosome. One model for the control of olfactory receptor (OR) expression invokes the existence of 1000 different sensory neurons, each expressing a unique combination of regulatory factors that governs the choice of a different OR gene. This deterministic model predicts that all OR genes will contain different *cis*-regulatory sequences that are recognized by unique sets of transcription factors. An alternative, stochastic model of receptor gene selection suggests that all odorant receptor genes within a zone contain the same *cis*-regulatory information and are controlled by the same set of transcription factors. In this model a special mechanism must exist to assure that only one receptor gene is chosen. Moreover, once a specific receptor is chosen for expression, this transcriptional choice must be stable for the life of the cell because receptor switching after stable synapse formation would seriously perturb odor discrimination.

A series of transgene experiments performed by Ben Shykind in my own laboratory, as well as in other labs, provide evidence for a mechanism of receptor choice that is stochastic (36,37). We have generated mice in which the endogenous P2 allele has been replaced with the P2-IRES-tau-lacZ allele. We have also introduced a randomly integrated P2-IRES-GFP transgene into the chromosome of this strain. In a deterministic model, we predict that a unique combination of transcription factors would activate both the endoge-

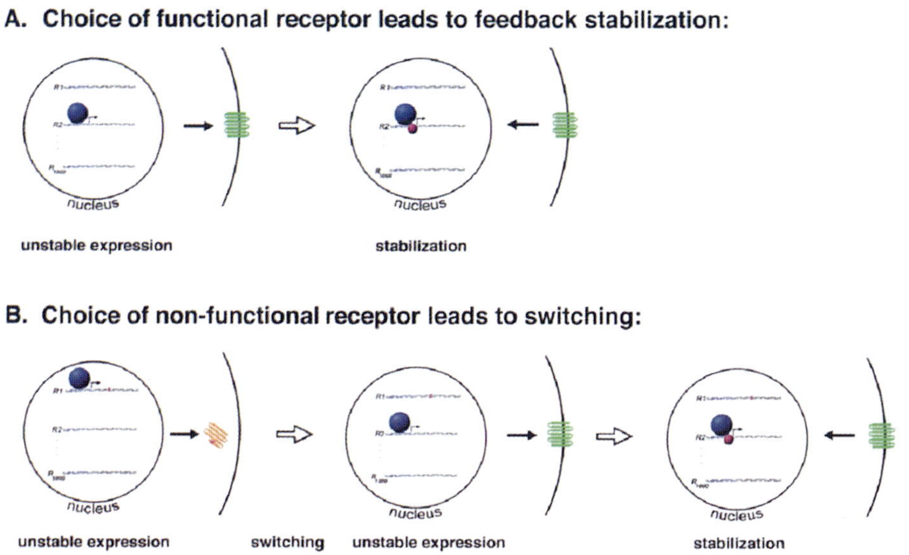

Figure 5. A Feedback Model Assuring the Stable Expression of a Functional Receptor. (A) The transcriptional machinery represented by a blue sphere expresses only one of 1000 odorant receptor genes (in this instance, R2). R2 encodes a functional receptor that elicits a feedback signal that leads to the stabilization of receptor choice (symbolized by a red sphere). (B) If the transcriptional machinery chooses the non-functional receptor, R1, which is not competent to mediate feedback stabilization, switching occurs. The transcriptional machine is then free to select a second receptor for expression that will ultimately mediate feedback stabilization. This model provides a mechanism to assure that a neuron expresses a functional odorant receptor.

nous and transgenic P2 alleles such that cells that express lacZ from the endogenous P2-IRES-tau-lacZ allele should also express GFP from the P2 transgene. Examination of the sensory epithelium in these mice, however, reveals a singularity of P2 expression. Cells that express the endogenous P2 allele never express the transgene. In a conceptually similar experiment, we generated transgenic mice that harbor an integrated array of multiple P2 transgenes that include P2-IRES-tau-lacZ and P2-IRES-GFP linked at the same chromosomal locus. In these strains, we also observe a singularity of transgene expression. Neurons that express the P2-IRES-tau-lacZ transgene do not express the linked P2-IRES-GFP gene. Taken together, these experiments provide support for a model in which receptor choice is not deterministic, rather it is stochastic.

Once a single receptor gene is chosen for expression, this transcriptional choice must be stable for the life of the cell because receptor switching after stable synapse formation would seriously perturb odor discrimination. In recent experiments, Ben Shykind in my lab along with the Reed and Sakano labs devised genetic strategies that permit the analysis of the stability of receptor choice (38,39,40). We have employed a lineage tracer to map the fate of sensory neurons that express either an intact or a nonfunctional deletion of the MOR28 gene. Mature neurons that express an intact MOR28 receptor, but have not yet formed stable synapses in the brain, can switch receptor expression, albeit at low frequency. Thus, we observe that switching is an inherent property of wild type receptor gene choice. Neurons that choose to express a mutant MOR28 receptor subsequently extinguish its expression and switch at high frequencies to express alternate receptors such that a given neuron stably transcribes only a single receptor gene. These observations suggest a mechanism of OR gene choice in which a cell selects only one receptor allele but can switch at low frequency. Expression of a functional receptor would then elicit a signal that suppresses switching and stabilizes odorant receptor expression. Neurons that initially express a mutant receptor fail to receive this signal and switch genes until a functional receptor is chosen (Fig. 5).

The mouse genome contains 340 OR pseudogenes, whereas the human genome contains 550 pseudogenes, several of which continue to be transcribed (12,16). Expression of a pseudogene would result in the generation of sensory neurons incapable of odor recognition. A mechanism that allows switching provides a solution to the pseudogene problem such that if pseudogenes are chosen, another transcriptional opportunity is provided assuring that each neuron expresses a functional receptor. This model of serial monogamy assures that neurons will express a single receptor throughout their life. This feedback model in which expression of a functional odorant receptor suppresses switching to other OR genes is reminiscent of one mechanism of allelic exclusion in T and B lymphocytes.

CLONING A MOUSE FROM AN OLFACTORY SENSORY NEURON

What mechanism assures that a single receptor gene is chosen stochastically in a sensory neuron? One model invokes DNA recombination of odorant receptor genes at a single active expression site in the chromosome. DNA recombination provides *Saccharomyces cerevisiae* (41), trypanosomes (42) and lymphocytes (43) with a mechanism to stochastically express one member of a set of genes that mediate cellular interactions with the environment. One attractive feature shared by gene rearrangements in trypanosomes and lymphocytes is that gene choice is a random event, a feature of receptor gene selection in olfactory sensory neurons. However, efforts to demonstrate a recombination event involving OR genes have been seriously hampered by the inability to obtain populations of neurons or clonal cell lines that express the same receptor. Kristin Baldwin in my laboratory, in a collaboration with Rudy Jaenisch, Kevin Eggan and Andy Chess at MIT, addressed this problem by generating ES cell lines and cloned mice derived from the nuclei of olfactory sensory neurons expressing the P2 receptor (Fig. 6) (44). The generation of cloned mice from cells of the nose derives from an initial insight of Woody Allen in his 1978 futuristic comedy, *Sleeper.* In this film, efforts are made to resurrect a totalitarian leader by cloning from his only surviving body part, his nose. Twenty-five years later, science successfully imitated art with the

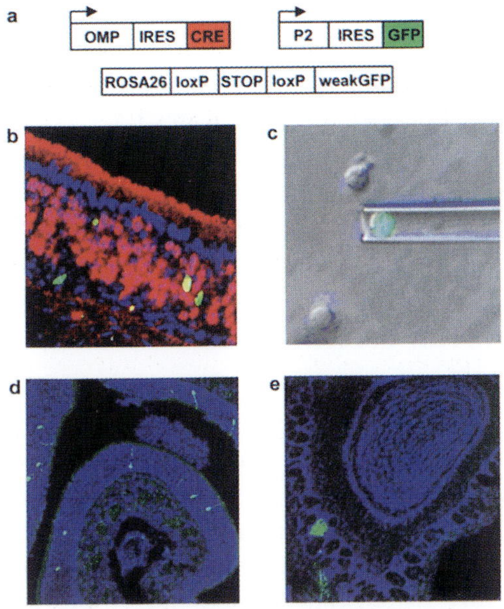

Figure 6. Cloning a Mouse from Olfactory Sensory Neurons Expressing the P2 Odorant Receptor. (a) A genetic strategy to label P2-expressing sensory neurons with GFP as well as to mark olfactory sensory neurons by virtue of a unique deletion in DNA. (b) The olfactory epithelium of a mouse with the genetic modifications described above. A single nucleus expressing the P2 odorant receptor gene was picked and introduced into an enucleated oocyte. The epithelium was stained with antibody to Cre recombinase (red) to mark sensory neurons and GFP (green) to identify P2-expressing cells. (c) A green neuron expressing P2-IRES-GFP was picked from dissociated olfactory epithelium of donor animals. (d) The olfactory epithelium from a mouse cloned from a nucleus expressing the P2 receptor shows the normal distribution of P2-expressing cells. Axons from these neurons converge on a single glomerulus in the olfactory bulb (e). All nuclei are stained with TOTO-3 blue. The observation that mice cloned from a nucleus expressing the P2 receptor gene do not preferentially express this gene in the sensory epithelium suggests that DNA recombination events do not accompany receptor gene choice. Adapted with permission from 44. Reprinted from *Nature* 428, 44-49, 2004, Eggan *et al.*, with permission from *Nature.*

generation of mice cloned from a single sensory neuron from the nose.

We would predict that if DNA recombination accompanies receptor gene choice then the olfactory epithelium from cloned mice derived from a sensory neuron expressing the P2 gene should be clonal with respect to receptor expression, such that all cells transcribe the rearranged P2 allele. Analysis of the sequence and organization of the DNA surrounding the P2 allele expressed in cloned mice revealed no evidence for either gene conversion or local transposition at the P2 locus. In addition, the pattern of receptor gene expression in the sensory epithelium of cloned mice was normal. Multiple odorant receptor genes are expressed without preference for the P2 allele transcribed in the donor nucleus (Fig. 6). These data, along with similar experiments by Peter Mombaerts (45), demonstrate that the mechanism responsible for the choice of a single odorant receptor gene does not involve irreversible changes in DNA. In a broader context, the generation of fertile cloned mice that are anatomically and behaviorally indistinguishable from wild type indicates that the genome of a postmitotic, terminally differentiated olfactory neuron can re-enter the cell cycle and be reprogrammed to a state of totipotency after nuclear transfer. The stochastic choice of a single OR gene is therefore not accomplished by DNA recombination but rather by a rate limiting transcriptional process, perhaps involving a single transcriptional machine capable of stably accommodating only one OR gene.

OLFACTION IN THE FLY: A FUNCTIONAL MAP IN THE ANTENNAL LOBE

The identification of an anatomic map in the olfactory bulb immediately poses the question as to whether this map provides a meaningful representation of odor quality that is translated into appropriate behavioral output. Recently, we have become interested in how the olfactory world is represented in the brain of the fruit fly. *Drosophila* provides an attractive system to understand the logic of olfactory perception. Fruit flies exhibit complex behaviors controlled by an olfactory system that is anatomically and genetically simpler than that of vertebrates. Genetic analysis of olfaction in *Drosophila* may therefore provide a facile system to understand the mechanistic link between behavior and the perception of odors. The recognition of odors in *Drosophila* is accomplished by sensory hairs distributed over the surface of the third antennal segment and the maxillary palp. Olfactory neurons within sensory hairs send projections to one of the multiple glomeruli within the antennal lobe of the brain (46,47). Leslie Vosshall and Allan Wong showed that most sensory neurons express only one of about 80 odorant receptor genes. Neurons expressing the same receptor project with precision to one or rarely two spatially invariant glomeruli in the antennal lobe, the anatomic equivalent of the olfactory bulb of mammals (48,49,50) (Fig. 7).

The anatomic organization in *Drosophila* is therefore remarkably similar to that of the olfactory system of mammals, suggesting that the mechanism of odor discrimination has been shared despite the 600 million years of evolution separating insects from mammals. This conservation may reflect the maintenance of an efficient solution to the complex problem of recognition and discrimination of a vast repertoire of odors in the environment. In both flies and mice, the convergence of like axons into discrete glomerular structures provides a map of receptor activation in the first relay station for olfactory information in the brain, such that the quality of an odorant may be reflected by spatial patterns of activity, first in the antennal lobe or olfactory bulb and ultimately in higher olfactory centers.

An understanding of the logic of odor perception requires functional analysis to identify odor-evoked patterns of activity in neural assemblies and ultimately the relevance of these patterns to odor discrimination. We have performed two-photon calcium imaging to examine the relationship between the anatomic map and the functional map in the antennal lobe (51). Jing Wang and Allan Wong in my lab developed an isolated *Drosophila* brain preparation that is

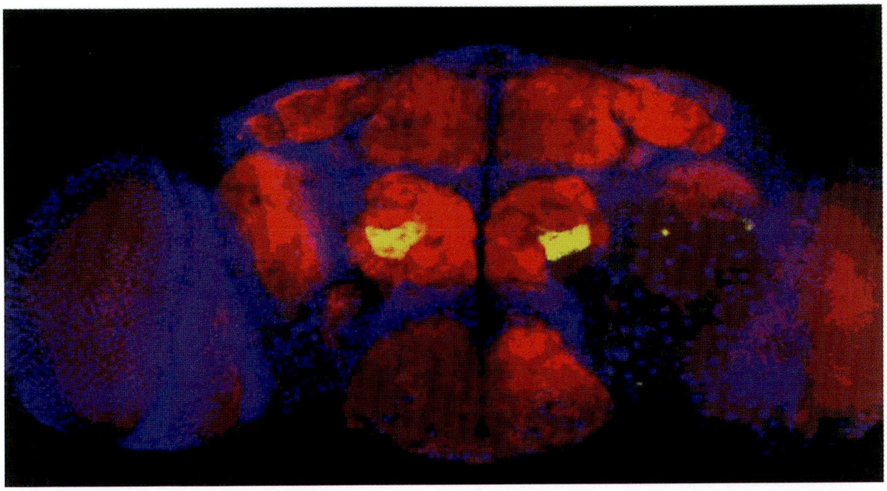

Figure 7. An Olfactory Sensory Map in the Fly Antennal Lobe. Neurons expressing the odorant receptor, OR47b, also express the transgene, synaptobrevin GFP, revealing convergence on a single spatially invariant glomerulus that is bilaterally symmetric in the antennal lobe.

amenable to two-photon imaging and is responsive to odor stimulation for up to five hours. We expressed the calcium-sensitive fluorescent protein G-CaMP in primary olfactory sensory neurons and projection neurons. G-CaMP consists of a circularly permuted EGFP flanked at the N-terminus by the calcium-binding site of calmodulin and at the C-terminus by the M13 fragment of myosin light chain kinase (52). In the presence of calcium, calmodulin interacts with the M13 fragment eliciting a conformation change in EGFP. The resulting elevations in fluorescent intensity reflect changes in the intracellular calcium concentration, a presumed mirror of electrical activity. Moreover, the ability to express G-CaMP in genetically defined populations of neurons allowed us to determine with certainty the locus of neural activity. Odor-evoked changes in fluorescence intensity within the antennal lobe are monitored by a laser-scanning two-photon microscope (53).

This imaging technique has allowed us to measure the responsivity of 23 glomeruli to 16 different odors (51). A number of interesting features of the glomerular response to odors are revealed by these experiments. First, different odors elicit different patterns of glomerular activation and these patterns are conserved among different animals (Fig. 8). At odor concentrations likely to be encountered in nature, the map is sparse and glomeruli are narrowly tuned.

Second, the patterns of activity are insular, such that neighboring glomeruli do not necessarily respond together to a given odor. Each glomerulus visualized anatomically appears to be a functional unit. Third, the patterns of glomerular activity are qualitatively similar upon imaging either sensory or projection neurons. These observations suggest the faithful transmission of sensory input to higher brain centers. Fourth, we have coupled genetic experiments with

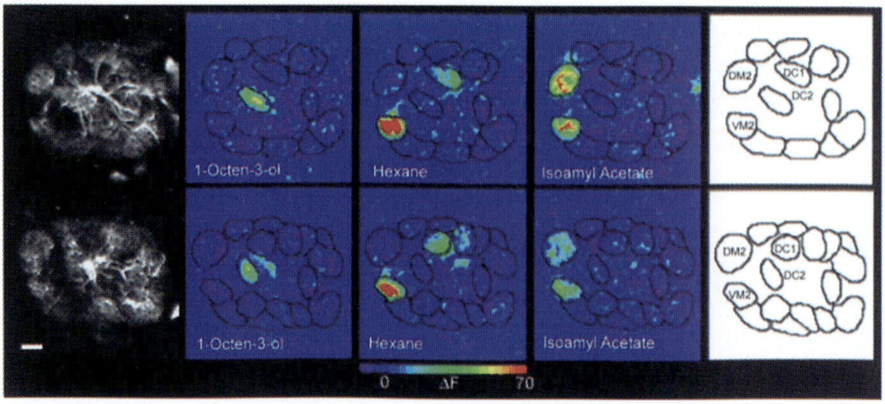

Figure 8. Different Odors Elicit Different Patterns of Glomerular Activation that are Conserved Among Different Organisms. Two different flies (upper and lower panels) bearing the GH146-Gal4 and UAS-G-CaMP transgenes were exposed to three odors. Glomerular responses reveal different patterns of activity for the different odors that are conserved in different animals. The panels to the left show the pre-stimulation images that reveal glomerlar structure and the panels to the right identify the specific glomeruli schematically. Reprinted from *Cell*, Vol 112, 2003, pp 271-282, Wang *et al.*, with permission from Elsevier.

imaging to demonstrate that the odor-evoked profile for a given glomerulus directly reflects the responsivity of an individual odorant receptor. This finding is consistent with prior molecular and anatomic studies that reveal that neurons that express only a single receptor in like axons converge on a single glomerulus. Thus these studies, along with other imaging approaches in insects (54,55), demonstrate that the anatomic map is indeed functional and suggests that each odor elicits a sparse pattern of glomerular activation that may confer a signature for different odors in the brain. Imaging experiments in vertebrates similarly reveal a functional representation of the anatomic map (56,57,58).

SPATIAL REPRESENTATIONS AND INNATE BEHAVIOR

All animals exhibit innate behaviors in response to specific sensory stimuli that are likely to result from the activation of developmentally programmed circuits. Allan Wong and Jing Wang in my lab, in collaboration with Greg Suh, David Anderson and Seymour Benzer at Caltech, asked whether we can relate patterns of glomerular activity elicited by an odor to a specific behavior (59). Some time ago Benzer observed that *Drosophila* exhibits robust avoidance to odors released by stressed flies. Gas chromatography and mass spectrometry identified one component of this "*Drosophila* stress odorant (DSO)" as CO_2. Exposure of flies to CO_2 alone also elicits an avoidance behavior at levels of CO_2 as low as 0.1% (Fig. 9).

We therefore performed imaging experiments with the calcium-sensitive fluorescent indicator G-CaMP and two-photon microscopy to ask whether we

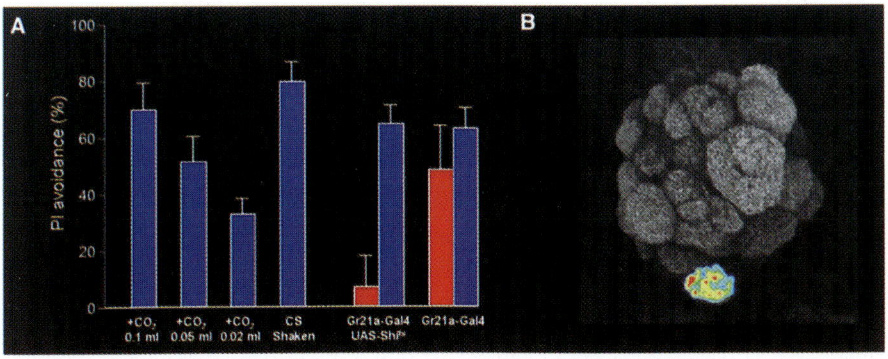

Figure 9. CO_2 Activates a Single Glomerulus and Elicits Avoidance Behavior. (A) Avoidance of air from stressed flies (CS) as well as of increasing concentrations of CO_2. Inhibition of synaptic transmission in GR21A neurons that project to the V glomerulus using shi^{ts} blocks CO_2 avoidance. Red and blue bars indicate avoidance behavior at the nonpermissive (28°C) and permissive (21°C) temperatures, respectively. (B) Two-photon imaging in a strain harboring GR21A-Gal4 and UAS G-cAMP reveals robust activation of the V glomerulus.

could discern a pattern of glomerular activity in response to DSO and CO_2. We first examined flies in which the G-CaMP indicator is driven in all neurons by the pan-neural activator, Elav-Gal4. DSO activates only two glomeruli, DM2 and the V glomerulus, whereas CO_2 activates only the V glomerulus. Activation of the V glomerulus was detected at CO_2 levels as low as 0.05% and this glomerulus was not activated by any of 26 other odorants tested (Fig. 9).

We demonstrated that axonal projections to V originate from sensory neurons expressing the receptor, GR21A (50). We therefore performed calcium imaging with flies in which the UAS G-CaMP reporter was driven by a GR21A promoter Gal4 activator. CO_2, as well as DSO activated GR21A sensory termini in the V glomeruli. We next asked whether the GR21A sensory neurons are necessary for the avoidance response to CO_2. Inhibition of synaptic transmission in the GR21A sensory neurons that innervate the V glomerulus, using a temperature-sensitive *shibire* gene, shi^{ts} (60), blocks the avoidance response to CO_2 (Fig. 9). Inhibition of synaptic release in the vast majority of other olfactory sensory neurons or in projection neurons other than those that innervate the V glomerulus, had no effect on this behavior.

The identification of a population of olfactory sensory neurons innervating a single glomerulus that mediates robust avoidance to a naturally occurring odorant provides insight in the neural circuitry that underlies this innate behavior. These observations suggest that a dedicated circuit that involves a single population of olfactory sensory neurons mediates detection of CO_2 in *Drosophila.* The simplicity of this initial olfactory processing offers the possibility of tracing the circuits that translate odor detection into an avoidance response.

HOW IS THE MAP READ?

Our experiments indicate that different odors elicit different patterns of glomerular activity within the antennal lobe and moreover that defined patterns of activity can be associated with specific behaviors. We can look at the pattern of activity in the fly antennal lobe with a two-photon microscope and discern, with a reasonable degree of accuracy what odorant the fly has encountered in nature. Thus we can with our eyes and our brain determine what odors the fly has encountered, but how does the fly brain read the sensory map?

A topographic map in which different odors elicit different patterns of activity in the antennal lobe suggests that these spatial patterns reflect a code defining odor quality. However, the mere existence of a map, whether anatomic or functional, does not prove that spatial information is the underlying parameter of an odor code. It has been suggested, for example, that the quality of an odor is reflected in temporal dynamics of a distributed ensemble of projection neurons (61,62). In this model, a given odor might activate a small number of glomeruli and a large ensemble of projection neurons (PNs) such that different odors elicit different temporal patterns of activity in the same PN. This temporal hypothesis in its simplest form postulates that the brain exploits circuit dynamics to create spatiotemporal patterns of neuronal activation to achieve a larger coding space. Whatever the code, patterns of activity in the antennal lobe must be translated by higher sensory centers to allow the discrimination of complex olfactory information. If odor quality is encoded by spatial patterns, we might expect that a representation of the glomerular map is retained in the protocerebrum.

We have begun to address the question of how the map in the antennal lobe is represented in higher olfactory centers by examining the pattern of projections of the neurons that connect the glomeruli to the protocerebrum. Allan Wong and Jing Wang randomly labeled individual projection neurons to visualize their processes that connect defined glomeruli with their targets in the mushroom body and protocerebrum. We have used an enhancer trap line in which Gal4 is expressed in a subpopulation of projection neurons along with the FLP-out technique, to label single projection neurons with a CD8-GFP reporter (63). A similar experimental approach has been used to determine the lineage relationship of individual PNs and to examine their pattern of axonal projections (64,65). We observe that most PNs send dendrites to a single glomerulus. Projection neurons that receive input from a given glomerulus extend axons that form a spatially invariant pattern in the protocerebrum (Fig. 10). PNs from different glomeruli exhibit patterns of axonal projections that are distinct, but often interdigitated (Fig. 11). Our data reveal a striking invariance in the spatial patterns of axon arbors of PNs that innervate a given glomerulus, a precision of connectivity that assures the specificity of information transfer.

The precision of projections of PNs reveals a spatial representation of glomerular activity in higher brain centers but the character of the map differs from that observed in the antennal lobe. Axon arbors in the protocerebrum are

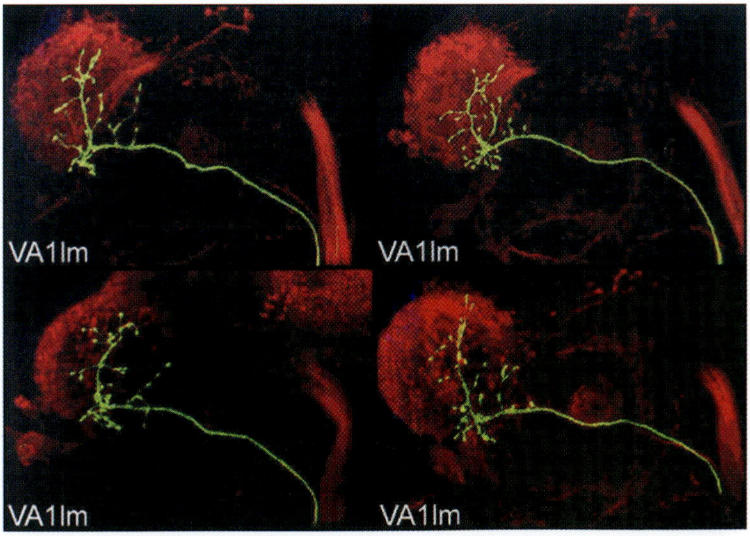

Figure 10. Projection Neurons that Innervate to the Same Glomerulus have Similar Axonal Projection Patterns. Individual projection neurons that connect to the VA1 LM glomeruli are visualized in the protocerebrum in different flies. These images reveal a striking constancy in the projection pattern among PNs that project to a given glomerulus. These observations reveal an invariant topographic map in the protocerebrum that differs in character from the map in the antennal lobe (with permission from 63). Reprinted from *Cell*, Vol 109, 2002, pp 229-241, Wong *et al.*, with permission from Elsevier.

diffuse and extensive, often extending the entire dimension of the brain hemisphere (Fig. 10,11). This is in sharp contrast to the tight convergence of primary sensory axons, whose arbors are restricted to a small 5–10 μm spherical glomerulus. As a consequence, the projections from different glomeruli, although spatially distinct, often interdigitate. Thus, the point-to-point segregation observed in the antennal lobe is degraded in the second order projections to the protocerebrum. This affords an opportunity for the convergence of inputs from multiple different glomeruli essential for higher order processing. Third order neurons in the protocerebrum might synapse on PNs from multiple distinct glomeruli, a necessary step in decoding spatial patterns to allow the discrimination of odor and behavioral responses.

CONCLUDING REMARKS

These data suggest a model in which the convergence of information from deconstructed patterns in the antennal lobe are reconstructed by "cardinal cell assemblies" that sit higher up in a hierarchical perceptual system in the protocerebrum. Olfactory processing will initially require that the structural

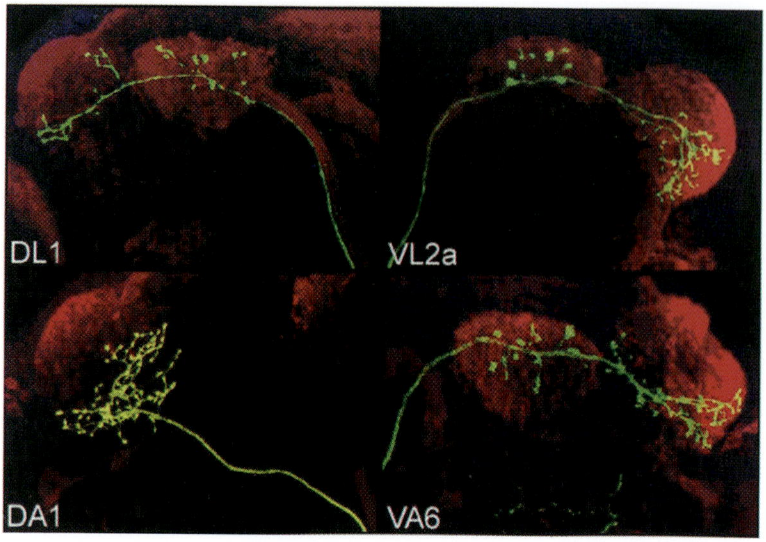

Figure 11. Axonal Patterns from Projection Neurons that Innervate to Different Glomeruli are Distinct. Axonal projections from single PNs can be visualized as they branch in the mushroom body and ultimately arborize in the protocerebrum. Projections neurons that connect to different glomeruli exhibit different patterns of axonal projections. The axon arbors in the protocerebrum are dispersed unlike the insular segregated arbors in the glomerulus, affording the possibility for integration in higher olfactory centers (with permission from 63). Reprinted from *Cell*, Vol 109, 2002, pp 229-241, Wong *et al.*, with permission from Elsevier.

elements of an odor activate an unique set of receptors that in turn result in the activation of a unique set of glomeruli. The odorous stimuli must then be reconstructed in higher sensory centers that determine which of the numerous glomeruli have been activated. The identification of a spatially invariant sensory map in the protocerebrum that is dispersive affords an opportunity for integration of multiple glomerular inputs by higher odor neurons.

The elucidation of an olfactory map in both the olfactory bulb or antennal lobe and in higher olfactory centers leaves us with a different order of problems. Though we may look at these odor-evoked images with our brains and recognize a spatial pattern as unique and can readily associate the pattern with a particular stimulus, the brain does not have eyes. Who in the brain is looking at the olfactory image? Who reads the map? How are spatially defined bits of electrical information in the brain decoded to allow the perception of an olfactory image? We are left with an old problem, the problem of the ghost in the machine.

Finally, how do we explain the individuality of olfactory perception? The innately configured representation of the sensory world, the olfactory sensory maps that I have described, must be plastic. Our genes create only a substrate upon which experience can shape how we perceive the external world. Surely the smell of a madeleine does not elicit in all of us that "vast structure of recollection" it evoked for Marcel Proust. For Proust, smell is the evocative sense, the sense that brings forth memory and associations with a richness not elicited by other sensory stimuli. Nowhere is this more apparent than in the eloquent words recalling the madeleine incident from *"Remembrance of Things Past"* (66).

"But when from a long distant past nothing subsists, after the people are dead, after the things are broken and scattered, still alone, more fragile but with more vitality, more unsubstantial, more persistent, more faithful, the smell and taste of things remain, poised a long time, like souls ready to remind us, waiting and hoping for their moment, amid the ruins of all the rest; and bear unfaltering in the tiny and impalpable drop of their essence, the vast structure of recollection."

ACKNOWLEDGEMENTS

This lecture encompasses the efforts in my laboratory over the past thirteen years to provide further insight into the molecular logic of olfactory sensory perception. I wish to thank the Howard Hughes Medical Institute, the National Institutes of Health, and the Mathers Foundation for their continued gracious support of our research. Howard Hughes Medical Institute provided an opportunity to interface molecular biology with neuroscience and has consistently encouraged and supported the efforts of the laboratory in novel directions. It is this work for which Linda Buck and I share the profound honor and good fortune of having been awarded the Nobel Prize in Physiology or Medicine. This award was not made to me as a man but for my work, a science that derives from the efforts of many brilliant students and from the incisive teachings of my colleagues. I take equal pride in the science that has been accomplished in the laboratory and in the scientists that have trained with me and have contributed to our efforts. I therefore feel that I accept this prize in trust as a representative of a culture of science in my laboratory and at Columbia University. I am deeply grateful for this culture. Over the past thirty years, Columbia has provided an atmosphere that fosters intellectual rigor and creativity and at the same time is imbued with a spirit of warmth and collaboration.

REFERENCES

1. Kant, I. [1781/1787] (1961) in: *Critique of Pure Reason,* N.K. Smith (transl), MacMillan, London.
2. Kandel, E., Schwartz, J.H., and Jessell, T.M. (eds) (2000) *Principles of Neural Science* (4th ed.), McGraw Hill, New York.
3. Buck, L. and Axel, R. (1991) A novel multigene family may encode odorant receptors: A molecular basis for odor recognition. *Cell* 65:175–187.
4. Pace, U., Hanski, E. Salomon, Y., and Lancet, D. (1985) Odorant-sensitive adenylate cyclase may mediate olfactory reception. *Nature* 316:255–258.
5. Sklar, P.B., Anholt, R.R.H., and Snyder, S.H. (1986) The odorant-sensitive adenylate

cyclase of olfactory receptor cells: differential stimulation by distinct classes of odorants. *J. Biol. Chem.* 261:15536–15543.

6. Jones, D.T. and Reed, R.R. (1989) G_{olf}, an olfactory neuron-specific G-protein involved in odorant signal transduction. *Science* 244:790–795.
7. Breer, H., Boekhoff, L., and Tarelius, E. (1990) Rapid kinetics of second messenger formation in olfactory transduction. *Nature* 345:65–68.
8. Zhao, H., Ivic, L., Otaki, J.M., Hashimoto, M., Mikoshiba, K., and Firestein, S. (1998) Functional expression of a mammalian odorant receptor. *Science* 279:237–242.
9. Krautwurst, D., Yau, K.-W., and Reed, R.R. (1998) Identification of ligands for olfactory receptors by functional expression of a receptor library. *Cell* 95:917–926.
10. Malnic, B., Hirono, J., Sato, T., and Buck, L.B. (1999) Combinatorial receptor codes for odors. *Cell* 96:713–723.
11. Touhara, K., Sengoku, S., Inaki, K., Tsuboi, A., Hirono, J., Sato, T., Sakano, H., and Haga, T. (1999) Functional identification and reconstitution of an odorant receptor in single olfactory neurons. *Proc. Natl. Acad. Sci. USA* 96:4040–4045.
12. Zhang, X. and Firestein, S. (2002) The olfactory receptor gene superfamily of the mouse. *Nat. Neurosci.* 5:124–133.
13. Godfrey, P.A., Malnic, B., and Buck, L.B. (2004) The mouse olfactory receptor gene family. *Proc. Natl. Acad. Sci. USA* 101:2156–2161.
14. Glusman, G., Yanai, I., Rubin, I, and Lancet, D. (2001) The complete human olfactory subgenome. *Genome Res.* 11:685–702.
15. Zozulya, S., Echeverri, F., and Nguyen, T. (2001) The human olfactory receptor repertoire. *Genome Biol.* 2:0018.1–0018.12.
16. Young, J.M. and Trask, B.J. (2002) The sense of smell: genomics of vertebrate odorant receptors. *Hum. Mol. Genetic.* 11:1153–1160.
17. Clyne, P.J., Warr, C.G., Freeman, M.R., Lessing, D., Kim, J., and Carlson, J.R. (1999) A novel family of divergent seven-transmembrane proteins: candidate odorant receptors in *Drosophila. Neuron* 22:327–338.
18. Gao, Q. and Chess, A. (1999) Identification of candidate *Drosophila* olfactory receptors from genomic DNA sequence. *Genomics* 60:31–39.
19. Vosshall, B.L., Amrein, H., Morozov, P.S., Rzetsky, A., and Axel, R. (1999) A spatial map of olfactory receptor expression in the *Drosophila* antenna. *Cell* 96:725–736.
20. Troemel, E.R., Chou, J.H., Dwyer, N.D., Colbert, H.A., and Bargmann, C.I. (1995) Divergent seven transmembrane receptors are candidate chemosensory receptors in *C. elegans. Cell* 83:207–218.
21. Sengupta, P., Chou, J.H, and Bargmann, C.I. (1996) *odr-10* encodes a seven transmembrane domain olfactory receptor required for responses to the odorant diacetyl. *Cell* 84:899–909.
22. Wald, G., Brown, P.K., and Smith, P.H. (1955) Iodopsin. *J. Gen. Physiol.* 38:623–681.
23. Nathans, J., Thomas, D., and Hogness, D.S. (1986) Molecular genetics of human color vision: the genes encoding blue, green, and red pigments. *Science* 232:193–202.
24. Chess, A., Simon, I., Cedar, H., and Axel, R. (1994) Allelic inactivation regulates olfactory receptor gene expression. *Cell* 78:823–834.
25. Ressler, K.J., Sullivan, S.L., and Buck, L.B. (1993) A zonal organization of odorant receptor gene expression in the olfactory epithelium. *Cell* 73:597–609.
26. Vassar, R., Ngai, J., and Axel, R. (1993) Spatial segregation of odorant receptor expression in the mammalian olfactory epithelium. *Cell* 74:309–318.
27. Ressler, K.J., Sullivan, S.L., and Buck, L.B. (1994) Information coding in the olfactory system: evidence for a stereotyped and highly organized epitope map in the olfactory bulb. *Cell* 79:1245–1255.
28. Vassar, R., Chao, S.K., Sitcheran, R., Nunez, J.M., Vosshall, L.B., and Axel, R. (1994) Topographic organization of sensory projections to the olfactory bulb. *Cell* 79:981–991.
29. Mombaerts, P., Wang, F., Dulac, C., Chao, S.K., Nemes, A., Mendelsohn, M., Edmondson, J., and Axel, R. (1996) Visualizing an olfactory sensory map. *Cell* 87:675–686.
30. Fritzsch, B. (2003) Development of inner ear afferent connections: forming primary

neurons and connecting them to the developing sensory epithelia. *Brain Res. Bulletin* 60:423–433.

31. McLaughlin, T., Hindges, R., and O'Leary, D.M. (2003) Regulation of axial patterning of the retina and its topographic mapping in the brain. *Current Opinion in Neurobiol.* 13:57–69.
32. Barnea, G., O'Donnell, S., Mancia, F., Sun, X., Nemes, A., Mendelsohn, M., and Axel, R. (2004) Odorant receptors on axon termini in the brain. *Science Brevia* 304:1468.
33. Wang, F., Nemes, A., Mendelsohn, M., and Axel, R. (1998) Odorant receptors govern the formation of a precise topographic map. *Cell* 93:47–60.
34. Feinstein, P., Bozza, T., Rodriguez, I., Vassali, A., and Mombaerts, P. (2004) Axon guidance of mouse olfactory sensory neurons by odorant receptors and the b2 adrenergic receptor. *Cell* 117:833–846.
35. Feinstein, P. and Mombaerts, P. (2004) A contextual model for axonal sorting into glomeruli in the mouse olfactory system. *Cell* 117:817–831.
36. Serizawa, S., Ishii, T., Nakatani, H., Tsuboi, A., Nagawa, F., Asano, M., Sudo, K., Sakagami, J., Sakano, H., Iijiri, T., *et al.* (2000) Mutually exclusive expression of odorant recepor transgenes. *Nat. Neurosci.* 3:687–693.
37. Vassali, A., Rothman, A., Feinstein, P., Zapotocky, M., and Mombaerts, P. (2002) Minigenes impart odorant receptor-specific axon guidance in the olfactory bulb. *Neuron* 35:681–696.
38. Serizawa, S. Miyamichi, K., Nakatani, H., Suzuki, M., Saito, M., Yoshihara, Y., and Sakano, H. (2003) Negative feedback regulation ensures the one receptor-one olfactory neuron rule in mouse. *Science* 302:2088–2094.
39. Lewcock, J.L. and Reed, R.R. (2004) A feedback mechanism regulates monoallelic odorant receptor expression. *Proc. Natl. Acad. Sci.* 10.
40. Shykind, B., Rohani, C., O'Donnell, S., Nemes, A., Mendelsohn, M., Sun, Y., Axel, R., and Barnea, G. (2004) Gene switching and the stability of odorant receptor gene choice. *Cell* 117:801–815.
41. Hicks, J., Strathern, J.N., and Klar, A.J. (1979) Transposable mating type genes in *Saccharomyces cerevisiae. Nature* 282:478–483.
42. Pays, E.,Van Assel, S., Laurent, M., Darville, M., Vervoort, T., Van Meirvenne, N., and Steinart, M. (1983) Gene conversion as a mechanism for antigenic variation in trypanosomes. *Cell* 34:371–381.
43. Brack, C., Hirama, M., Lenhard-Schuller, R., and Tonegawa, S. (1978) A complete immunoglobulin gene is created by somatic recombination. *Cell* 15:1–14.
44. Eggan, K., Baldwin, K., Tackett, M., Osborne, J., Gogos, J., Chess, A., Axel, R., and Jaenisch, R. (2004) Mice cloned from olfactory sensory neurons. *Nature* 428:44–49.
45. Li, J., Ishii, T., Feinstein, P., and Mombaerts, P. (2004) Odorant receptor gene choice is reset by nuclear transfer from mouse olfactory sensory neurons. *Nature* 428:393–399.
46. Stocker, R.F. (1994) The organization of the chemosensory system in *Drosophila melanogaster:* a review. *Cell Tissue Res.* 275:3–26.
47. Laissue, P.P., Reiter, C., Hiesinger, P.R., Halter, S., Fischbach, K.F., and Stocker, R.F. (1999). Three-dimensional reconstruction of the antennal lobe in *Drosophila melanogaster. J. Comp Neurol.* 405:543–552.
48. Gao, Q., Yuan, B., and Chess, A. (2000) Convergent projections of *Drosophila* olfactory neurons to specific glomeruli in the antennal lobe. *Nat. Neurosci.* 3:780–785.
49. Vosshall, L.B., Wong, A,M., and Axel, R. (2000) An olfactory sensory map in the fly brain. *Cell* 102:147–159.
50. Scott, K., Brady, R., Jr., Cravchik, A. Morozov, P., Rzhetsky, A., Zuker, C., and Axel, R. (2001) A chemosensory gene family encoding candidate gustatory and olfactory receptors in *Drosophila. Cell* 104:661–673.
51. Wang, J.W., Wong, A.M., Flores, J., Vosshall, L.B., and Axel, R. (2003) Two-photon calcium imaging reveals an odor-evoked map of activity in the fly brain. *Cell* 112:271–282.
52. Nakai, J., Ohkura, M., and Imoto, K. (2001) A high signal-to-noise Ca^{2+} probe composed

of a single green fluorescent protein. *Nat. Biotechnol.* 19:137–141.
53. Denk, W., Strickler, J,H., and Webb, W.W. (1990) Two photon laser scanning fluorescence microscopy. *Science* 248:73–76.
54. Joerges, J., Juttner, A., Galizia, C.G., and Menzel, R. (1997) Representations of odours and odour mixtures visualized in the honeybee brain. *Nature* 387:285–287.
55. Ng, M., Roorda, R.D., Lima, S.Q., Zemelman, B.V., Morcillo, P., and Miesenbock, G. (2002) Transmission of olfactory information between three populations of neurons in the antennal lobe of the fly. *Neuron* 36:463–474.
56. Rubin, B.D. and Katz, L.C. (1999) Optical imaging of odorant representations in the mammalian olfactory bulb. *Neuron* 23:499–511.
57. Uchida, N., Takahashi, Y.K., Tanifuji, M., and Mori, K. (2000) Odor maps in the mammalian olfactory bulb: domain organization and odorant structural features. Nat. *Neurosci.* 3:1035–1043.
58. Meister, M. and Bonhoeffer, T. (2001) Tuning and topography in an odor map on the rat olfactory bulb. *J. Neurosci.* 21:1351–1360.
59. Suh, S.B., Wong, A.M., Hergarden, A.C., Wang, J.W., Simon, A., Benzer, S., Axel, R., and Anderson, D.J. (2004) A single population of olfactory neurons mediates an innate avoidance behavior in *Drosophila. Nature* 431:854–859.
60. Kitamoto, T. (2001) Conditional modification of behavior in *Drosophila* by targeted expression of a temperature-sensitive *shibire* allele in defined neurons. *J. Neurobiol.* 47:81–92.
61. Laurent, G. (1999) A systems perspective on early olfactory coding. *Science* 286:723–728.
62. Wilson, R.I. and Laurent, G. (2004) Transformation of olfactory representations in the *Drosophila* antennal lobe. *Science* 303:366–370.
63. Wong, A.M., Wang, J.W., and Axel, R. (2002). Spatial representation of the glomerular map in the *Drosophila* protocerebrum. *Cell* 109:229–241.
64. Jefferis, G.S., Marin, E.C., Stocker, R.F., and Luo, L. (2001) Target neuron prespecification in the olfactory map of *Drosophila. Nature* 414:204–208.
65. Marin, E.C., Jefferis, G.S., Komiyama, T., Zhu, H., and Luo, L. (2002) Representation of the glomerular olfactory map in the *Drosophila* brain. *Cell* 109:243–255.
66. Proust, Marcel (1913) Vol. I. *Swann's Way* in: *Remembrance of Things Past,* Random House, New York, pg. 50–51.
67. Dunipace, L., Meister, S., McNealy, C., and Amrein, H., (2001) Spatially restricted expression of candidate taste reseptors in the *Drosophila* qustatory system. *Corr. Biol* 11: 822–835.

LINDA B. BUCK

SEATTLE

I was born in 1947 in Seattle, Washington, a city surrounded by mountains, forests, and the sea. My mother was the daughter of Swedish immigrants who had come to the US in the late nineteenth century while my father's family had Irish roots on one side and ancestors extending back to the American Revolution on the other. I was the second of three children, all girls. My mother was a homemaker who was exceptionally kind and witty and loved word puzzles. My father was an electrical engineer who, at home, spent much of his time inventing things and building them in our basement. It may be that my parents' interest in puzzles and inventions planted the seeds for my future affinity for science, but I never imagined as a child that I would someday be a scientist.

During my childhood, I did the things that girls often do, such as playing with dolls. I was also curious and easily bored though, so I frequently embarked on what were to me new adventures. Aside from school and music lessons, my life was relatively unstructured and I was given considerable independence. I learned to appreciate music and beauty from my mother and my father taught me how to use power tools and build things. I spent a lot of time with my maternal grandmother, who told me magical stories about her girlhood in Sweden and, to my delight, taught me how to sew clothes for my dolls. I was fortunate to have wonderfully supportive parents who told me that I had the ability to do anything I wanted with my life. They taught me to think independently and to be critical of my own ideas, and they urged me to do something worthwhile with my life, in my mother's words, to "not settle for something mediocre". I realize now that I internalized those lessons and that they have influenced my work as a scientist.

I received my undergraduate education at the University of Washington, which was only a few miles from our home. I had always wanted to have a career in which I would help others, so I initially decided to major in Psychology, thinking that I would become a psychotherapist. Over time, my interests expanded and I entertained a variety of different career possibilities. However, none seemed ideal and I was reluctant to embark on something that might prove to be inappropriate. Over the next several years, I intermittently traveled, lived on a nearby island, and took more classes in Seattle. I finally found my direction when I took a course in immunology, which I found fascinating. I would be a biologist.

DALLAS

In 1975, I began graduate school in the Microbiology Department at the University of Texas Medical Center in Dallas. The department had recently undergone an expansion in the area of immunology, making it a major center in this still young area and a stimulating place to learn. I had done a small amount of research at the University of Washington, first in psychology with Walter Makous and then in immunology with Ursula Storb, but it was in Texas that I truly learned to be a scientist. I had a wonderful thesis advisor, Ellen Vitetta, who demanded excellence and precision in research, habits that I believe are important to learn as a student. For my thesis, I compared the functional properties of subsets of B lymphocytes that differed in the class of cell surface immunoglobulin that they used as antigen receptors. In this work and much of my subsequent work, I thought in terms of molecules and the molecular mechanisms underlying biological systems, and sought to gain insight into those mechanisms in my experiments.

NEW YORK

In 1980, I moved to Columbia University in New York City to do postdoctoral work in immunology with Benvenuto Pernis. As a graduate student, I had become fascinated with the unexplained requirement for major histocompatibility complex (MHC) proteins in immune responses, a mystery that was later solved. I decided to explore this puzzle, focusing on class II MHC proteins found on the surface of B lymphocytes. I found that, contrary to expectation, the MHC proteins rapidly accumulated inside these cells when they were activated. My further experiments indicated that they were being internalized from the cell surface and were probably being recycled to it. It was known that antigen is endocytosed with antigen receptors and then degraded. One possibility raised by the internalization and apparent recycling of MHC molecules was that, following internalization, they might be targeted to a specialized microenvironment where they could interact with degraded antigen. The MHC-antigen complexes might then be exported to the cell surface for corecognition by T helper cells.

By this time, it had become clear to me that to study molecular mechanisms underlying biological systems, which is what interested me, I needed to learn the recently developed techniques of molecular biology. To this end, I moved to the laboratory of Richard Axel at Columbia University. Richard had begun to work in the area of neuroscience several years earlier through collaboration with Eric Kandel, who was also at Columbia. Their collaboration had focused on molecular studies of the nervous system of *Aplysia*, a sea snail. This was the model organism that Eric had used in many of his studies of learning and memory, for which he received a Nobel Prize in 2000. Perhaps not surprisingly, I was interested in searching for genes encoding neuronal cell surface receptors. However, at that time, Richard wanted to continue studying *Aplysia*, so I agreed to a project in which I would try to develop a technique for cloning genes expressed in one *Aplysia* neuron, but not another. After spending a short time learning molecular techniques from Jim Roberts, a student in the lab, I started my *Aplysia* project. Eric Kandel's group showed me how to isolate giant *Aplysia* neurons that had been assigned names and could be identified by their locations and, within a relatively short time, I began to uncover genes that were differentially expressed among *Aplysia* neurons.

While studying a neuropeptide gene expressed in neuron number R15, I discovered that the gene was also expressed in some other neurons, but that its primary transcript was alternatively spliced in different neurons to give different polyproteins. The two polyproteins could generate two different combinations of peptides in different neurons, suggesting a way to produce physiological or behavioral programs with partially overlapping components. While working on the neuropeptide gene, I encountered numerous technical challenges that increased my knowledge of molecular biology and honed my abilities. During this period, I learned a lot of molecular biology from Richard and other members of his lab. I also got to know Eric Kandel, who has continued to be a wonderful source of inspiration and encouragement for me over the years.

From my first introduction to neuroscience, I had been fascinated by the brain's cellular and connectional diversity. In parallel with my *Aplysia* experiments, I sporadically tried to find a way to scan the genome for genes that had undergone gene rearrangement or gene conversion in neurons, thinking that genes that showed this characteristic might be involved in the generation of neuronal diversity. One method that I devised showed promise in *Drosophila*, but was not sensitive enough for the much larger genome of a mammal, which is what interested me. Nonetheless, these efforts were a great source of creative enjoyment for me as I proceeded with the more mundane task of searching for minute alternative exons in the *Aplysia* genome.

I was grateful that Richard was tolerant of my high-risk endeavors. He was an unusual mentor in that he gave people in his lab extensive independence in charting their own course once they had established themselves. During this time, I had many colleagues at Columbia with whom I enjoyed long discussions about science. Among these were George Gaitanaris, who has remained a close friend over the years, and Tom Jessell and Jane Dodd, neuroscientists from whom I learned a great deal about neural development.

As I was nearing the end of my *Aplysia* project, I read a paper that changed my life. It was a 1985 publication from Sol Snyder's group that discussed potential mechanisms underlying odor detection. This was the first time I had ever thought about olfaction and I was fascinated. How could humans and other mammals detect 10,000 or more odorous chemicals, and how could nearly identical chemicals generate different odor perceptions? In my mind, this was a monumental puzzle and an unparalleled diversity problem. It was obvious to me that the first step to solving the puzzle was to determine how odorants are initially detected in the nose. This meant finding odorant receptors, a class of molecules that had been proposed to exist, but had not been found. I decided that this is what I had to do as soon as my neuropeptide work was completed.

In 1988, I embarked on a search for odorant receptors, staying on in Richard's lab for this purpose. In a recent commentary in the journal *Cell*, I described what was known about odor detection at that time and the approaches that I tried in the quest to find the elusive odorant receptors. In short, it was known that odorants depolarize, and thereby activate, olfactory sensory neurons in the nose. Although there were varied proposals as to what kind of molecules might interact with odorants, there was compelling evidence that olfactory transduction involved G-protein induced increases in cAMP. After trying several different approaches, I identified the odorant receptor family by designing experiments based on three assumptions. First, since odorants vary in structure and can be discriminated, there would be a family of varied, but related odorant receptors, which would be encoded by a multigene family. Second, odorant receptors would be at least distantly related to the relatively small set of G protein coupled receptors whose sequences were known at that time. And finally, odorant receptors would be selectively expressed in the olfactory epithelium, where olfactory sensory neurons are located. It took some time to devise and develop the methods I used in my search, but in the end they succeeded. Looking at the first sequences of odorant receptors obtained from rat, I was moved by Nature's marvelous invention. This work showed that the rat has a multigene family that codes for in excess of one hundred different odorant receptors, all related, but each one unique. The unprecedented size and diversity of this family explained the ability of mammals to detect a vast array of diverse chemicals as having distinct odors. In 1991, Richard Axel and I published the identification of odorant receptors.

BOSTON

In 1991, I departed for Boston to be an assistant professor in the Neurobiology Department at Harvard Medical School. There, I was immersed in an environment in which I could broaden my understanding of the nervous system. I received excellent support from my chairman, Gerry Fischbach, as I set up my lab. I also developed many excellent colleagues, including David Hubel, whose pioneering studies of the visual system with Torsten Wiesel, for which they received a Nobel Prize in 1981, had always been an inspiration to me. In 1994, I became an Investigator of the Howard Hughes Medical Institute, which has generously supported our work for the past eleven years. Over the next decade, I remained at Harvard, gradually rising through the ranks to become associate and then full professor. In 1994, I met Roger Brent, a marvelous intellect and fellow scientist who has been my partner and an important part of my life ever since.

The discovery of odorant receptors had explained how the olfactory system detects odorants. My next goal was to learn how signals from those receptors are organized in the brain to generate diverse odor perceptions. I was joined in this endeavor by a series of excellent students and postdoctoral fellows. The discoveries on the organization of the olfactory system that were cited by the Nobel Foundation were made over a period of ten years, during which I was a faculty member at Harvard.

The first question we asked was how odorant receptors (ORs) are organized in the olfactory epithelium of the nose. This work was begun by Kerry Ressler, an M.D./Ph.D. student who came to the laboratory for a few months just as the equipment and supplies I had ordered began to arrive in January 1992. I had decided to switch from rat to mouse as a model organism because of the advantage of using isogenic inbred strains for dissecting a multigene family, and the possibility of generating transgenic mice. After cloning and sequencing a series of mouse OR genes, Kerry did our first *in situ* hybridization experiments to examine patterns of OR gene expression. By June, Kerry had returned as a full time student and Susan Sullivan had joined the lab as a postdoctoral fellow. At this point, we began to precisely analyze OR expression patterns and to compare them in different individuals. Prior to the present era of digital photographs that can be stored and analyzed on a computer, this was painstaking work that involved displaying photographic slides on a desktop viewer and recording, on transparencies, the locations of individual labeled cells in different animals. Our studies showed that each OR gene is expressed in about 1/1000 olfactory sensory neurons, that the olfactory epithelium has several spatial zones that express nonoverlapping sets of OR genes, and that neurons with the same OR are randomly scattered throughout one zone. This indicated that signals derived from different ORs are segregated in different sensory neurons and in the information they transmit to the brain. It further indicated that, in the olfactory epithelium, neurons that detect the same odorant are dispersed and those that detect different odorants are interspersed. Thus, there is a broad organization of sensory information into several zonal sets in the epithelium, but, overall, information is encoded in a highly distributed manner. We published these findings in 1993. Similar observations in rat by Richard Axel and his colleagues were also reported that year.

Having determined how inputs from different ORs are organized in the nose, we asked how they are arranged at the next structure in the olfactory pathway, the olfactory bulb. In the bulb, the axons of olfactory sensory neurons synapse in about 2000 spherical structures, called glomeruli. Kerry began to use retroviral vectors to investigate how the axons of neurons expressing specific ORs are organized in the bulb, but then we inadvertently found another way to address the question. While using *in situ* hybridization to identify a number of OR genes expressed in each epithelial zone for chromosomal mapping studies, Susan found that, in one tissue section, an OR probe labeled a single spot in the bulb, which proved to be a glomerulus. Using probes that recognized single OR genes rather than subfamilies of related OR genes, we found that each probe labeled OR mRNAs in sensory axons that were confined to one or a few glomeruli at only two sites, one on either side of the bulb. Different OR probes labeled different glomeruli and those glomeruli had virtually identical locations in different individuals. I still remember a meeting with Kerry and Susan in my office in which I asked

Kerry how many sections separated different labeled glomeruli in different bulbs. All of us were stunned by his answer, because it provided the first hint that the bulb might have a stereotyped map of OR inputs and we could not imagine how this could be generated given the organization of OR gene expression in the epithelium. This mystery still has not been solved. These studies indicated that while thousands of neurons expressing the same OR are highly dispersed in the epithelium, their axons all converge in a few specific olfactory bulb glomeruli. The result is a stereotyped map of OR inputs in which signals derived from different ORs are segregated in different glomeruli and in the bulb projection neurons whose dendrites innervate those glomeruli. Remarkably, Bob Vassar in Richard Axel's lab had concurrently found that different OR probes labeled different glomeruli in the rat bulb. Our two groups published these findings in 1994.

Several years later, we began to investigate how the OR family and the patterning of OR inputs encode the identities of different odorants. Using single cell RT-PCR (reverse transcriptase-polymerase chain reaction), Bettina Malnic, a fellow in the lab, had been comparing gene expression in single olfactory sensory neurons. Her work demonstrated that each neuron expresses only a single OR gene, something that we had previously suspected, but that needed to be verified. Bettina was initially focused on the identification of genes that might be involved in OR gene choice or axon targeting in the bulb, but we decided to change course when Takaaki Sato visited our lab and told us about his calcium imaging studies of odor responses in the olfactory epithelium. This was the beginning of a highly successful collaboration in which Takaaki used calcium imaging to define the odor response profiles of individual neurons and Bettina then used RT-PCR to identify the OR expressed by each responsive neuron. These studies demonstrated that the OR family is used in a combinatorial manner. Different neurons are recognized, and thereby encoded, by different combinations of ORs, but each OR is used as one component of the combinatorial receptor codes for many different odorants. As discussed in my Nobel Lecture, these studies also provided explanations for several intriguing features of human odor perception, including how a slight change in the structure of an odorant can dramatically change its perceived odor quality.

As soon as we had determined how OR inputs are organized in the olfactory bulb, we began to explore how they are arranged at the next structure in the olfactory pathway, the olfactory cortex. Lisa Horowitz, an M.D./Ph.D. student in the lab, initially investigated connections between the bulb and cortex using classical anatomical techniques. By depositing different tracers in the dorsal and ventral bulb, she determined that these areas project axons to the same regions of the cortex. In agreement with previous findings, this indicated that there could not be a point-to-point patterning of connections between the bulb and cortex. We decided to abandon traditional approaches and to instead ask whether we could chart neural pathways genetically by expressing a gene encoding a transneuronal tracer in olfactory sensory neurons. Lisa found that this was indeed possible. When she made transgenic mice that expressed barley lectin in all olfactory sensory neurons, the lectin crossed two synapses, labeling second-order neurons in the bulb and then third-order neurons in the cortex. This work, which we published in 1999, opened the way to investigating a wide array of questions concerning neural circuits, including those that carry olfactory information.

We then went on to use the genetic tracer to examine how inputs from individual types of ORs are organized in the olfactory cortex. To do this we used gene targeting to generate mice that coexpressed barley lectin with a single OR gene. Lisa, together with a fellow in the lab, Jean-Pierre Montmayeur, prepared the DNA constructs for gene targeting. Zhihua Zou, another fellow, then made and analyzed mice that coexpressed the tracer with different OR genes. The approach worked, but was difficult, with Zhihua investing almost a year in perfecting the conditions needed to detect minute amounts of the tracer in cortical neurons. These studies revealed that the olfactory cortex has a stereotyped map of OR inputs, but one that is radically different from that in the bulb. As I discussed in my Nobel Lecture, the segregation of OR inputs in different glomeruli and neurons in the bulb gives way in the cortex to a complex array of OR inputs in which signals from different ORs partially overlap and single cortical neurons appear to receive signals from combinations of different ORs. This offers a means by which the individual components of an odorant's receptor code could be integrated at the level of single neurons. This could serve as an initial step in the reconstruction of an odor image from its deconstructed features, which are conveyed by the OR elements of the receptor code. We published our findings on the cortex in 2001.

During the ten year period at Harvard in which we did the work described above, my laboratory also investigated a number of other questions. These included studies of the chromosomal organization of OR genes and the evolution of the OR gene family by Susan Sullivan, studies of the development of OR gene expression patterns by Susan and Staffan Bohm, and bioinformatic studies by Bettina Malnic and Paul Godfrey that defined and compared the OR gene repertoires of human and mouse. We also conducted a series of studies on the detection of pheromones in the vomeronasal organ, including studies by Emily Liman and Anna Berghard that revealed differences between transduction molecules involved in odor versus pheromone detection, the discovery of zonal patterns of transduction molecules likely to be involved in pheromone detection by Anna, analyses of vomeronasal responses to pheromones and odorants by Mehran Sam, and the discovery, by Hiroaki Matsunami, of a family of candidate pheromone receptors. During the latter part of this period, Hiroaki Matsunami, Jean-Pierre Montmayeur, and Stephen Liberles also began to explore the mechanisms underlying taste detection, in the process discovering candidate receptors for both bitter and sweet tastes, both of which were also found by other groups at about the same time.

SEATTLE

In 2002, I returned to Seattle to be a Member of the Division of Basic Sciences at Fred Hutchinson Cancer Research Center and Affiliate Professor of Physiology and Biophysics at the University of Washington. I had always intended to someday return to the West Coast and had already stayed longer in Boston than I had anticipated. When Mark Groudine, then Director of the Basic Sciences Division at Fred Hutchinson, offered me a faculty position there, I gladly accepted. The Hutchinson Center had a reputation for cutting edge science as well as a high level of collegiality, both of which were important to me. In addition, by moving to Seattle, I would be closer to my partner, Roger, who lived in Berkeley, and to my family and friends in Seattle.

In Seattle, we are continuing to explore the mechanisms underlying odor perception as well as the means by which pheromones elicit instinctive behaviors. We have also become interested in the neural circuits that underlie innate behaviors and basic drives, such as fear, appetite, and reproduction. We are currently developing molecular techniques to uncover those circuits and to define their composite neurons and the genes they express. In a different vein, we have developed a high throughput approach in which we are using chemical libraries to identify genes that control aging and lifespan, our chief interest being whether there might be a central mechanism that determines lifespan and regulates the aging of cells throughout the body.

LOOKING BACK

Since Richard Axel and I published the discovery of odorant receptors in 1991, it has been immensely satisfying for me to see many laboratories using these receptors in a large scale effort to dissect the mechanisms that underlie the sense of smell and the developmental processes that shape the organization of the olfactory system. Molecular approaches to studying olfaction have extended to other vertebrates as well as to invertebrate species, with Cori Bargmann's group discovering a large variety of chemosensory receptors in the nematode worm, *C. elegans*, and several groups, including Richard Axel's, identifying families of odorant and taste receptors in the fruit fly, *D. melanogaster*.

Looking back over my life, I am struck by the good fortune I have had to be scientist. Very few in this world have the opportunity to do everyday what they love to do, as I have. I have had wonderful mentors, colleagues, and students with whom to explore what fascinates me and have enjoyed both challenges and discoveries. I am grateful for all of these things and look forward to learning what Nature will next reveal to us.

As a woman in science, I sincerely hope that my receiving a Nobel Prize will send a message to young women everywhere that the doors are open to them and that they should follow their dreams.

UNRAVELING THE SENSE OF SMELL

Nobel Lecture, December 8, 2004

by

LINDA B. BUCK

Howard Hughes Medical Institute, Fred Hutchinson Cancer Research Center, 1100 Fairview Avenue North, Seattle, WA 98109–1024, USA.

INTRODUCTION

The subject of my lecture is the sense of smell, one of the five senses through which we perceive the world. Through the sense of smell, humans and other mammals can perceive a vast number and variety of chemicals in the external world. It is estimated that humans can sense as many as 10,000 to 100,000 chemicals as having a distinct odor. All of these "odorants" are small, volatile molecules. However, they have diverse structures and somehow those different structures are perceived as having different odors (Figure 1).

The sense of smell is mediated by the olfactory system, a system that is characterized by exquisite sensitivity and discriminatory power. Even a slight change in the structure of an odorant can change its perceived odor. For example, the close relative of a chemical that is perceived as pear can have the scent of an apple. In addition to odorants, the olfactory system detects pheromones, chemicals that are released from animals and act on members of the same species, stimulating hormonal changes or instinctive behaviors, such as mating or aggression. The olfactory system also detects predator odors, which can elicit innate fear responses.

Over the past 16 years, our work has focused on two questions. First, how do mammals detect so many different environmental chemicals? And second, how does the brain translate those chemicals into diverse odor perceptions and behaviors?

Odorants are initially detected by olfactory sensory neurons, which are located in the olfactory epithelium lining the nasal cavity (Figure 2). These neurons transmit signals to the olfactory bulb of the brain, which then relays those signals to the olfactory cortex. From there, olfactory information is sent to a number of other brain areas. These include higher cortical areas thought to be involved in odor discrimination as well as deep limbic areas of the brain, which are thought to mediate the emotional and physiological effects of odors. In contrast to odorants, pheromones are detected primarily in the vomeronasal organ, or VNO, a separate olfactory structure in the nasal septum.

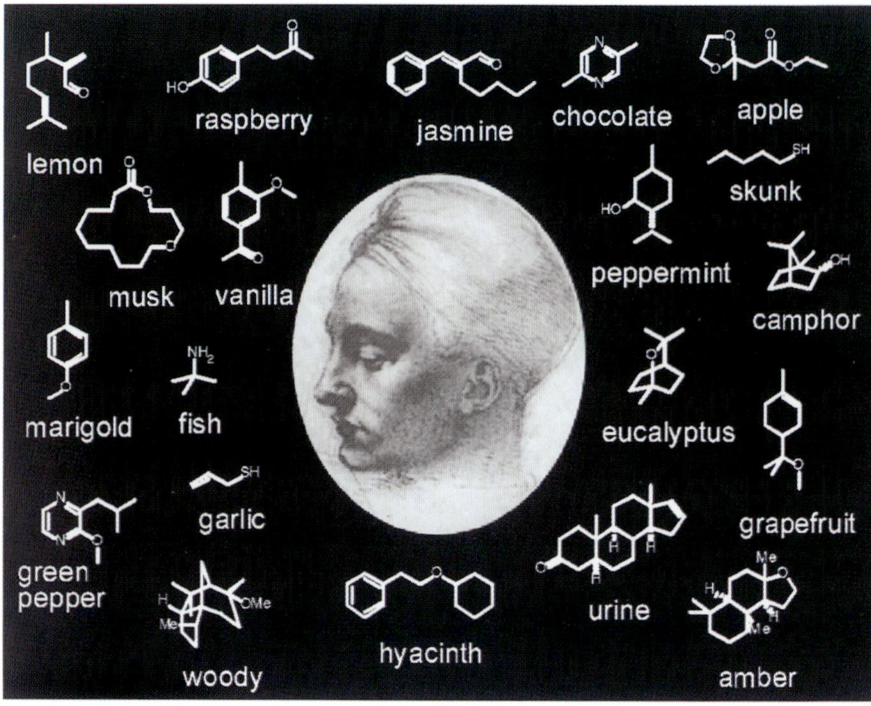

Figure 1. Humans and other mammals perceive a vast number of chemicals as having distinct odors.

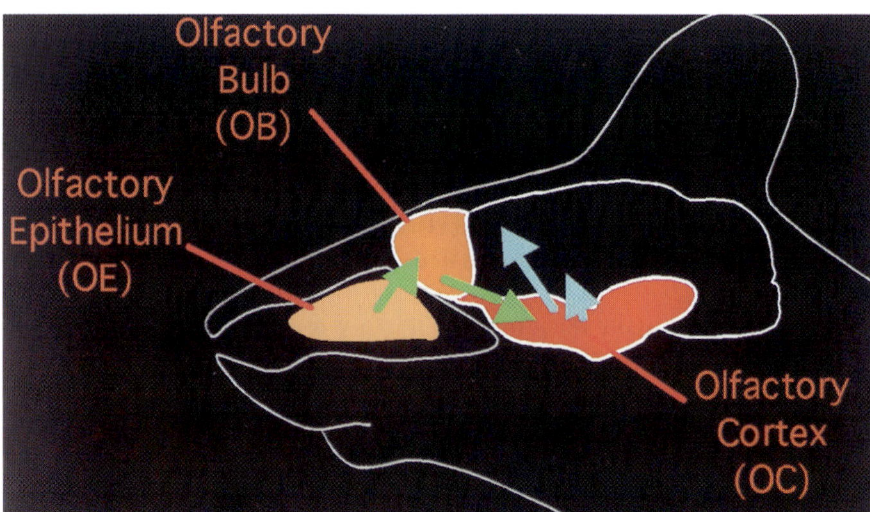

Figure 2. The olfactory pathway. Odorants are detected by olfactory sensory neurons in the olfactory epithelium. Signals generated in those neurons are relayed through the olfactory bulb to the olfactory cortex and then sent to other brain areas.

From VNO neurons, signals are relayed through the accessory bulb to the medial amygdala and then the hypothalamus, areas implicated in hormonal and behavioral responses to pheromones.

The olfactory epithelium contains millions of olfactory sensory neurons. It also contains supporting cells and a basal layer of stem cells. Olfactory sensory neurons are short-lived cells that are continuously replaced from the stem cell layer. At the surface of the epithelium, each neuron extends cilia into the nasal lumen, allowing it to come in contact with odorants dissolved in the nasal mucus. Each neuron communicates with the brain via a single axon that it extends to the olfactory bulb.

ODORANT RECEPTORS

In our initial experiments, Richard Axel and I asked how it is that these neurons detect odorants. Beginning in 1965 with the work of Robert Gesteland (Gesteland *et al.*, 1965), numerous electrophysiological studies had shown that different olfactory sensory neurons are depolarized, or activated, by different odorants. John Amoore proposed that these neurons had odorant receptor proteins that varied in their affinity for different odorants (Amoore, 1970; Amoore, 1977). In the mid 1980s, hints started to emerge about signal transduction in the cilia of the olfactory neurons. Doron Lancet and Sol Snyder and their colleagues showed that odorants induce GTP-dependent increases in adenylyl cyclase activity in the cilia, suggesting the involvement of intracellular G proteins (Pace *et al.*, 1985; Sklar *et al.*, 1986), and Randy Reed identified Gαolf, a G protein that could mediate this response and was highly expressed in olfactory sensory neurons (Jones and Reed, 1989).

In 1988, Richard Axel and I embarked on a search for odorant receptors. The strategy we devised was based on three assumptions. First, odorant receptors would be selectively expressed in the olfactory epithelium. Second, since odorants vary in structure, there would be a family of varied, but related receptors, and those receptors would be encoded by a multigene family. And third, odorant receptors would be related to other types of receptors that interact with intracellular G proteins. By 1989, molecular cloning had revealed the structures of about 20 of these G protein-coupled receptors, or "GPCRs". All of these receptors had seven potential transmembrane domains and they shared a few amino acid sequence motifs.

On the basis of these assumptions, we set out to search for a family of GPCRs expressed in the rat olfactory epithelium (Buck and Axel, 1991). To do this, we first used PCR (the polymerase chain reaction) to look for receptors expressed in the olfactory epithelium that were related to known GPCRs. We designed 11 degenerate oligonucleotide primers that matched amino acid sequences in transmembrane domains 2 and 7 of known GPCRs. We then used these primers in all 30 pairwise combinations to amplify related sequences in cDNA prepared from rat olfactory epithelium RNA. From the 30 PCR reactions, we obtained 64 different PCR products in the appropriate size range. Each of these appeared as a distinct band using agarose gel electrophoresis.

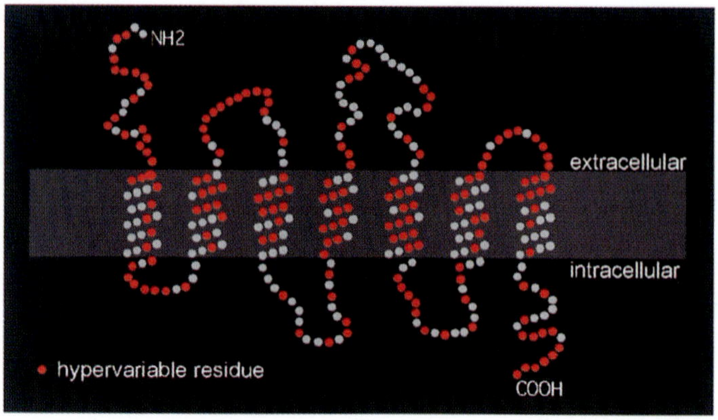

Figure 3. Topology of an odorant receptor in the membrane. Individual amino acid residues are indicated by balls. Red balls indicate residues that were hypervariable among ten odorant receptors. Adapted from Buck and Axel, 1991.

We then asked whether any of the 64 PCR products contained multiple members of a multigene family. To do this we cut the DNA in each PCR product with a restriction enzyme. Most of the bands were cut into a small number of fragments that added up to the original in size. However, one band, #13, was cut into a large number of fragments, suggesting that it might contain multiple members of a multigene family. When we cloned and sequenced five of the DNAs in #13, we found what we had been looking for. All five encoded novel proteins with the hallmarks of GPCRs. Moreover, all five were related, but each one was unique.

Using these DNAs as probes, we isolated a series of related cDNAs from an olfactory epithelium cDNA library. We initially examined the proteins encoded by ten of the cDNAs. All ten proteins had the seven potential transmembrane domains characteristic of GPCRs. In addition, they had several amino acid sequence motifs seen in other GPCRs. However, the ten receptors all shared sequence motifs not seen in any other GPCRs, indicating that they were members of a novel receptor family.

Figure 3 shows a model of one of these receptors in the membrane with individual amino acids represented as balls. Red balls indicate amino acids that were especially variable among the ten receptors. Importantly, though related, the ten olfactory receptors varied extensively in amino acid sequence. This hypervariability was consistent with an ability of the receptors to interact with odorants with different structures.

Consistent with the selective expression of these receptors in the olfactory epithelium, a mixed olfactory receptor DNA probe hybridized to RNA from the olfactory epithelium, but not other tissues. Moreover, enriching for olfactory sensory neurons also enriched for receptor RNAs, suggesting that the receptors were expressed predominantly or exclusively by olfactory sensory neurons.

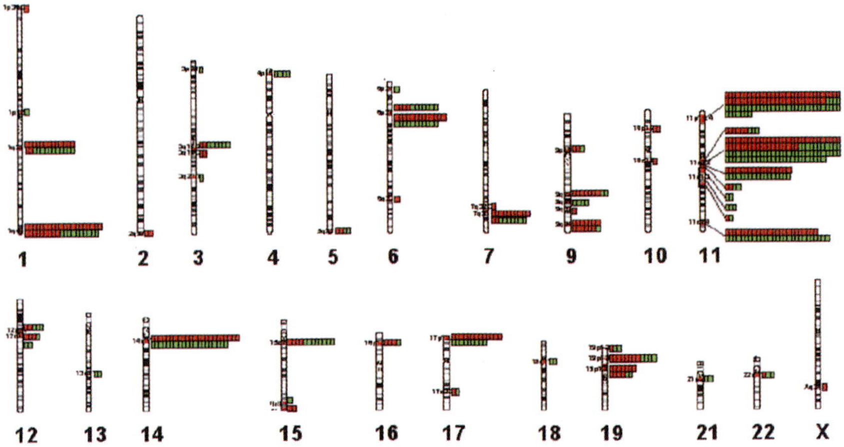

Figure 4. The chromosomal distribution of human odorant receptor genes. Intact receptor genes are shown in red and pseudogenes in green. Adapted from Malnic *et al.*, 2004.

On Southern blots of genomic DNA, single receptor probes hybridized to multiple bands, and a mixed receptor probe hybridized to a large number of bands. This further indicated that the receptors we had found were encoded by a large multigene family. Genomic library screens indicated that the multigene family contained in excess of 100 members. In later studies, we obtained evidence for about 1000 different olfactory receptor genes in mouse.

On the basis of these results, we concluded that the receptor family we had identified coded for odorant receptors (ORs) expressed by olfactory sensory neurons in the nose (Buck and Axel, 1991). Subsequent studies showed that homologous families of odorant receptors are present in vertebrate species ranging from fish to humans (reviewed in Mombaerts, 1999). In 1991, after publishing our work on odorant receptors (Buck and Axel, 1991), I left Richard Axel's lab to join the faculty of Harvard Medical School.

A decade later, the sequencing of human and mouse genomes made it possible to determine the number of OR genes in these species. This was done by Lancet and Zozulya for human (Glusman *et al.*, 2001; Zozulya *et al.*, 2001) and by Firestein and Trask for mouse (Young and Trask, 2002; Zhang and Firestein, 2002), and in my lab it was done for both species by Bettina Malnic and Paul Godfrey (Godfrey *et al.*, 2004; Malnic *et al.*, 2004). These studies indicate that humans have about 350 different ORs and mice have about 1000. This indicates that roughly 1–5% of the genes in the genome are devoted to the detection of odorants. OR genes are highly distributed across the genome. In our studies of the human genome, we found OR genes on 21 different chromosomes and at 51 different chromosomal loci, where they are found singly or in clusters (Figure 4) (Malnic *et al.*, 2004).

In the mid 1990s, two additional families of receptors were found in the olfactory system. These receptors, called V1Rs and V2Rs, are unrelated to ORs in protein sequence, but both types have the characteristic seven transmembrane structure of GPCRs. V1R and V2R genes are selectively expressed in the VNO, suggesting that they might be pheromone receptors. Both receptor families have more than 100 members. The V1R family was identified in 1995 by Dulac and Axel (Dulac and Axel, 1995), and the V2R family was identified in 1997 by Hiroaki Matsunami in my lab and also by the laboratories of Catherine Dulac and Nicholas Ryba (Herrada and Dulac; Matsunami and Buck, 1997; Ryba and Tirandelli, 1997).

ORGANIZATION OF ODORANT RECEPTORS IN THE OLFACTORY EPITHELIUM

The discovery of odorant receptors explained how the olfactory system detects a vast array of chemicals in the external world. It also did something else that was important: it provided a set of molecular tools to explore how the nervous system translates chemical structures into odor perceptions. This is what we set out to do in my lab at Harvard.

In the mouse, we found evidence for as many as 1000 different OR genes. We first asked how information from different ORs is organized in the olfactory epithelium (Ressler *et al.*, 1993). In these experiments, Kerry Ressler, a graduate student in the lab, and Susan Sullivan, a postdoctoral fellow, hybridized labeled OR gene probes to sections through the mouse nose (Figure 5). These studies showed that the olfactory epithelium has distinct spatial zones that express nonoverlapping sets of OR genes (Figure 6). Each OR gene is expressed in about 1/1000 neurons and those neurons are randomly scattered within one zone. Similar findings were made in Richard Axel's lab in rat (Vassar *et al.*, 1993). The OR expression zones form stripes that extend along the anterior posterior axis of the nasal cavity.

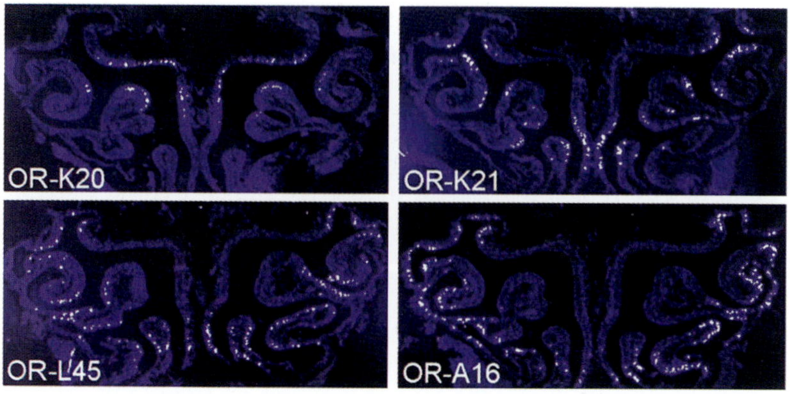

Figure 5. Expression patterns of odorant receptor genes in the mouse olfactory epithelium. Tissue sections through the mouse nose were hybridized to four different receptor gene probes. Adapted from Ressler *et al.*, 1993; Sullivan *et al.*, 1996.

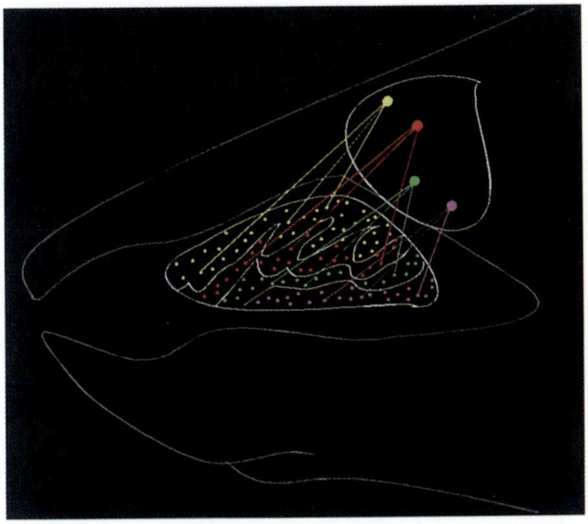

Figure 6. The organization of odorant receptor inputs in the olfactory epithelium and olfactory bulb. Sensory neurons expressing the same receptor are scattered within one epithelial zone, but their axons converge in specific glomeruli in the olfactory bulb. Adapted from Ressler *et al.*, 1993; Ressler *et al.*, 1994; Sullivan *et al.*, 1996.

These findings told us two important things. First, input from one type of OR is highly distributed in the epithelium. Therefore, neurons with receptors for one odorant, for example a strawberry odorant, must be interspersed with neurons that have receptors for another odorant, such as a lemon one. Second, each neuron may express only one OR gene. We later confirmed this by examining gene expression in single neurons (Malnic *et al.*, 1999). Thus, in the nose, inputs from different ORs are segregated in different neurons, and the information that each neuron transmits to the brain is derived from a single receptor type.

COMBINATORIAL RECEPTOR CODES FOR ODORS

In later studies, we asked how the OR family encodes the identities of different odorants. To explore this question, we searched for ORs that recognize specific odorants (Malnic *et al.*, 1999). This work was done by Bettina Malnic in the lab in collaboration with Takaaki Sato and Junzo Hirono at the Life Electronics Research Center in Japan. We first exposed single mouse olfactory sensory neurons to a series of odorants, using calcium imaging to visualize their responses. We then isolated each responsive neuron and used RT (reverse transcriptase)-PCR to determine the OR gene it expressed. In every case, we identified only one expressed OR per neuron, confirming that each neuron expresses a single OR gene.

For test odorants, we used four different classes of n-aliphatic odorants with different functional groups and carbon chains ranging in length from 4 to 9 carbon atoms. Each neuron was imaged as it was exposed sequentially to

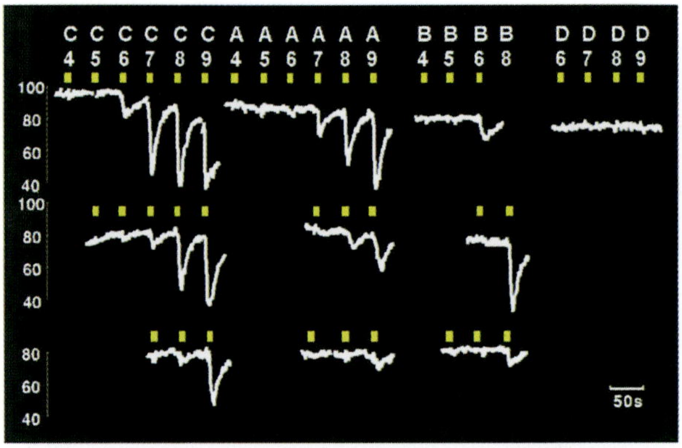

Figure 7. Responses of a single olfactory sensory neuron to different odorants. Fluorescence emission was monitored during sequential exposure of a Fura-2 containing neuron to a series of odorants (C4-D9). Responses to lower odorant concentrations are shown below. Adapted from Malnic *et al.*, 1999.

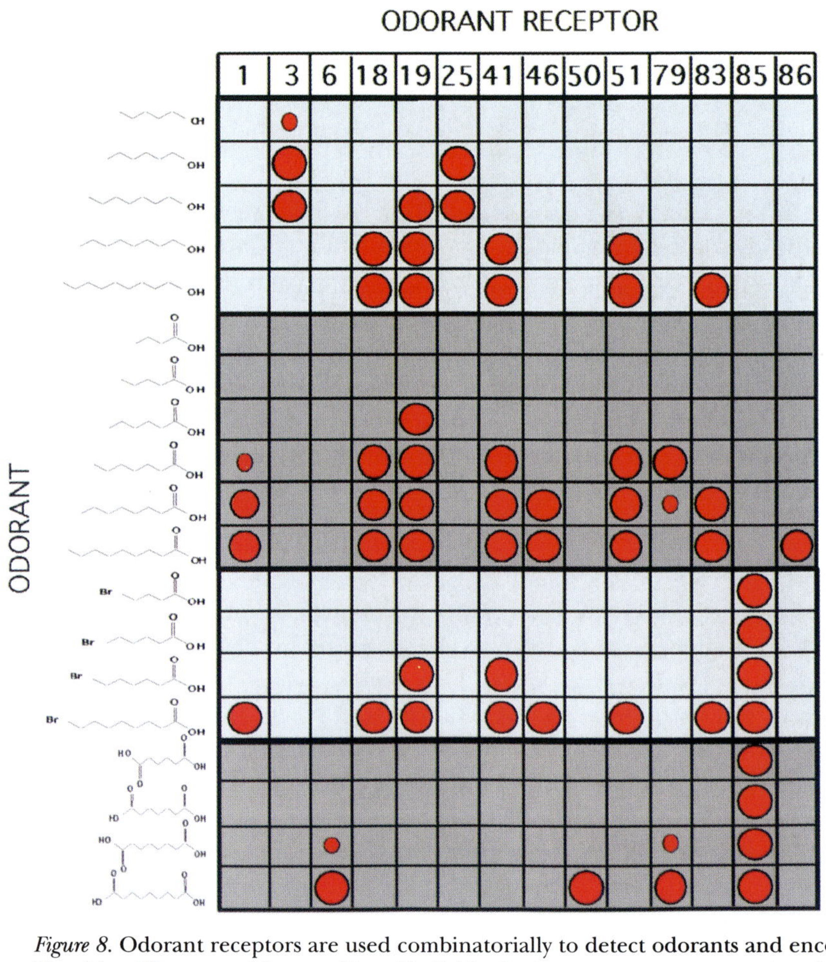

Figure 8. Odorant receptors are used combinatorially to detect odorants and encode their identities. The recognition profiles of individual odorant receptors to a series of odorants were determined by calcium imaging and single cell RT-PCR. The sizes of circles reflect response intensity. Adapted from Malnic *et al.*, 1999.

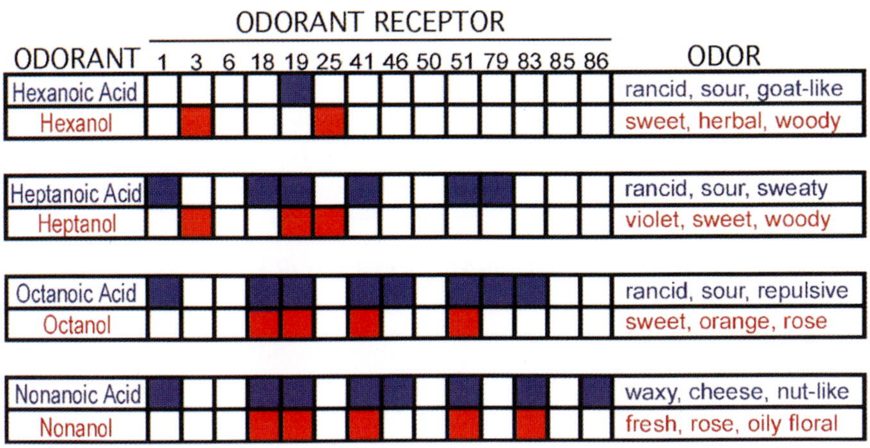

Figure 9. Closely related odorants with different perceived odors are detected by different combinations of receptors. Adapted from Malnic *et al.*, 1999.

different odorants (Figure 7). If a response was seen, the neuron was retested with a lower concentration of the same odorant.

Figure 8 shows the response profiles of 14 neurons and therefore the recognition properties of the ORs expressed in those neurons. These data make three important points. First, each OR can recognize multiple odorants, something previously shown by Stuart Firestein for one rat OR (Zhao *et al.*, 1998). Second, each odorant can be detected by multiple different ORs. And finally, and most importantly, different odorants are recognized by different combinations of ORs.

These results indicated that ORs are used combinatorially to encode odor identities (Malnic *et al.*, 1999). Different odorants are detected and thereby encoded by different combinations of ORs. However, each OR serves as one component of the codes for many odorants. Different odorants have different "receptor codes". Given the number of possible combinations of 1000 different ORs, this combinatorial coding scheme could allow for the discrimination of an almost unlimited number of odorants. Even if each odorant were detected by only three ORs, this scheme could potentially generate almost one billion different odor codes.

These studies also provided insight into several puzzling features of human odor perception (Malnic *et al.*, 1999). Changing the structure of an odorant even slightly can alter its perceived odor. Sometimes the change in odor can be dramatic. The aliphatic acids and alcohols that we used in our studies are excellent examples of this phenomenon (Figure 9). All of the acids have unpleasant odors, such as rancid, sour, or sweaty. In contrast, all of the alcohols have pleasant odors, such as herbal, woody, or orange. In our studies, pairs of acids and alcohols that differed by a single functional group invariably had different receptor codes (Figure 9). This indicates that even a slight change in the structure of an odorant can alter its receptor code, and thereby change its perceived odor.

Our studies showed that a change in the concentration of an odorant can also change its receptor code. At higher concentrations, additional ORs were invariably recruited into the odor response. This may explain why changing the concentration of an odorant can alter its perceived odor.

A STEREOTYPED MAP OF ODORANT RECEPTOR INPUTS IN THE OLFACTORY BULB

These studies indicated that, in the nose, different odorants are detected by different combinations of ORs and that the different OR combinations ultimately generate different odor perceptions. How is this accomplished? How does the brain translate an odorant's combinatorial receptor code into a perception?

Each olfactory sensory neuron in the olfactory epithelium sends a single axon to the olfactory bulb of the brain. Here the sensory axon enters a spherical structure called a glomerulus, where it synapses with the dendrites of bulb neurons. The mouse olfactory bulb has about 2000 glomeruli, each of which receives input from several thousand olfactory sensory neurons. Each sensory neuron synapses in only one glomerulus. Similarly, each mitral cell in the bulb receives input from a single glomerulus. Mitral cells are relay neurons that transmit signals to the olfactory cortex.

In the bulb, we found something very different from what we had seen in the nose (Ressler *et al.*, 1994). Here, single OR probes labeled OR mRNA in sensory axons in only a few glomeruli, and those glomeruli were located at only two spots, one on either side of the bulb (Figure 10A). We found that different OR probes labeled different glomeruli, and surprisingly, those glomeruli had nearly identical locations in different individuals (Figure 10B). These findings were made by Kerry Ressler and Susan Sullivan in my lab. In independent experiments in rat, Vassar and Axel obtained similar results, though individual OR probes generally labeled a larger number of glomeruli at more locations in the bulb in their studies (Vassar *et al.*, 1994).

Our studies in mouse indicated that the axons of thousands of sensory neurons with the same OR converge in only 2-4 glomeruli, each of which is likely to be dedicated to one OR (Figure 10) (Ressler *et al.*, 1994). They further indicated that sensory information that is broadly organized into 4 zonal sets in the nose is transformed in the bulb into a stereotyped sensory map (Figure 6). In this map, inputs from different ORs are targeted to different glomeruli and the bulb neurons associated with those glomeruli. Remarkably, this map is virtually identical in different individuals. The olfactory epithelium and bulb have one important thing in common, however. At both sites, inputs from different ORs are segregated. Each sensory neuron in the epithelium, and each glomerulus and relay neuron in the bulb, appears to be dedicated to only one type of OR.

The structure of the bulb map is likely to be important in at least two respects. First, it is likely to maximize sensitivity to low concentrations of odorants. Signals from 5000 or so neurons with the same OR converge on

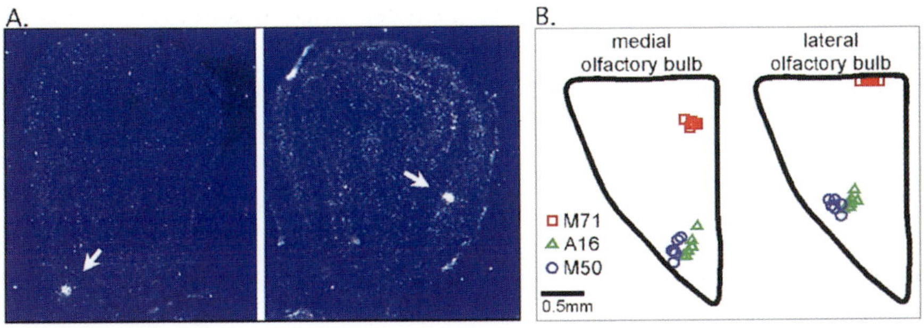

Figure 10. The olfactory bulb has a stereotyped map of OR inputs. A. A single OR gene probe hybridized to sensory axons in only 1-2 glomeruli on either side of the olfactory bulb. B. Different OR probes (A16, M50, M71) hybridized to different glomeruli and those glomeruli had similar locations in six different bulbs.

2–4 glomeruli and about 50 mitral cells, allowing a high degree of signal integration. Second, the bulb map is likely to be important for the stimulation of odor memories. Sensory neurons in the epithelium are short lived and are continuously replaced. However, the bulb map remains constant over time. Thus the neural code for an odor remains intact, assuring that odorants can elicit distant memories.

ODOR CODING IN THE OLFACTORY EPITHELIUM AND BULB

Given our finding that each odorant is recognized by a combination of ORs (Malnic *et al.*, 1999), these results imply that the code for an odor in the nose is a dispersed ensemble of neurons, each expressing one OR component of the odorant's receptor code (Figure 11). In the bulb, the code is a specific

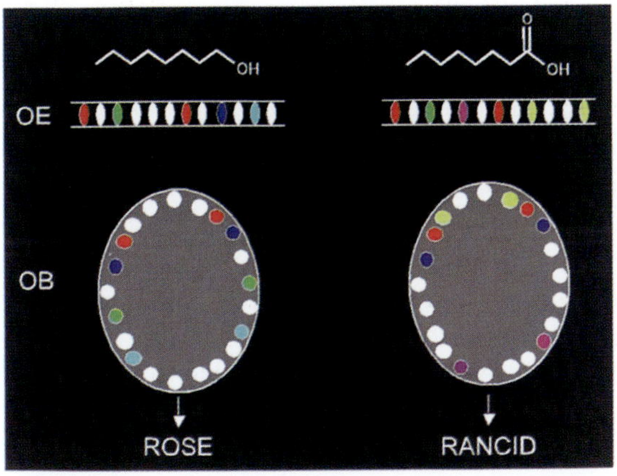

Figure 11. Odor coding in the olfactory epithelium and olfactory bulb. In this schematic, inputs from different ORs that recognize an odorant are indicated by different colors. In the olfactory epithelium (OE), the code for an odorant is a dispersed ensemble of neurons, each expressing one component of the odorant's receptor code. In the olfactory bulb (OB), it is a specific combination of glomeruli whose spatial arrangement is similar among individuals. Partially overlapping combinations of OR inputs generate distinct odor perceptions.

combination of glomeruli that receive inputs from those ORs and have a similar spatial arrangement in different individuals. This arrangement is consistent with many studies of odor-induced activity in the epithelium and bulb, beginning in the 1950s with the studies of Lord Adrian, who discovered that different mitral cells in the rabbit bulb respond to different sets of odorants (Adrian, 1950; Adrian, 1956; Buck, 1996).

STEREOTYPY, DIVERGENCE, AND CONVERGENCE IN OLFACTORY CORTEX

What happens to this information at higher levels of the nervous system to ultimately generate diverse odor perceptions?

Mitral cell relay neurons in the bulb extend axons to the olfactory cortex, a large area that stretches along the ventral lateral part of the brain. The olfactory cortex is composed of a number of distinct anatomical areas, at least some of which are likely to have different functions. The largest area is the piriform cortex, which itself has morphologically distinct anterior and posterior halves.

In the 1980s, Lewis Haberly and others showed that a tracer placed in one small region of this cortex would back-label mitral cells in many parts of the bulb (Haberly, 1998). This clearly indicated that the organization of sensory information in the olfactory bulb could not be recapitulated in the cortex. But how olfactory information is organized in the olfactory cortex was a mystery.

We were initially interested in three questions regarding the olfactory cortex. First, do different areas of the olfactory cortex, which may have different functions, receive signals derived from different subsets of ORs or, alternatively, does each area receive input from the entire OR repertoire? Second, is input from one OR scattered in the cortex, as in the nose, is it targeted to unique, stereotyped sites, as in the bulb, or is it organized in some other way? And finally, given that each odorant is recognized by multiple ORs, are inputs from different ORs combined in individual cortical neurons, or are they segregated in different neurons as in the nose and bulb?

To determine how OR inputs are organized in the cortex, we first asked whether it would be possible to trace neural circuits genetically. In those studies, Lisa Horowitz, a graduate student in the lab, made transgenic mice that expressed a plant protein, barley lectin, in all olfactory sensory neurons in the nose. In those mice, the expression of barley lectin, or BL, was controlled by the promoter of the OMP gene, a gene that is selectively expressed by olfactory sensory neurons.

Using BL-specific antibodies, we detected BL in olfactory sensory neurons in the olfactory epithelium, glomeruli and relay neurons in the bulb, and also in neurons in the olfactory cortex (Horowitz *et al.*, 1999). This indicated that BL produced by olfactory sensory neurons in the nose could travel across two synapses to label connected neurons first in the olfactory bulb and then in the olfactory cortex.

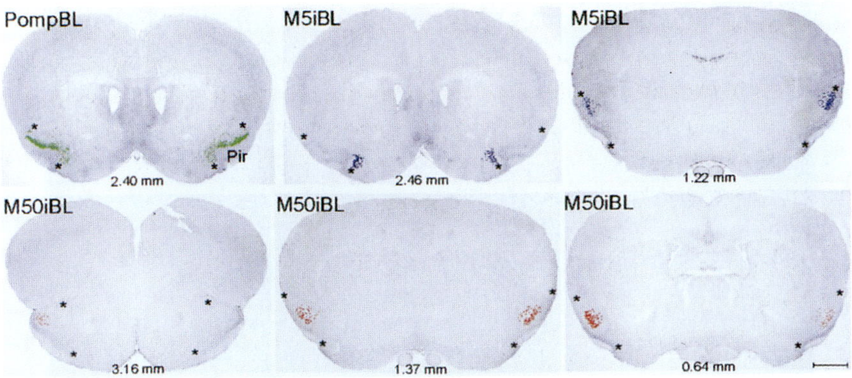

Figure 12. Inputs from one OR are targeted to 2–3 clusters of neurons in the anterior piriform cortex. Coronal sections through the anterior piriform cortex of mice in which barley lectin was expressed in all olfactory sensory neurons (PompBL) or only in neurons expressing the M5 (M5iBL) or M50 (M50iBL) odorant receptor. Asterisks indicate the outer limits of the piriform cortex (Pir). Distance from an anterior-posterior landmark is shown in mm. (Adapted from Zou *et al.*, 2001).

Having developed a genetic method for charting neural circuits, we were able to move to the next step. That was to ask how inputs from individual ORs are organized in the cortex.

Our goal was to coexpress BL with only one of the 1000 different OR genes. To do this, Lisa Horowitz and Jean-Pierre Montmayeur in our lab altered individual OR genes by inserting, just 3' to their coding regions, an IRES sequence followed by a BL coding sequence. Using gene targeting in embryonic stem cells, Zhihua Zou, a postdoctoral fellow in the lab then made "knockin" mice that contained an altered allele of either the M5 or M50 OR gene (Zou *et al.*, 2001). In these knockin mice, BL was produced only in neurons that expressed the M5 or M50 OR.

In the olfactory cortex, the axons of bulb neurons branch and form synapses in layer Ia with the dendrites of pyramidal neurons located in layers II and III. In PompBL mice, which express BL in all olfactory sensory neurons (Horowitz *et al.*, 1999), we saw labeled neurons in layers II and III throughout the olfactory cortex (Figure 12). In the M5 and M50 knockin mice, we also detected labeled cortical neurons, but they were located in distinct clusters (Figure 12) (Zou *et al.*, 2001). Moreover, the clusters appeared to have similar locations in different individuals.

In each knockin strain, we detected 2–3 clusters of labeled neurons in the anterior piriform cortex (Figure 12). Most of these clusters were bilaterally symmetrical in the left and right brain. We also found clusters of labeled neurons in several other areas of the olfactory cortex. In each cluster, the high-

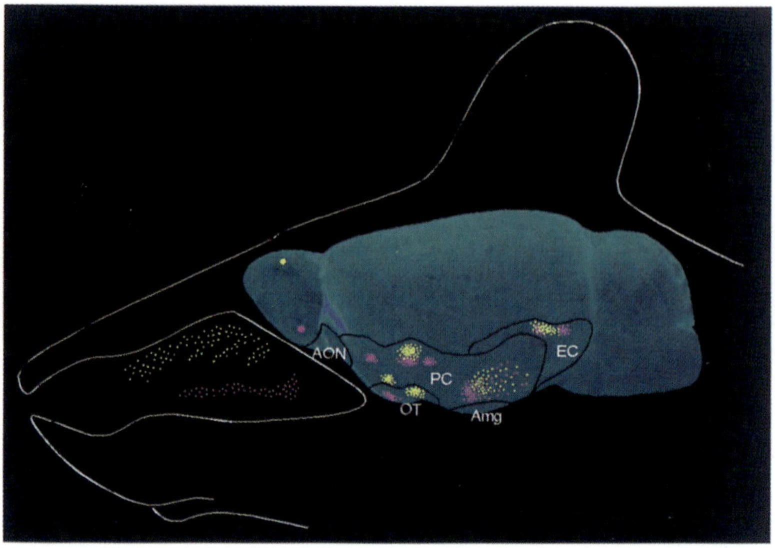

Figure 13. The olfactory cortex has a stereotyped map of OR inputs. Organization of inputs from the M5 (yellow) and M50 (pink) odorant receptors in the olfactory epithelium, bulb, and cortex. Black lines and abbreviations indicate different areas of the olfactory cortex. AON, anterior olfactory nucleus; PC, piriform cortex; OT, olfactory tubercle; Amg, olfactory nuclei of amygdala; EC, lateral entorhinal cortex.

est density of labeled neurons was in the center, but even in the center, only about half of the resident pyramidal neurons were labeled for the BL tracer.

Detailed analysis of the clusters in the anterior piriform cortex revealed that they had similar locations and similar dimensions in different individuals and, in most cases, they were bilaterally symmetrical (Zou *et al.*, 2001). The clusters had different locations in the two knockin strains, but one of the M5 clusters appeared to partially overlap with one of the M50 clusters.

These results showed that the olfactory cortex has a stereotyped map of OR inputs (Figure 13). In this map, signals derived from one type of OR are targeted to several loose clusters of cortical neurons. The clusters of neurons that receive input from a particular OR are found at specific locations, which are virtually identical among individuals.

These studies clearly indicated that input from one OR diverges to multiple areas of the olfactory cortex. This divergence of OR inputs may allow a parallel processing of OR signals in which signals from the same ORs are combined or modulated in different ways prior to transmission to other brain regions that have different functions.

We found that in the anterior piriform cortex, the clusters of neurons that receive input from one OR occupy about five percent of the total area along the anterior-posterior and dorsal-ventral axes (Zou *et al.*, 2001). In PompBL mice, which express BL in all olfactory sensory neurons, there were about

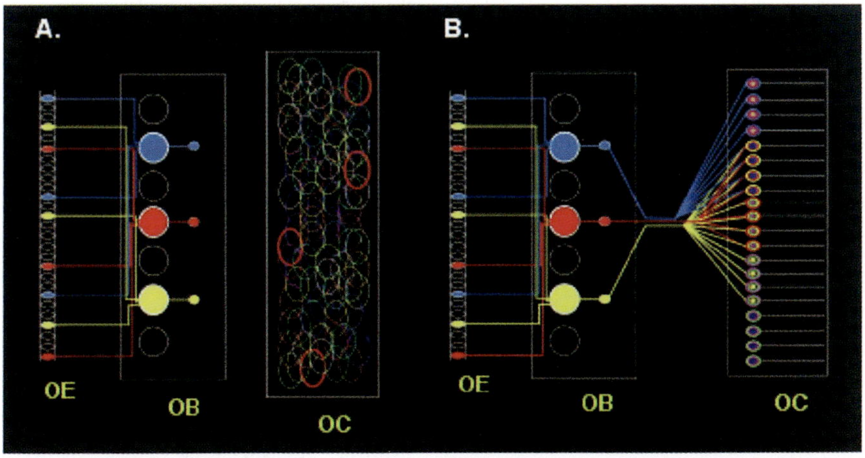

Figure 14. Schematic diagrams showing the organization of odorant receptor inputs in the olfactory epithelium (OE), olfactory bulb (OB), and olfactory cortex (OC). Inputs from different ORs are segregated in different neurons and glomeruli in the OE and OB. In contrast, it appears that different receptor inputs overlap extensively in the OC (A) and that single cortical neurons receive signals from a combination of receptors (B).

180,000 BL-labeled neurons in the anterior piriform cortex. If each cortical neuron received input from only one of 1000 different ORs, one might expect to see about 180 labeled neurons in this area each knockin mouse. However, we detected about 4000–6000 BL-labeled neurons in the anterior piriform cortex in each knockin strain (Zou *et al.*, 2001).

These results indicated that the map of OR inputs in the olfactory cortex is markedly different from that in the olfactory bulb. First, while inputs from different ORs are spatially segregated in different glomeruli in the bulb, they are likely to overlap extensively in the cortex (Figure 14A). Second, while signals from different ORs are segregated in different neurons in both the nose and bulb, each cortical neuron is likely to receive signals derived from multiple different ORs (Figure 14B). Since each odorant is recognized by a combination of ORs, this may permit an initial integration of multiple components of an odorant's receptor code that is critical to the generation of diverse odor perceptions.

These findings raise the possibility that neurons in the olfactory cortex function as coincidence detectors that are activated only by correlated combinatorial inputs from different ORs. For example, in a simple model, signals from different ORs that recognize vanillin would be targeted to partially overlapping locations in the cortex, but the only neurons activated by vanillin would be those that receive coincident signals derived from more than one of the vanillin ORs.

In sensory systems, environmental stimuli are deconstructed and then reconstructed in the brain to create perceptions. The organization of OR inputs seen in the olfactory cortex may serve as an initial step in the reconstruction of an odor image from its deconstructed features.

ACKNOWLEDGMENTS

I would like to acknowledge the very talented students and postdoctoral fellows in my lab who did the experiments that I discussed today. Kerry Ressler and Susan Sullivan did all of the early studies on OR inputs in the olfactory epithelium and bulb. Hiroaki Matsunami identified and characterized the V2R family of candidate pheromone receptors. Bettina Malnic conducted the studies of OR specificities in collaboration with Takaaki Sato and Junzo Hirono at the Life Electronic Research Center in Japan. Bettina and Paul Godfrey defined the OR gene repertoires of human and mouse. Lisa Horowitz developed the genetic method for tracing neural circuits. And Zhihua Zou, in collaboration with Lisa and Jean-Pierre Montmayeur, conducted the studies of OR inputs in the olfactory cortex.

REFERENCES

Adrian, E. D. (1950). Sensory discrimination with some recent evidence from the olfactory organ. *Br Med Bull* 6, 330–333.

Adrian, L. (1956). The action of the mammalian olfactory organ. *J Laryngol Otol 70,* 1–14.

Amoore, J. E. (1970). Molecular basis of odor (Springfield, Charles C. Thomas).

Amoore, J. E. (1977). Specific anosmia and the concept of primary odors. *Chemical Senses and Flavor 2,* 267–281.

Buck, L. B., and Axel, R. (1991). A novel multigene family may encode odorant receptors: a molecular basis for odor recognition. *Cell 65,* 175–187.

Buck, L. B. (1996). Information coding in the vertebrate olfactory system. *Annu Rev Neurosci 19,* 517–544.

Dulac, C., and Axel, R. (1995). A novel family of genes encoding putative pheromone receptors in mammals. *Cell 83,* 195–206.

Gesteland, R. C., Lettvin, J. Y., and Pitts, W. H. (1965). Chemical transmission in the nose of the frog. *J Physiol 181,* 525–559.

Glusman, G., Yanai, I., Rubin, I., and Lancet, D. (2001). The complete human olfactory subgenome. *Genome Res 11,* 685–702.

Godfrey, P. A., Malnic, B., and Buck, L. B. (2004). The mouse olfactory receptor gene family. *Proc Natl Acad Sci USA 101,* 2156–2161.

Haberly, L. B. (1998). The olfactory cortex. In The synaptic organization of the brain, G. M. Shepherd, ed. (New York, Oxford University Press), pp. 377–416.

Herrada, G., and Dulac, C. (1997). A novel family of putative pheromone receptors in mammals with a topographically organized and sexually dimorphic distribution. *Cell 90,* 763–773.

Horowitz, L. F., Montmayeur, J. P., Echelard, Y., and Buck, L. B. (1999). A genetic approach to trace neural circuits. *Proc Natl Acad Sci USA 96,* 3194–3199.

Jones, D. T., and Reed, R. R. (1989). Golf: an olfactory neuron specific-G protein involved in odorant signal transduction. *Science 244,* 790–795.

Malnic, B., Godfrey, P. A., and Buck, L. B. (2004). The human olfactory receptor gene family. *Proc Natl Acad Sci USA.*

Malnic, B., Hirono, J., Sato, T., and Buck, L. B. (1999). Combinatorial receptor codes for odors. *Cell 96,* 713–723.

Matsunami, H., and Buck, L. B. (1997). A multigene family encoding a diverse array of putative pheromone receptors in mammals. *Cell 90*, 775–784.

Mombaerts, P. (1999). Molecular biology of odorant receptors in vertebrates. *Annu Rev Neurosci 22*, 487–509.

Pace, U., Hanski, E., Salomon, Y., and Lancet, D. (1985). Odorant-sensitive adenylate cyclase may mediate olfactory reception. *Nature 316*, 255–258.

Ressler, K. J., Sullivan, S. L., and Buck, L. B. (1993). A zonal organization of odorant receptor gene expression in the olfactory epithelium. *Cell 73*, 597–609.

Ressler, K. J., Sullivan, S. L., and Buck, L. B. (1994). Information coding in the olfactory system: evidence for a stereotyped and highly organized epitope map in the olfactory bulb. *Cell 79*, 1245–1255.

Ryba, N. J., and Tirindelli, R. (1997). A new multigene family of putative pheromone receptors. *Neuron 19*, 371–379.

Sklar, P. B., Anholt, R. R., and Snyder, S. H. (1986). The odorant-sensitive adenylate cyclase of olfactory receptor cells. Differential stimulation by distinct classes of odorants. *J Biol Chem 261*, 15538–15543.

Sullivan, S. L. Adamson, M. A. Ressler, K. J. Kozak, C. A. and Buck, L. B. (1996) The chromosomal distribution of mouse odorant receptor genes. *Proc Natl Acad Sci USA* 93:884–888.

Vassar, R., Chao, S. K., Sitcheran, R., Nunez, J. M., Vosshall, L. B., and Axel, R. (1994). Topographic organization of sensory projections to the olfactory bulb. *Cell 79*, 981–991.

Vassar, R., Ngai, J., and Axel, R. (1993). Spatial segregation of odorant receptor expression in the mammalian olfactory epithelium. *Cell 74*, 309–318.

Young, J. M., and Trask, B. J. (2002). The sense of smell: genomics of vertebrate odorant receptors. *Hum Mol Genet 11*, 1153–1160.

Zhang, X., and Firestein, S. (2002). The olfactory receptor gene superfamily of the mouse. *Nat Neurosci 5*, 124–133.

Zhao, H., Ivic, L., Otaki, J. M., Hashimoto, M., Mikoshiba, K., and Firestein, S. (1998). Functional expression of a mammalian odorant receptor. *Science 279*, 237–242.

Zou, Z., Horowitz, L. F., Montmayeur, J. P., Snapper, S., and Buck, L. B. (2001). Genetic tracing reveals a stereotyped sensory map in the olfactory cortex. *Nature 414*, 173–179.

Zozulya, S., Echeverri, F., and Nguyen, T. (2001). The human olfactory receptor repertoire. *Genome Biol 2*, RESEARCH0018.

Physiology or Medicine 2005

BARRY J. MARSHALL and J. ROBIN WARREN

"for their discovery of the bacterium Helicobacter pylori and its role in gastritis and peptic ulcer disease"

THE NOBEL PRIZE IN PHYSIOLOGY OR MEDICINE

Speech by Professor Staffan Normark of the Nobel Assembly at Karolinska Institutet.
Translation of the Swedish text.

Your Majesties, Your Royal Highnesses, Ladies and Gentlemen,

Napoleon Bonaparte was not poisoned, but died of a stomach ulcer that had been transformed into cancer. Author James Joyce, still disappointed at the poor reception of his latest novel *Finnegans Wake,* died of a perforated ulcer. Ulcers don't strike only the famous. They are one of the most common afflictions of humanity. For a long time, ulcers were regarded as being a result of stress and improper diet. Barry Marshall's and Robin Warren's discovery that ulcers are caused by a bacterial infection was therefore completely revolutionary and was initially met by great skepticism.

Robin Warren, in his professional role as a clinical pathologist, observed spiralshaped bacteria in tissue specimens from the stomachs of many patients who had undergone gastroscopy. These bacteria were abundantly present, associated with the epithelium of the stomach and shielded from dangerous hydrochloric acid by a thick mucous layer. Warren saw that the underlying mucous membrane always showed signs of inflammation and therefore suggested that the spiral-shaped bacteria caused gastritis.

Barry Marshall became interested in Warren's findings and decided to culture the bacterium. Many fruitless attempts were made. During the Easter 1982 holiday, the agar plates were left in the incubator by mistake, and when they were inspected after the holiday they contained numerous colonies consisting of the same bacterium that Warren had observed under his microscope. Soon it became clear that a whole new bacterial genus had been isolated. It was eventually named *Helicobacter pylori.*

Marshall and Warren now conducted a large clinical study, which showed that *Helicobacter pylori* was found in the stomach of most patients with ulcers either in the duodenum or the stomach and that the bacterium was always associated with inflammation of the mucous membrane.

In order to prove that the bacterium that had been isolated caused a disease, Marshall tried to fulfill Koch's Fourth Postulate. This implies that an infectious agent that has been isolated must cause the same disease in an experimental animal as in humans. In the absence of a suitable animal model, Marshall decided to drink a bacterial culture containing *Helicobacter* and he suffered a severe inflammation of the stomach.

Before the connection between *Helicobacter* infection and ulcer had become clear, the disease was often chronic with frequent relapses. Marshall and Warren were among the first to show that ulcer patients can be cured for good if the *Helicobacter* is eliminated from the stomach with the help of antibiotics.

Today we know that most ulcers are caused by *Helicobacter* infection. A chronic infection that involves the entire stomach also increases the risk of stomach cancer, the second most common of all forms of cancer. People are infected in their early childhood years and then carry the infection throughout their life. Half of humanity is infected, but luckily most people have no symptoms of the infection. *Helicobacter pylori* lives only in the human stomach. From the standpoint of both the bacterium and its human host, ulcer disease, cancer and death are to be regarded as a failure in an otherwise long and relatively harmonious relationship.

Barry Marshall and Robin Warren,

Against prevailing dogmas, you discovered that one of the most common and important diseases of mankind, peptic ulcer disease, is caused by a bacterial infection of the stomach. Your discovery has meant that this frequently chronic and disabling condition can now be permanently cured by antibiotics to the benefit of millions of patients. Your pioneering work has also stimulated research all around the world to better understand the link between chronic infections and diseases such as cancer. On behalf of the Nobel Assembly at Karolinska Institutet, I wish to convey to you our warmest congratulations, and I now ask you to step forward to receive the Nobel Prize from the hands of His Majesty the King.

BARRY J. MARSHALL

I was born in 1951 in Kalgoorlie, a prosperous mining town 370 miles east of Perth, Western Australia. Kalgoorlie was a gold rush town which sprang up in the desert after the Irishman paddy Hannan struck gold there in 1892.

At the time I was born my father was 19 years old and in the final year of his apprenticeship as a fitter and turner. My mother quit her nursing training to have me at the age of eighteen years.

We moved quite a bit through my early childhood. After my father finished his apprenticeship, my parents decided to go and work in the new Uranium mine in Rum Jungle in the Northern Territory. They drove their Model A Ford up Australia's west coast about 1000 miles but stopped at Carnarvon when the car broke down. The whaling station at Carnarvon was also offering excellent wages for good tradesmen and my father was one of the best. We lived near the whaling station while I grew from two to four years and my brother William was born there.

My first memories are of life in Carnarvon. I recall a boat trip back to Perth on one occasion and a DC3 aeroplane flight to Perth on another. Our house was on Babbage Island about 100 yards from the beach. We had electricity, an outhouse toilet, dirt floors in parts of the house, a telephone, refrigerator, a car, a cat and a dog. Nearby was a derelict steam engine on a railway siding. We had neighbours close by and other kids to play with.

By then, my grandparents had the license on the Tower Hotel in Kalgoorlie and periodically we would return there to live. In Kalgoorlie I remember doing all kinds of things as a six and seven year old including making bows and arrows, slingshots and lighting crackers after school.

After a period back in Kalgoorlie, my mother decided to move the family to Perth where my second brother, Andrew, was born in 1958. I was seven years old at the time. I suppose my mother could see the young boys in Kalgoorlie leaving school at 16 and going down the mines to work. It was an attractive proposition for them. They earned high salaries and had a wild social life drinking and partying on their off days. She wanted more for her children and hoped we would study and enter a profession. Moving to the city was the first step. We are lucky she made that decision. My two brothers and sister all went through University and have highly successful careers and happy lives.

In school I sporadically hit the top of the class but mostly did not work hard enough to stay up there. At home I had plenty of interesting reading material. Dad always explained the car engine when he repaired it and he had many technical books so I was making electromagnets by age eight as well as reading my mothers medical and nursing books. I suspect I was born with a boundless curiosity and this was encouraged through my childhood.

Being the eldest of four children, I was expected to be the responsible one and often found myself controlling two younger brothers who shared my exuberant and inquisitive nature. I still feel guilty about the time I advised my younger brother to jump out of a tree and he broke his arm.

My first exposure to fame came at age twelve when I was left in charge of the younger siblings while my mother attended to the grocery shopping. I had a history of responsible baby-sitting by this time so nothing should have gone wrong. During the morning my 18 month-old sister found a milk bottle half full of kerosene and drank some, perhaps also aspirating a little so that my brothers and I found her choking but did not know why. I called the emergency services and an ambulance arrived about fifteen minutes later. During the wait, as I had learned some basic CPR at the Royal Lifesaving Society Swimming training, I tried to perform mouth to mouth resuscitation on my little sister. I know now that it was pointless because she was actually still breathing. However, my close mouth contact enabled me to smell the kerosene and make the diagnosis of poisoning. I featured in the newspaper a few days later, with my fully recovered little sister on my lap. It was a good story about how to call the emergency number, and why you should not put poison into drink containers. Very kindly my mother did not leak to the press the fact that it was I who had left the kerosene within reach of young Marie!

In our dad's shed, my brothers and I had access to all the tools needed to build or dismantle anything. I frequently got into trouble doing both. My favourite book as a child was an old Newne's Children's Encyclopaedia which my grandfather had bought just before World War II and donated to our family after seeing how interested we were in it. Each volume had special chapters called "Things Boys can Do". My brothers and I would pick out interesting projects. As the years went by, and I grew up, I recall building a slingshot, a crystal set, a Morse-code set, various guns, a hydrogen generator for balloons, electric devices and minor explosives. In those days fireworks had been banned, but chemicals were easily available from pharmacies and chemical suppliers so, in the tradition of Alfred Nobel, we would create various explosive mixtures and make firecrackers and bombs. This started rather benignly with simple gunpowder but graduated to more dangerous oxidising agents after a few years. Many times we were in trouble after disturbing the neighbours, but were fortunate never to cause serious injury. I often found myself in trouble with my parents when someone was hurt, but despite the minor punishments, I know my parents were quite proud of my ingenuity.

Occasionally my father, Bob, was on the receiving end of my "brilliant" work. Observing a fraying cord on his electric drill, I repaired it but accidentally swapped the neutral and earth wire. He jumped rather high when he tried to use it a few days later while standing on wet grass. On another occasion my brothers decided to fly lighter-than-air balloons for our team at the school sports carnival. Since helium was not available, we built a device which pressurised domestic house gas and filled the balloons. Our technology was rather primitive however and these balloons contained quite bit of air as well, but they did float satisfactorily. My father recognised this and warned us that they might be a little dangerous if they came in contact with an open flame. As an example, he demonstrated the risk by touching a lighted cigarette to one of the balloons as it floated under the back patio. He was enveloped in a ball of flame and his eyebrows were singed off. This did not worry us very much because we had seen him in this state before as he often seemed to be washing engine parts in gasoline and then testing the spark plugs of engines nearby.

After high school, at Newman College, although interested in science and mathematics, I felt that my mathematical ability was not strong enough to do electrical engineering, so I chose medical school as an alternative which was at least as interesting, and which did not require daily exposure to calculus! In addition the opportunity to study biological sciences was an attraction, particularly biochemistry which was not available in high school.

I met my wife Adrienne, a psychology student, at the University of Western Australia and we married in 1972 while I was doing my fifth year in medicine. I graduated from the University MBBS (Bachelor of Medicine, Bachelor of Surgery) in 1975 and thereafter performed internship and residencies in internal medicine at the Queen Elizabeth II Medical Centre (Sir Charles Gairdner Hospital). In those days I had no definite goals in medicine, but was interested in all aspects of clinical medicine including geriatrics, oncology and rheumatology. I was more interested in an academic career combining research with clinical medicine in a University hospital environment. I began my training as a specialist physician in 1978. In 1979 I moved to Royal Perth Hospital in order to become more experienced with cardiology and open heart surgery, which was only performed at that hospital in Perth.

Although we didn't appreciate it at the time, my wife Adrienne and I must have been very busy during those years. We had four children, Luke, born in 1973, Bronwyn in 1975, Caroline in 1978 and Jessica in 1981. Adrienne was finishing the honours year of her psychology degree as Luke was being born. She was working in-between babies as a counsellor with the Education Department. My non-medical time was spent delivering children to various child-minding facilities, renovating our house, and indulging in my hobby of computers and electronics.

In the second half of 1981, my rotation took me to the gastroenterology division. It was there that I met Robin Warren. As part of my training I was encouraged to perform a clinical research project each year. I was already totally engrossed in a study of heat stroke in "fun runners" and might have progressed to sports or environmental medicine from there. However, I asked my boss, Dr Tom Waters, if there was a gastroenterology project I could start. He told me that Robin Warren had given him a list of patients with curved bacteria present on their stomach biopsies and needed someone to follow-up the patients to see what clinical diseases they had. I was especially interested because one of the people on Robin's list was a woman I had seen in my ward, who had severe stomach pain but no diagnosis. In desperation we had referred her to a psychiatrist and commenced antidepressant medication for want of a better treatment. The only abnormal finding had some redness in the stomach and Robin's bacteria on the stomach biopsy.

So I called Robin in the basement area of Royal Perth Hospital where the Pathology department resided. It was to be the first of many afternoon visits in the next year. In those days, Robin used to drink strong black coffee and smoke small cigars, "cigarillos" I believe they were called. I too used to indulge occasionally, and would try out one of Robin's cigars from time to time during our meetings. In our first meeting, Robin showed me slides of the curved bacteria he had seen, and explained the histopathology of the gastric mucosa to me.

I am often asked what made me listen to Robin and take up the research with him. Clearly this was an interesting thing to study, previously undescribed bacteria living in the acid-filled stomach. But I may have had other advantages compared with colleagues Robin had approached over the previous two years.

I was undifferentiated in that I wasn't coming from a background in gastroenterology so that my knowledge and ideas were founded in general medical basic science rather than the dogma one was required to learn in specialist medicine. As a trainee general physician with broader training, I was comfortable with the notion of infectious disease and antibiotic therapies. I am told by others that I have a lateral thinking broad approach to problems, sometimes to my detriment. In school my grades always suffered because I was continually mucking about with irrelevant side issues which I often found to be more interesting.

At around that time also, I was aware of publications in the literature describing *Campylobacter jejuni* as a newly discovered common cause of foodborne gastroenteritis and colitis. Thus, I had seen pictures of campylobacters and could identify that Robin's organisms appeared to be quite similar. In retrospect, one advantage of doing this research in Perth was that, as a modern Western society, *H. pylori* was already in decline by 1981, so that rather than 80% of persons having the *CLO*, bacteria were only present in 30–50%. Thus, in any biopsy collection taken that year, Robin could see both infected specimens with inflammation (gastritis) and uninfected specimens which hardly ever had gastritis i.e. a "control group". A further advantage I did have in 1981 was the new connection we had from the medical library to the National Library of Medicine at the NIH (*Medline*). Perhaps because of my interest in computer programming, this resource appealed to me and enabled me to enlist the librarians at Royal Perth Hospital to extensively search the past and current literature on gastric bacteria.

By the end of that first afternoon with Robin I was very interested. Over the next six months I followed the literature from book chapters, to their references, to deeper references, to material in library archives. I found that spiral gastric bacteria had been reported again and again but passed over. I could see an interesting paper being produced, perhaps in an obscure microbiological journal, but had no idea at the time of what we were really about to discover.

At the end of 1981, my gastroenterology term had almost finished and my term allocations for 1982 had been chosen. In the midst of all this time consuming and interesting research work I was still a physician in training. I was fitting in the research around education and patient commitments. In the first 6 months of 1982, I was to be a haematology registrar looking after the bone marrow transplant patients. In the second 6 months I was to be the physician at Port Hedland Hospital, a rotation to a point 2,000 kilometres north of Perth which attracted "hardship bonus", i.e. $5,000 extra over 6 months. By then, I was very excited about the spiral bacteria. I had developed a degree of confidence in our methodology, and believed that we could safely carry out a study on 100 or so patients. I was able to keep the work going, continuing the research by fitting it around my other duties.

In November 1981, Adrienne had delivered our fourth child, Jessica. I was beginning a project which would occupy every minute of my spare time for the next 6 months. Adrienne was on maternity leave and was full time at home. It meant I could leave much of the parenting to her. We never did find the time to complete our home renovations and at the last minute in 1986 took out an extra home loan to pay someone to finish it for us. We had to have it in a rentable condition while we were in the USA.

I was fortunate to have a partner who was as enthusiastic about the work as I was. She also enjoys a challenge and shares my sense of adventure. Adrienne's background in Psychology and experimental research was invaluable and she was always around to discuss the design of studies and the results of various other research works I had found. Over the years we took lots of chances. I took jobs on inadequate pay for many years. As my contemporaries were making their careers and achieving success I seemed to be falling further behind. I always had Adrienne's full support. When she urged caution or vetoed some of my excesses, I knew it was time to really listen and re-evaluate. As time went on she became by my unofficial editor. All my early papers were edited by her and she helped with much of the discussion. Her liberal arts background means she is a more fluent writer than I. Over the past 25 years she has also helped to write and edit most of my books and speeches. All of the talks and speeches given in Stockholm were written with her substantial help.

My hobby of electronics was also an important aid in my research. In the evenings during 1981, I continued with my hobby of computing and electronics, so that by the end of that year I had completed the construction of a home computer capable of word processing. I was able to type grant proposals, consent forms and protocols. I was always on the leading edge of technology and my communication with overseas researchers was efficient because of that. It also meant I was able to access information not readily available. By 1981 I could function better as a single unfunded scientist than many units with multiple support staff.

The family moved to Port Hedland in July 1982 and I took all my references and textbooks with me. It was an important period. I had time to do an extensive literature search by correspondence and also had time to digest the results of our study and write it up for presentation. It was a great time for the family too. Winter in Port Hedland was beautiful, every day sunny with a temperature in the 80's. We had a bit of extra money and we spent many weekends travelling in remote communities and camping with the kids under the stars.

In October 1982, I presented the preliminary findings from our study to the local College of Physicians meeting, where it received a mixed response. I found that my contract at Royal Perth would not be renewed the following year. I had successfully completed my training as a physician and now wanted to work in gastroenterology or microbiology to continue the work. These jobs at Royal Perth were not available.

Fortune stepped in when I was approached by Drs. Norm Marinovich and Ian Hislop at Fremantle Hospital who suggested they would find me a senior registrar position and fund me to continue. Fremantle is the third and smallest of the teaching Hospitals in Perth and has a tradition of openness and experimentation. In the next two years at Fremantle I had an enthusiastic group of people working with me. Dr Ian Hislop, Norm Marinovich, Harvey Turner, David McGechie, Ross Glancy, Neil Noakes, Graeme Francis, Peter Rogers, Neil Stingemore, as well great support from the Medical Superintendent, Peter Smith. The only downside of the appointment was that I was forced to halt my collaboration with Robin Warren. Robin did not have an appointment at Fremantle so the pathologist there Ross Glancy joined the team.

They were happy and very productive years. I was able to confirm very quickly that our observations of the bacteria at Royal Perth Hospital also applied in other parts of the city, the majority of peptic ulcer patients having the organism. I was still officially unfunded. The hospital was picking up all the costs of my work. It was at Fremantle in those two years that the first effective treatments were devised. I solved the conundrum of why bismuth has been such an effective stomach treatment for the past 200 years. I did my famous self experimentation and the early urease tests were developed.

A great piece of luck in early 1983 was finding Dr Martin Skirrow in the UK. I got his phone number from David McGechie. Skirrow arranged for the first presentation at the European Campylobacter Meeting in September 1983. Harvey Turner arranged a travel grant to take me to Brussels and the Gist Brocades Company, helped so that I could extend the trip, visit Martin in the UK and Guido Tytgat's group in Amsterdam.

In September 1983, I visited Martin Skirrow in Worcester England, and attended an endoscopy session at the Worcester Infirmary. Martin's registrar, Cliodna McNulty, was able to successfully isolate the organism 3 days later, showing that the spiral bug was not merely an Australian phenomenon but was present in ulcer patients in the UK as well. Martin Skirrrow in Britain and Adrian Lee in Sydney were enormously encouraging, helping me with the microbiology in those early years.

In 1984 therefore, there were several groups around the world obtaining results which paralleled those of our group in Perth. In Australia, Adrian Lee in Sydney with Stuart Hazel and Hazel Mitchell also Nick Talley, John Lambert and Tom Borody were early researchers who made significant advances in the *H. pylori* work. After the Brussels meeting, a core of researchers in Europe immediately picked up the research and much of the most important work on HP has been done by that group: my old friends, Mario Quina in Portugal, Tony Axon and Ashley Price in the UK, Francis Megraud in France, Peter Malfertheiner in Germany, Manuel Lopez Brea and Jose Pajares Garcia in Spain, Penti Sipponen in Finland, Dino Vaira and Giovanni Gasparini in Italy, Colm O'Morain in Ireland, Leif Andersen in Denmark, Alexander Hirschl in Austria, Guido Tytgat, Ernst Kuipers and Erik Rauws in The Netherlands, Michel Deltenre in Belgium, Pierre Michetti in Switzerland, Torkel Wadstrom and Lars Engstrand in Sweden. We became a closely knit group. The European group grew out of the campylobacter group I had met in Brussels in 1983 and today I count the members of that group amongst my closest friends. We have shared a remarkable story together.

In the USA David Graham, Pete Peterson and Martin Blazer began as critics. They set out to disprove the hypothesis but quickly became leaders in the field of HP research in the USA. With Tadetaka (Tachi) Yamada, although he was not directly involved in the HP research, they played an important role in moving various bodies such as the NIH towards action and acceptance of HP as an ulcer cause. In Asia, Takashi Shimoyama, Ken Kimura, Susumu Okabe, Yoshihiro Fukuda, Toshio Fujioka, Bow Ho, and K.L. Goh, were doctors who I was in contact with through the 1980s. They were developing their own HP research and supporting mine. In Asia, the *H. pylori* research was taken up very quickly and I made my first visit to Japan in 1985 to present my work. There are too many others to list here. Needless to say, reports that I was alone in the promotion of HP as a pathogen are somewhat exaggerated.

But 1984 was a difficult year. I was unsuccessfully attempting to infect an animal model. There was interest and support from a few but most of my work was rejected for publication and even accepted papers were significantly delayed. I was met with constant criticism, that my conclusions were premature and not well supported. When the work was presented, my results were disputed and disbelieved, not on the basis of science but because they simply could not be true. It was often said that no one was able to replicate my results. This was untrue but became part of the folklore of the period. I was told that the bacteria were either contaminants or harmless commensals.

At the same time I was successfully experimentally treating patients who had suffered with life threatening ulcer disease for years. Some of my patients had postponed surgery which became unnecessary after a simple 2 week course of antibiotics and bismuth. I had developed my hypothesis that these bacteria were the cause of peptic ulcers and a significant risk for stomach cancer. If I was right, then treatment for ulcer disease would be revolutionized. It would be simple, cheap and it would be a cure. It seemed to me for the sake of patients this research had to be fast tracked. The sense of urgency and frustration with the medical community was partly due to my disposition and age. However, the primary reason was a practical one. I was driven to get this theory proven quickly to provide curative treatment for the millions of people suffering with ulcers around the world.

Becoming increasingly frustrated with the negative response to my work I realized I had to have an animal model and decided to use myself. Much has been written about the episode and I certainly had no idea it would become as important as it has. I didn't actually expect to become as ill as I did. I didn't discuss it with the ethics committee at the hospital. More significantly, I didn't discuss it in detail with Adrienne. She was already convinced about the risk of these bacteria and I knew I would never get her approval. This was one of those occasions when it would be easier to get forgiveness that permission. I was taken by surprise by the severity of the infection. When I came home with my biopsy results showing colonization and classic histological damage to my stomach, Adrienne suggested it was time to treat myself. I had a successful infection, I had proved my point.

At the end of 1984 I was funded by the Australian Medical Research Council to conduct a prospective double blind trial to see if antibiotics could cure duodenal ulcers. It was conditional on getting a large number of patients into the study so I decided to move back to Royal Perth Hospital where the patient load is far higher. It meant I would be leaving my Fremantle colleagues and it was with some reluctance that I moved. When I returned to Australia in 1996, I was asked to be Patron of the Fremantle Hospital Research Foundation and I take great pride in having that position. At Royal Perth I was again working with Robin, John Armstrong, Len Matz, John Pearman, Stewart Goodwin, Doug Annear and Helen Royce.

Even though I was not officially collaborating with Robin when I was working at Fremantle Hospital in 1983–84, we still met to discuss the papers we were writing for the Lancet and would meet for dinner with our wives. We had one of these dinners only a few weeks after my self experimentation experiment. I was enthusiastic about the results and the severity of my illness. It was also the first confirmation of infection with documented results. I was eager to share the news with Rob and he was equally excited about it. Early the next morning he had a call from a journalist in the USA at 5 am who had his timing totally off. No one is ever able to figure out what time it is in Perth. When asked the usual question about "How do you know it's a pathogen and not a harmless commensal?" Rob blabbed the results of my still unreleased work with "I know because Barry Marshall has just infected himself and damm near died"; a slight exaggeration, but it made for good copy. What he didn't know was that the journalist he was speaking to was from the "Star" newspaper a tabloid that often features with stories about alien babies being adopted by Nancy Reagan. This was right up their alley. The next day the story appeared, "Guinea-pig doctor discovers new cure for ulcers… and the cause."

This became one of the serendipitous, life changing events in my life and I have Rob's temper to thank for it. Firstly, I was contacted by a continuous line of patients in the USA who read the story and were desperate for treatment. I was able to help. I was treating patients by proxy in the USA as early as 1984.

Ten years later this became important in a dispute with another doctor who claimed to be the first. I still had the records from some of these patients and was able to get in touch with them to prove my claim to be first.

The second result was that it was read by Mike Manhart, a microbiologist working for Proctor and Gamble in the USA. He tracked down my published letters and realized the economic potential for P & G who made a bismuth drug and set up a business relationship. P & G later patented much of my work and also helped me with patents on my diagnostics. There was little money in any of this for the first 10 years but after 1995 it became a significant income for us. P & G funded a fellowship for me in the USA to replicate and push the research there. We departed Australia, believing that it shouldn't take more than 2–3 years to convince the world that antibiotics would cure most gastric diseases.

It also brought Bruce and Claudette McCarty into my life. Bruce was head of Health and Personal Care products at P & G and became an important mentor. He arranged support funding to set up a lab at University of Virginia. Bruce became a good friend and a keen advocate for H. Pylori research in the USA. He taught me a lot about how business works best in a trusting and responsible way to benefit everyone. It also seemed to me that he and Claudette spent lots of time in their life just having fun with family and friends. Tragically, Bruce died in 2004. It was a great sadness for me and Adrienne that he and Claudette were not there in Sweden to see me receive the Prize. He always believed in me and his faith in the work and great enthusiasm never failed.

The ten years spent at University of Virginia, were a chance to extend my research, particularly in the area of treatment and diagnostics. I became an advocate for treatment though many called me a zealot. They were often hard years for the family particularly the first few years when we were on a financial shoestring. They were rewarding as well. I had continuous stream of letters from patients who had been treated and freed from a lifetime of pain and disruption. I worked with a great team at UVA. Richard McCallum was head of Gastroenterology and Dick Guerrant in Infectious Diseases. McCallum gave me free rein and sponsored my academic rise in the USA. I also did great collaborative work with Dick over the years. David Peura, a long-time H. Pylori enthusiast from the US Army, moved to UVA in 1992. My team included nurse Susie Hoffman, nuclear technician Michael Plankey, post-doc Matthew Coombs, data manager Linda Mosen, programmer Sherry Boyd, assistant Nancy Noblette and many others. We were regarded as being outside the mainstream but were a great enthusiastic group and became lifelong friends.

I also met Bill and Sandy Fry in 1987. Bill owned a company Tri-Med along with Phil Patterson and Kevin Dye. Bill Fry bankrolled a USA study for my CLOtest diagnostic and launched it in the USA. Later he was to also pick up the C14-Urea Breath Test and shepherded it through the FDA at a cost of several million dollars. I count Bill amongst my closest friends, a brilliant salesman and an example of a team leader. No matter how black things looked,

Bill could always find a silver lining for us even though I am certain he was secretly concerned about our chances of success. Bill's credo which he lives by is that "Good things happen to good people". We have done a lot of good stuff together and had many great times.

Patients often wanted to make a donation to the work so I set up a foundation to use the money for patient and doctor education about the research. On one occasion there had been a story about the cure in the Sunday papers across the USA. In the following weeks we received 30,000 letters all with donations of a dollar or two to pay for postage and photocopying of information. We had to hire in students to handle it all.

Over the years the journalists who covered the story helped significantly in educating the public to ask for and later demand the new treatments from unwilling doctors. Suzanne Chazin in the Readers Digest, Terry Monmany in the New Yorker, Mark Ragg in the Bulletin and Larry Altman in the New York Times all wrote detailed reviews of the work that became important sources of information. The BBC show Ulcer Wars produced by Michael Mosley is still shown around the world.

The tide of acceptance began to turn in the early 1990s and by 1992 I could go to meetings and receive as much praise as criticism. 1994 was a watershed year for us. In February 1994 the NIH held a consensus meeting in Washington DC which ended after 2 days with the statement to the effect that the key to treatment of duodenal and gastric ulcer was detection and eradication of Helicobacter pylori.

I had been waiting for ten years for this day and I felt a combination of relief and satisfaction that I had achieved what I set out to do. Years before, I had developed the hypothesis, tested it, proved it and now it had reached official acceptance.

The next year proved to be harder. I began to receive awards and recognition. At the hospital, I was still carrying a full load of patient care and research. However, I was increasingly dissatisfied. Much of my time was being spent attending meetings and travelling. I think the pressure of the previous 10 years was beginning to show. Because I had been so involved in the exponential rise of Helicobacter, I had been unable to update my training in new areas of molecular biology which by then were coming to represent a large proportion of the Helicobacter publications.

In typical fashion Adrienne took over the decision-making and at the end of 1994, I took a year of leave from the university. We cashed in my superannuation and decided to live on that for a year to figure out what would be next in our lives. In that year I still travelled and lectured but my primary work was with Tri-Med, getting the breath test through the FDA regulatory process. I am proud of my diagnostics tests, the CLOtest and PYtest. They are often my forgotten children, eclipsed by my work on treatment. Although less glamorous than high impact papers, reliable cheap and available diagnostics are just as important in medicine as treatments. They don't always get the same recognition. After 1994 my business interests became more important. The diagnostics were starting to earn an income. In Australia, close friend

Rod Blechynden took on the role of managing it for me. Rod and Adrienne take care of the business aspects of my work. Their work frees me to focus on my research.

Once I had completed that project, Adrienne decided it was time for us all to go home. I was still unsure but it has turned out to be the best decision for me and the family. We moved back to Perth in 1996. I was awarded the McFarlane Burnet Fellowship which funded my lab at the University of WA for a 5 year period. In 1998, Tri-Med USA bought the manufacturing rights to CLOtest. I was keen to keep the manufacturing in Western Australia. It has been a long term ambition of mine to develop industry here in Perth. I set up a new manufacturing facility here but, sadly, it didn't last. Tri-Med in the US was later sold and the new owners moved all the manufacturing back to the USA. Tri-Med in Perth continues in a small way doing R&D and selling medical products.

Before finishing I want to acknowledge all those scientists who failed to recognize HP. Without them I would have had a very different career. Some of their stories are described in my book "Helicobacter Pioneers". I also want to thank Irvin Modlin for the foreword he wrote for it. He is a great guy and was able to say things about the joy of scientific research that I never could.

One of the truly great things about winning the Nobel Prize in 2005 was that I was living and working back home. I got to share it and celebrate with those who had been involved in the initial work at Royal Perth and Fremantle Hospital.

I continue to live in Perth Western Australia. I have an appointment at The University of Western Australia and still see patients at the gastroenterology department at Sir Charles Gairdner Hospital. My other interests continue. I take an active role in Tri-Med and in 2005 began a new project with vaccine company Ondek.

There were many occasions when luck played a role in my life; meeting with Robin, the first culture of the bacteria and chance meetings with many people who helped me and collaborated with me. I look back and am grateful to the many friends and family who helped me along the way, most importantly, my wife Adrienne, and my children, their partners and my grandchildren.

HELICOBACTER CONNECTIONS

Nobel Lecture, December 8, 2005

by

BARRY J. MARSHALL

NHMRC Heliobacter pyroli Research Laboratory, QEII Medical Centre, Nedlands, WA 6009, Australia.

SUMMARY

After preliminary studies in 1981, Marshall and Warren conducted a study in which the new bacterium, *Helicobacter pylori,* was cultured. In that series, 100% of 13 patients with duodenal ulcer were found to be infected. The hypothesis that peptic ulcer was caused by a bacterial infection was not accepted without a fight. Most experts believed that *Helicobacter* was a harmless commensal infecting people who had ulcers for some other reason. In response, Marshall drank a culture of *Helicobacter* to prove that the bacteria could infect a healthy person and cause gastritis. The truth behind peptic ulcers was revealed; i.e. very young children acquired the *Helicobacter* organism, a chronic infection which caused a lifelong susceptibility to peptic ulcers. Marshall developed new treatments for the infection and diagnostic tests which allowed the hypothesis to be evaluated and proven. After 1994 *Helicobacter* was generally accepted as the cause of most gastroduodenal diseases including peptic ulcer and gastric cancer. As a result of this knowledge, treatment is simply performed and stomach surgery has become a rarity.

ACKNOWLEDGEMENTS

My wife Adrienne for encouraging me in this work and reviewing my manuscript, and Robin Warren for showing me the spiral bacteria and explaining the meaning of gastritis.

INTRODUCTION

The title, "*Helicobacter* Connections" refers to the two components of our discovery. Firstly, we were able to associate a new bacterium with peptic ulcer disease. Secondly, we could see that the new bacteria could explain many phenomena observed by other gastric researchers over the previous 100 years. By connecting this literature with our own observations, we were able to confirm our hypothesis rather quickly. As a result, other researchers were often dismayed at our supreme confidence that these new bacteria were serious pathogens and that antibiotics would provide a cure for peptic ulcer.

To quote historian Daniel Boorstin: "The greatest obstacle to knowledge is not ignorance; it is the illusion of knowledge". The relevance of his quotation is that in 1982 the cause of peptic ulcer was "already known". Ulcers were caused by excessive amounts of acid secondary to personality, stress, smoking, or an inherited tendency. The successful introduction of H2-receptor-antagonists (H2RA) five years earlier seemed to confirm this idea because nearly all ulcers could be healed by lowering stomach acid secretion with these drugs. Thus, when *Helicobacter* was revealed, doctors were not looking for a new cause of peptic ulcer, that territory had already been taken by the illusion of knowledge.

BACKGROUND

Figure 1 shows photographs of peptic ulcers taken at endoscopy. The diagram indicates the usual locations on a map of the stomach. The most common type of peptic ulcer is located in the duodenum, a few centimetres past the pyloric sphincter which controls the outlet from the stomach. In the photograph, a dark area on the duodenal ulcer shows that this has already eroded into a blood vessel and bleeding has occurred. Vomiting of blood is one of the major symptoms of an ulcer, sometimes with a fatal outcome if a large enough artery is eroded. Duodenal ulcers usually cause some stomach pain, typically during the night after the evening meal is digested. This is not always true however and some people can suffer a fatal ulcer complication without any warning that an ulcer is present.

The second photograph is of a gastric or stomach ulcer. This shows the common appearance with a scarred white ulcer base, kept rather clean looking by the digestive juices. However, it can easily be seen how this ulcer, if

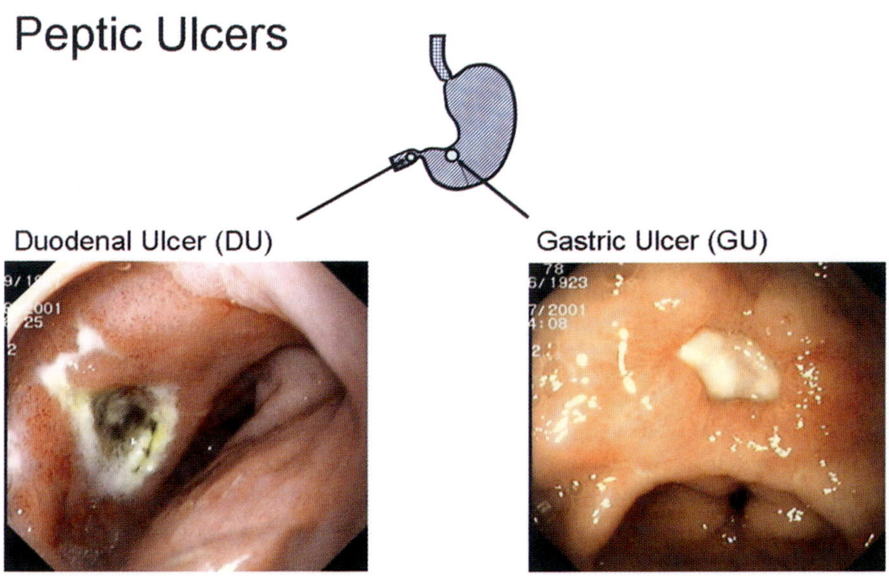

Figure 1. Typical Appearance of Peptic Ulcer.

Cost of Peptic Ulcer
Centers for Disease Control, Atlanta, USA 1993

- One out of ten Americans during lifetime.
  - 1 million hospitalizations
  - 6500 deaths per year.
- Annual health care costs $6 billion:
  - $3 billion in hospitalization costs,
  - $2 billion in physician office visits, and
  - $1 billion in decreased productivity and days lost from work.

Table 1. Cost of Peptic Ulcer USA 1993.

it was deeper, could penetrate all the way through the wall allowing gastric contents to leak into the peritoneal space causing fatal peritonitis.

Although peptic ulceration can occur at any age, it typically develops in adulthood with a peak incidence above the age of forty. Ulcers are more common in men and cigarette smokers, and tend to run in families. Once begun, ulcer disease lasts many years, with an unpredictable tendency towards healing and recurrence. From post mortem studies, peptic ulcer was known to affect 10% of persons at some time in their lives. According to data from the Centers for Disease Control in Atlanta, the cost of peptic ulcer in the USA in 1993 was close to 6 billion dollars per annum (Table 1).

It just seems impossible to imagine these days that ulcer sufferers lived their lives with the possibility of suddenly being struck down with a potentially fatal illness. This explains why it was possible to sell very expensive treatment to people with ulcers once an effective treatment was marketed. This treatment became available with the discovery of the H2RA drugs, the first two of which were cimetidine (Tagamet) and ranitidine (Zantac).

By 1983, the Smith Kline and French company was making a billion dollars per year from Tagamet. Zantac, the second drug in the H2RA class, was destined to sell more than 3 billion dollars per year for most of the 1980's. The only other way that ulcers could be controlled medically was with white chalky antacid and the amount needed to heal an ulcer reliably was about a bucketful taken over four weeks. Cure of ulcer disease required removal of the lower third of the stomach by surgery. However, about 10% of patients treated with surgery became "gastric cripples", unable to enjoy food for the rest of their lives, with chronic gastrointestinal symptoms and difficulty maintaining a normal body weight.

In an article from that era in *Fortune Magazine,* Joel Dreyfuss called Tagamet "the pale green pill that cures ulcers", but this was an overstatement because the ulcers almost always recurred once the drug was stopped. Thus cimetidine was a treatment, not a cure. By 1983 it was clear that for most ulcer patients, lifelong treatment was going to be necessary.

THE PILOT STUDY, 1981

Beginning in about August 1981, I took over the clinical studies for a list of patients with the new bacteria. An initial chart review of Robin Warren's 27 best cases did not reveal any obvious associations between the bacteria and clinical disease. However I did notice an old patient of mine, a 50 year old woman with undiagnosed abdominal pain in whom the bacteria had been the only abnormal finding.

Then, assisted by colleagues Tom Waters, Chris Sanderson and gastrointestinal nurse Dorothy Heys, I collected gastric biopsy specimens from patients attending for endoscopy. Since the histology of ulcer borders was often disturbed and always inflamed, Robin instructed us to sample the stomach wall (mucosa) a few centimeters away from ulcers or local gastric lesions, so that tissue was representative of the antral mucosa in general and it could be assessed for presence of gastritis. In retrospect these specimens were quite different from specimens in most other studies because if gastritis was present it could not have been attributed to a nearby ulcer. One biopsy was taken for histology and one for microbiology. Robin had special stains performed on his histology specimen and John Pearman, our microbiologist, supervised attempts to Gram stain the tissue and culture the organism. Except for the early days of our pilot study, I did not send clinical data with the tissue specimens. Thus, in most cases, histological scoring was performed without any knowledge of the endoscopy findings. Conversely, I did not see the individual results of the laboratory analyses until much later.

PREVIOUS LITERATURE ON SPIRAL BACTERIA

While these activities were in progress, I searched the literature in more depth. Following some initial leads Robin had given me, I rediscovered several reports of spiral bacteria in animals and man. It was apparent that the new spiral organism was not just a strange infection occurring in Western Australia, but was the same as the "spirochaete" which had been described in the literature several times in the previous 100 years. I was particularly interested in old reports from the USA. In 1940, Stone Freedberg from Harvard Medical School had seen spirochaetes in 40% of patients undergoing stomach resection for ulcers or cancer. About 10 years later, the leading US gastroenterologist, Eddie Palmer at Walter Reid Hospital, had performed blind suction biopsies on more than 1000 patients but had been unable to find the bacteria. His report concluded that bacteria did not exist except as post mortem contaminants.

Most of the old references to gastric bacteria had not connected spiral organisms to any significant disease process. The best example of such an observation was an electron micrograph taken by Susumu Ito and included in his chapter for The Handbook of Physiology (1). Reproduced in Figure 2, Ito's illustration shows a detailed view of *Helicobacter pylori*, with flagella present on one end. I later discovered that to obtain this specimen Ito had

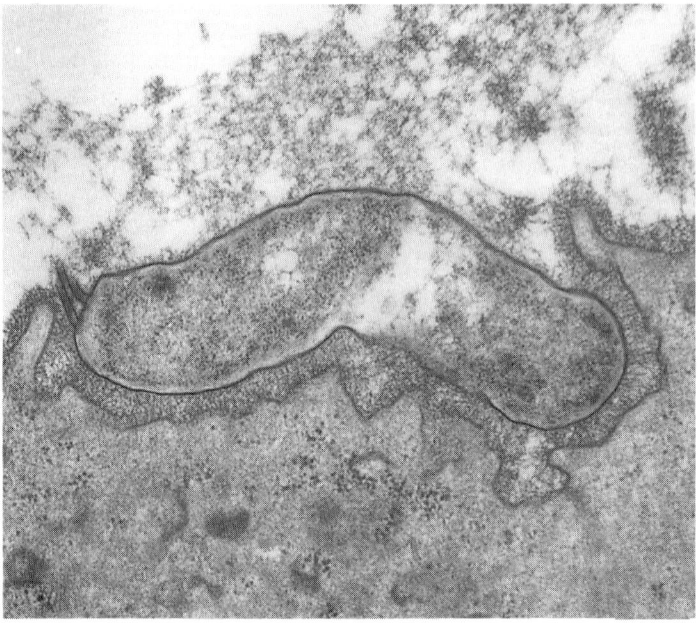

Figure 2. Susumu Ito, an anatomist in Boston, swallowed a suction biopsy instrument and sampled his own stomach to reveal the organism shown above. The smooth bacterial cell wall, sheathed flagella and proximity to the epithelial cells identifies it as *Helicobacter pylori.* Ito believed that spiral bacteria in the stomach were normal for humans, as they appeared to be in cats and dogs. He could not have known that in the 1960's almost all Japanese adults carried *H.pylori.*

performed a blind suction biopsy of his own stomach. He had seen similar organisms in cats where they were almost universally present without any associated pathology. As a result, he assumed that the human organism from his own stomach was also a commensal. Ito could not have known that in his generation almost all Japanese were infected with *Helicobacter.* According to many other studies around that time, gastritis was so common as to be a "normal" appearance in Japanese, the race which also suffered from the world's highest rate of gastric cancer. From Ito's and Freedberg's reference lists, other reports of gastric spirochaetes in animals and man were obtained, to as far back as that of Bizzozero in 1892 (2).

In our initial series taken during the latter half of 1981, we could easily see the bacteria on Gram stained smears of gastric tissue but we were unable to culture them. My gastroenterology rotation was due to finish on December 31st but my colleagues supported the idea of a prospective study in which further attempts could be made to culture the bacteria and look for disease associations.

PROSPECTIVE STUDY, 1982

Towards the end of 1981 I wrote the protocol for a prospective study of 100 consecutive elective endoscopy patients. The documents were submitted to the Royal Perth Hospital Human Ethics Committee at the end of that year so that the study could begin before March 1982.

I chose 100 patients simply because in the days before computer spreadsheets it allowed percentages to be easily calculated when we wrote the paper. The aims of the study were to determine the prevalence of the bacteria in an endoscopy population, to try to culture the organism, to see what diseases were associated with it and to detect an infection source if there was one.

During the first half of 1982 I was actually a medical registrar in the hematology service but I was able to fit in the study activities around my new duties. In addition, I was also conducting a study of heatstroke in marathon runners, so I was very busy.

For the new project, I would stay at the hospital each evening to interview inpatients who were due for endoscopy the next morning starting at 7:30 A.M. In the morning I would arrive early so that I could interview the new outpatients at 8 A.M. as they were being prepared for their endoscopies.

For each patient I was required to explain the study and then have them sign a consent form so that biopsy samples could be taken. I then asked 30 or so questions related to lifestyle, pets, travel, occupation, medications, dental hygiene, gastrointestinal symptoms and medical conditions. Finally, I looked in their mouths to briefly assess the state of their oral hygiene and dentition.

I considered many explanations for the apparent commonness of the new bacterium. How did these bacteria get into the stomach? Could it be that people were taking cimetidine, lowering their acid level and then being infected? Did the bacteria live in the mouth as part of the normal flora? Could poor oral hygiene and periodontal disease be a risk factor? I asked every kind of question I could think of about dentition. I asked patients, "How many teeth do you have and how often do you clean your teeth?" and I heard some pretty extreme answers! By my reckoning, a person who has no teeth and never cleans his teeth has good dental hygiene – but his teeth are gone.

During the ensuing endoscopy, Dorothy Heys, an important ally of mine, would remind the gastroenterologist to take two extra antral biopsies, from a location away from any local lesion. The gastroenterologist would then complete the endoscopy report. During morning tea break and at lunch time I would collect the various biopsy specimens and deliver them to the pathology and microbiology labs for processing.

At the time I wrote the study protocol, I had no preconceived notion as to what diseases might be associated with the bacterium. Therefore the main goals of the study were to understand the histology and microbiology, rather than to discover the cause of peptic ulcer. However, in June, long after the 100 patients had been completed, I obtained all their endoscopy reports and coded these for the main endoscopic diagnoses. Diagnostic categories were simplified to include duodenal ulcer, gastric ulcer, gastritis, duodenitis, bile in the stomach, cancer, oesophageal disease and "other."

THE FIRST CULTURE: APRIL 8–13TH 1982 (EASTER)

At the time we started our studies, *Campylobacters* were very new. Harrison's Textbook of medicine only included *C. fetus* as a human pathogen although *C. jejuni,* a contaminant of fresh chicken carcasses, had been recently described in the English journals as a cause of gastroenteritis. So our culture methods focused around techniques for similar organisms, generically called "*Campylobacter like organisms*" (CLO's). We used the "Lee method" which is a microaerophilic culture necessary for *Campylobacter.* Professor Adrian Lee, a chicken specialist at the University of New South Wales in Sydney, had reported culturing spiral bacteria from the mouth and the colon of laboratory mice (3). In fact, with hindsight, we had chosen exactly the right technique from about one month after we started the work in 1981, but the months went by and we didn't culture the organism. This was particularly frustrating because we could see masses of bacteria on the Gram stained mucus smears which I delivered to the microbiology laboratory within a few minutes of the biopsy being taken.

The first successful culture was from a patient biopsied on the Thursday before Easter 1982. The patient, number 37, was a 70 year old male. He had a history of duodenal ulcer and gastric ulcer. He was anemic, with an artificial heart valve for which he required the anticoagulant, coumadin, in order to keep the metallic parts free of clot. So ulcer disease was a major problem for this patient's management and his life was continuously threatened by his duodenal ulcer disease. He did have a small duodenal ulcer at endoscopy but the research biopsies were still taken as they did not need to be near the ulcer.

It is my recollection that, at Royal Perth Hospital that month, a methicillin resistant *Staphylococcus aureus* had been detected. This "superbug", if it became widespread in Western Australia, would potentially cost the hospital about 10 million dollars per year in expensive antibiotic costs. To prevent this, some patients had been quarantined and surveillance cultures were being performed on all staff that had been anywhere near the affected ward. The microbiology lab was very busy and so there was no time to examine my research cultures on Easter Saturday as would normally have been done. Therefore, the culture plates remained in the incubator, untouched, from Thursday morning until Tuesday morning, five whole days. On the Tuesday after Easter small transparent colonies were present on the plates and these proved to be a rather pure culture of a Gram negative rod. John Pearman waited until he had a second culture before he called me to the lab and, grinning like a Cheshire cat, showed me the new organism. I was pleased, but unconvinced because the cultured bacteria did not have a very convincing spiral shape. However, I was in a good mood so it seemed an appropriate time for John to confess that the laboratory staff had been processing our research biopsies identically to the routine method used for throat-swab cultures. If nothing interesting was seen on the Petri dish at 48 hours, they had been discarding the specimens!

This might have been appropriate for throat cultures, because these carry many contaminating commensal organisms from the mouth causing the plates to be completely covered with irrelevant bacillus and fungal species after 48 hours. However, our research biopsies were actually rather clean. Typically, after the endoscope was passed through the patient's mouth into the stomach, any free stomach acid was sucked out through the biopsy channel. This meant that mouth organisms contaminating the endoscope were washed away and/or killed. The biopsy forceps were introduced down the channel with its cup-like jaws closed, and they were only opened in the stomach as the biopsy was taken. Then, with the tissue sample enclosed within the forceps, it was withdrawn through the endoscope and then opened so that the specimen could be removed with a sterile needle. This meant that gastric biopsy samples were often much cleaner than other "oral" specimens. Gastric tissue samples tended to grow nothing, or the new gastric organism. Even the non-selective blood agar plates produced almost pure cultures of *Helicobacter*, even after as long as the 4th or 5th day.

Prior to John's confession, I had no idea that the cultures were being discarded routinely at 48 hours. I had been wasting my time for six months! Now that the bacteria had been cultured however, a completely new line of research was open to me. What were these bacteria and how did they survive in the stomach? Was there a serological response to them? What antibiotics might I use? How were they transmitted? I still had no idea that they were important for anything more than gastritis because the study was prospective and blinded.

The 100 patient study was completed at the end of May 1982. Perhaps because of my enthusiasm, I had recruited 100 patients rather quickly, with only two declining to take part. In the School of Medicine Statistics division I found Norm Stenhouse who agreed to supervise the data analysis. This involved asking Robin and John Pearman to send their data tables separately to his student Rose Rendell.

I did the same with my demographic data and clinical questionnaire. I completed the process immediately before my family of six departed for Port Hedland, a mining town 1 900 km from Perth, in the North of Western Australia. In a frenzied weekend while my wife Adrienne was packing for the trip, I ducked out and spent all Saturday morning at the gastroenterology department photocopying the 100 endoscopy reports. Several weeks later, now the acting physician at Port Hedland, I scored patients for the presence or absence of the main visible endoscopic lesions and mailed that final datatable to Rose.

DATA ANALYSIS AND RESULTS

Back at Medical Statistics in Perth, Rose entered the data and then performed the analysis using SPPS. Eventually, a box of paper containing descriptive statistics and crosstabs of bacteria vs. everything else was delivered to me in September 1982.

Endoscopic Appearance	*Total*	*With Bacteria*	*p*
Gastric Ulcer	22	18 (77%)	0.0086
Duodenal Ulcer	13	13 (100%)	0.00044
All Ulcers	31	27 (87%)	0.00005
Oesophagus Abnormal	34	14 (41%)	0.996
Gastritis	42	23 (55%)	0.78
Duodenitis	17	9 (53%)	0.77
Bile in Stomach	12	7 (58%)	0.62
Normal	16	8 (50%)	0.84
Total	*100*	*58 (58%)*	

Table 2. Correlation between Endoscopic Findings and Bacteria

Note: Total number of patients in this table exceeds 100 because some patients had more than one diagnosis. Four patients had both gastric and duodenal ulcer, all were positive for bacteria.

On the first assessment I noted that most of the patients with ulcers were positive for the bacteria, as were about half of the patients without ulcers. This was interesting, but I tried not to get excited about what could have been sheer chance. I then went back to the endoscopy reports and double-checked the data. Rather than finding that I had over-read the number of duodenal ulcers, the opposite was true. I found that the one duodenal ulcer patient without bacteria had undergone surgery soon after her endoscopy and the bacteria were present when the far larger surgical specimen had been examined. I added her to the infected group. Later, Robin must have re-checked the samples and agreed that bacteria were present. I then double-checked various other fine details and submitted the revised data with more specific analysis requests back to Rose. While I awaited the final set of tables, I searched the literature for further reports of gastritis and gastric bacteria, especially as they might relate to an association with peptic ulcer.

The results of our study of 100 patients are shown in Table 2. The association of bacteria with endoscopic diagnoses was dramatic. Just over half of all the patients had bacteria, but all patients with duodenal ulcer had bacteria; 13 out of 13. Imagine that you're tossing a coin, how often do you get 13 heads in a row? The chance of 13 consecutive "heads" would be less than 1 in 1000. The actual P value for this association, using a two-tailed test, was 0.00044. So our finding was very, very unlikely to be by chance.

This finding appeared at a time when academic physicians were used to seeing hundreds of patients in clinical trials of peptic ulcer treatments. Typically, those studies were designed to demonstrate the differences between two acid lowering drugs. Large numbers of patients were necessary to differentiate an 85% cure rate from a 90% cure rate. So would gastroenterologists accept a revolutionary discovery, the main cause of peptic ulcer, on the basis of 13 patients from Perth, Western Australia? It was just not going to happen.

A second, extremely interesting, aspect of the data was that 18 out of 22 patients with gastric ulcer had the bacteria. Four patients had both types of ulcer and all were *Helicobacter* positive. So, with only a gastric ulcer, 77 % had the bacteria. But with duodenal ulcer, 100% had the bacteria. This difference was not statistically significant, but was very interesting if it held up. If our hypothesis was correct, why would duodenal ulcer be more tightly connected to the gastric bacteria than gastric ulcer? Why would ulcers occur down in the duodenum, when the type of mucosa there is different, intestinal type in fact, to which the bacteria did not attach? The varying connection between ulcer type and the bacteria seemed an unusual finding at first, but it rang a bell.

I remembered that I had read a paper about gastritis written by Magnus in 1952 (4). He studied accident victims in Minnesota, finding that quite a few had peptic ulcer disease. Interestingly, he noticed that where he found gastric ulcer, gastritis was present in 80% of cases, but if he found <u>duodenal</u> ulcer, gastritis was present in 100%. He could not explain why gastritis would be linked so strongly to the ulcer of the duodenum, rather than the stomach. Magnus discovered almost the exact same percentages that we had found for the link between bacteria and peptic ulcer. It was certainly a paradox and so everybody had ignored Magnus's findings because they did not fit in with what people thought would be the norm. When I presented our data in October 1982 at a meeting in Perth, a local gastroenterologist said to me; "Barry you've got that wrong, people with duodenal ulcers don't have gastritis. The stomach is usually normal." From what I had seen of Warren's biopsies, I could say "How do you know since nobody ever biopsies the stomach of duodenal ulcer patients?" In case I was wrong, I went back and checked my facts. By the end of 1982 I was certain that our data was actually quite consistent with other poorly-understood studies. The other interesting fact I knew from the literature was that when gastric ulcers developed in patients taking non steroidal anti-inflammatory drugs (NSAID's), the gastric mucosal histology was usually quite normal. i.e. gastritis was absent. This seemed to fit with the four patients in our study who had gastric ulcer, but normal histology. The questionnaire recorded that they were taking NSAID's.

This all seemed rather logical to me. In the stomach, anything you eat is directly applied to the mucosa. So, you could have *Helicobacter* causing ulcers associated with gastritis, and this would be the most common variety. Alternatively, even if *Helicobacter* were not present, the stomach wall could be corroded by anything else you might swallow. But whereas NSAID's could sit around in the stomach for many hours, it would be quite difficult for them to actually reach the duodenum in high enough concentrations to cause an ulcer. So you might expect that a purer form of peptic ulcer would exist down in the duodenum, where the influence of ingested drugs was much less. My hypothesis would be strengthened if *Helicobacter* were present in the duodenum. But how could they cause trouble down there, on intestinal type mucosa to which they could not stick?

MICROBIOLOGICAL STUDIES

During the second half of 1982, John Armstrong received some culture specimens and had them negatively stained to examine the morphology of the organism more exactly. He showed that it was 3.5 micrometers long, with 1.5 wavelengths of a spiral form. Usually, five or so flagella could be seen at one end of the bacterium. The flagella were sheathed, which meant they were more related to *Vibrio* and *Spirillum* species (i.e. cholera) than *Campylobacter* species, which have an unsheathed flagellum. An image of a dividing organism was chosen for our letters to *The Lancet* which were published in June 1983 (5).

Besides the morphology, we attempted to characterize the new bacterium according to the presence or absence of various biochemical markers, as shown in tables in Bergey's Manual of Determinative Microbiology. In 1982 there was not much else one could do to characterize newly discovered bacteria. In the days before polymerase chain reaction, techniques for analysis of DNA were rudimentary. The biochemical tests revealed a rather chemically inert organism, unable to produce acid from the metabolism of simple sugars. The new bacterium was catalase and oxidase positive and, at least in the hands of the technician at Royal Perth Hospital, it was urease negative. One can only speculate as to how the urease enzyme of *Helicobacter pylori* could have been missed.

So by the end of 1982 I was starting to get pretty excited about this. I finished my training at Royal Perth Hospital and was offered an endoscopy training post at Fremantle Hospital with Ian Hislop who had a background in gastritis from his days as a fellow at the Mayo Clinic. He said to me; "Barry this is intriguing data. I think you're wrong but it is a curious finding and we need to look into it."

Robin and I took two months to decide on a way to publish our first letters to *The Lancet.* Robin, quite rightly, could claim the initial observation and association with gastritis as his own work. However, I claimed that we now recognized the importance of his observation because of the clinical study and the linkage with peptic ulcer. If we were correct, then the discovery was worthy of the world's most widely read clinical medical journal, rather than a specialist pathology journal interested in gastritis. We called the editor of *The Lancet,* David Sharpe, who suggested we write two separate letters detailing Robin's initial findings in the first, and our joint work in the second.

In my June 1983 *Lancet* letter, I described a new species of bacteria, a cross between a *Vibrio* and a *Campylobacter.* I mentioned some of the microbiological data, but did not reveal any of the linkages with peptic ulcer disease. According to the extensive literature on gastritis, there were two major diseases which could be caused by the mucosal inflammation. These were peptic ulcer and gastric cancer. Although there were masses of papers on gastric cancer, the histology was described rather poorly in most of these and illustrations were never detailed enough to reveal bacteria. From our studies however, it appeared that nearly all gastritis was associated with the new bacteria. All the "other" types of gastritis seemed rather rare. Perhaps the

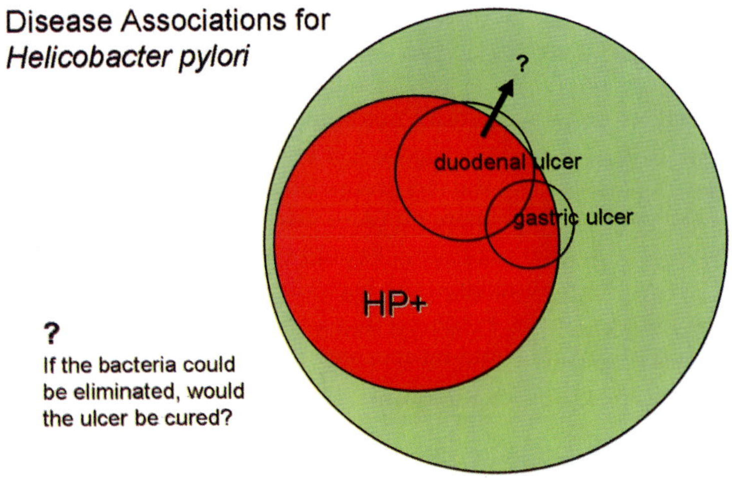

Figure 3. Disease Associations for *Helicobacter pylori.*

many different types of gastritis were just different names for the same thing, described at different stages of its natural history.

Since the new bacteria were associated with gastritis, my reading convinced me that this process was a launching pad for other important diseases. I expressed this hypothesis with the sentence "then they may play a role in other poorly understood gastric diseases such as peptic ulcer and gastric cancer." Somehow David Sharpe allowed it to reach the printers, I suppose he was tired of arguing.

More Hypotheses:

After sending our letters to *The Lancet,* Robin and I continued collaborating as we wrote a full paper to the same journal. There was quite a lot of data on those 100 patients so it was a complicated process to present it all in a concise and logical form. After several months of study we both knew every detail of every case by heart. I did have other activities but by 1983 had decided to focus my career on the new bacterium and see where it led me.

As we planned the full paper, and I studied more patients in my daily practice, several new conclusions dawned on me. From the 1982 data I could create the Venn diagram in Figure 3 which details a population similar to what we saw in our endoscopy patients. The patients in the large green circle did not have *Helicobacter* so they might be regarded as "normal." About half the patients we saw did have *Helicobacter* and they are shown as the large red inner circle. I knew that the patients with duodenal ulcer were almost all within this red group. In fact, there were almost no duodenal ulcer patients in the green group. It seemed almost impossible to get a duodenal ulcer if you didn't have the *Helicobacter.*

In addition, I was certain that some people with duodenal ulcer would have experienced *Helicobacter* eradication just by accident, as part of high-dose antibiotic treatment for other infectious diseases. So, assuming these people existed, I would have expected to see patients with *Helicobacter*-negative duodenal ulcer disease. Yet such persons were exceedingly rare, if they existed at all.

In a mental experiment I could extrapolate backwards from the above observations. Referring again to Figure 3, if I could eradicate the *Helicobacter*, I would move a patient from the red group, where a duodenal ulcer was possible, into the green group where an ulcer was nigh impossible. Therefore, just on the basis of logical reasoning, I could conclude that antibiotic treatment, if it permanently eliminated the bacteria, would also cure ulcers.

As I started work in 1983, I realized that a lot more data would be needed before the new bacterium could be accepted as an important pathogen. I set about to answer the following questions:

Q1. Do patients with bacteria have antibodies?

A. Yes. Laboratory technician Greg Wynn stained some *Helicobacter* smears with sera from my infected and non-infected patients. He could easily demonstrate the presence of IgG with anti-human fluorescent antibodies. This proved that the human immune system considered the bacteria to be pathogens.

Q2. Do antibacterial agents heal gastritis?

A. Yes. Robin and I had suggestive data from the patient we treated with tetracycline in 1982. Then, in 1983, I observed that bismuth, a time honored ulcer treatment mentioned by Kussmaul over 100 years earlier, killed *Helicobacter* in-vitro. In a single-blind prospective study of about 30 patients, I documented suppression of the bacteria and temporary healing of gastritis when patients took DeNol, an ulcer treatment containing bismuth. Regrettably, the bacterial infection usually relapsed, as did the gastritis and the ulcer. This experiment, although only partially successful, did encourage me to try other treatments in the ensuing 12 months. However, for the next 12 months, it did seem to many people that *Helicobacter* was just commensal flora associated with, but not causative for, peptic ulcers.

Q3. Had Koch's postulates been fulfilled for the new bacteria?

A. No. Koch's postulates are the time-honored way in which new bacteria are proven to be pathogens. There are four postulates as follows:

1. The bacteria must be present in every case of the disease.
2. The bacteria must be isolated from the host with the disease and grown in pure culture.
3. The specific disease must be reproduced when a pure culture of the bacteria is inoculated into a healthy susceptible host.
4. The bacteria must be recoverable from the experimentally infected host.

My attempt to fulfill Koch's postulates started in January 1984 with experiments on four piglets. I collaborated with Stewart Goodwin, Chief of Microbiology at Royal Perth Hospital, who had developed an interest in

Helicobacter. He and Robin found many spiral bacteria (mostly *Campylobacter*) on my piglets, but the *Helicobacter* I instilled did not take. Their stomach biopsies remained normal. Eventually the piglets grew so large that I was obliged to terminate the experiment and, like most failed experiments, it was never published. This was a rather frustrating time because, without an animal model, it was difficult to see if the new bacterium could cause disease.

Q4. What is the natural history of the disease process?

A. This was a major puzzle. Try as I might I could not elicit a history of an acute illness from my ulcer patients. Clearly they could not have been born with gastric bacteria. I had many adults with the bacteria but I had no clue as to where and when the bacteria had been acquired.

Q5. Was this disease confined to people with ulcers?

A. No. Interestingly, I saw many patients who had ulcer symptoms, but in whom no ulcer could be found. Many doctors believed that such patients had a psychosomatic illness. However, I soon collected many such "crazy people" in whom symptoms greatly improved during antibiotic treatment. I started to believe that it was not always necessary to have a visible ulcer in order to suffer from ulcer symptoms. Perhaps duodenal inflammation, a pre-ulcer condition, could cause pain. This concept had already been discussed by authorities in the field including Howard Spiro, author of a major gastroenterology textbook. According to Spiro, our bacteria might be the cause of duodenitis, an inflammation of the duodenum which was also called "Moynihan's disease."

Q6. How does Helicobacter survive in the stomach?

A. By hiding under the mucus layer where the pH is neutral. Initially, Robin and I could see that the bacteria were beneath and within the mucus layer, so they might not be exposed to acid in the chronic stage of the infection.

Q7. Do ulcer treatments which heal ulcers affect the bacteria?

A. Maybe. I had continued my search of the literature from 1890 to the current date, but with many new interpretations. Bismuth salts had been used to treat gastric diseases for about 200 years. It was well known that heavy metals were antibacterial to spirochaetes as bismuth, arsenic and mercury had all been used to treat syphilis. In Germany, bismuth had been a component in stomach therapy for 200 years and many antacid mixtures still contained bismuth salts. In Australia and Europe bismuth subcitrate was a proven ulcer treatment available sold under the brand name "De-Nol." This drug healed ulcers just as well as cimetidine, but without decreasing stomach acid. In fact, its mechanism of action was rather mysterious, perhaps related to some kind of coating action which protected the mucosa from acid. Of special interest to me was the fact that ulcer recurrence was less after bismuth treatment than after treatment with cimetidine.

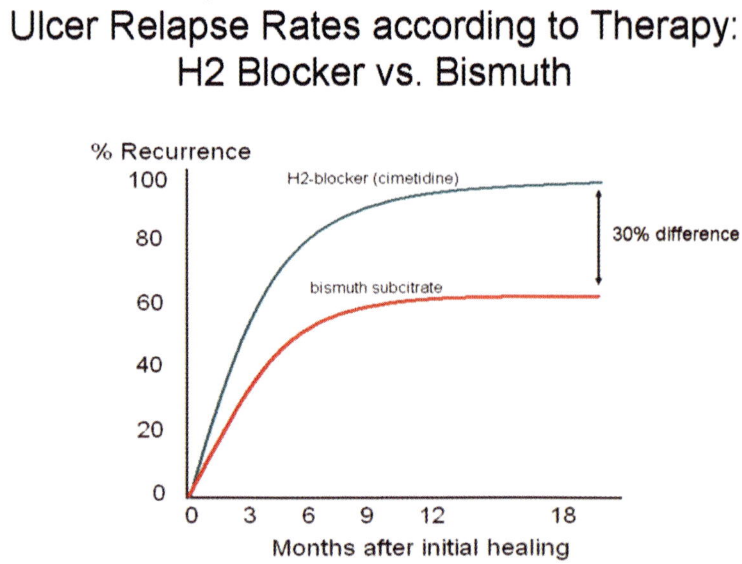

Figure 4. Relapse curves for Ulcers treated with Cimetidine or Bismuth.

A typical example of such a clinical trial is shown in Figure 4. Martin and Hollanders (6) treated duodenal ulcer patients with either cimetidine or bismuth. After the ulcers had healed, patients remained off all treatment until their ulcers recurred. At two years, about 90% of patients had suffered an ulcer recurrence in the cimetidine group, but significantly less patients, only 60%, had ulcer recurrence in the De-Nol group. The authors commented that "drug treatment given for a short period in duodenal-ulcer disease influences the progress of the disease." De-Nol seemed to heal the ulcer far better than cimetidine, but they could not understand why. Some commentators believed that the relapse was merely delayed and the follow-up period merely needed to be extended. However, in my interpretation, a remarkable thing had happened. Since relapse curves in both groups were more or less horizontal after two years, I concluded that no further relapses would be expected in either group. This meant that 30% of patients in the bismuth group had been completely cured. This could be explained if bismuth had eradicated the bacteria. Was ulcer treatment with bismuth acting as an "antibiotic"? At last I had a new hypothesis which I could test very simply.

Q8. Was bismuth an antibacterial?

A. Yes. My laboratory colleagues at Fremantle Hospital helped me carry out the simple experiment described below. 10 mm filter papers were dipped into De-Nol liquid and allowed to dry. *Helicobacter pylori* were then heavily inoculated onto individual blood agar culture plates. The discs were placed in the center of the plates which were cultured from Friday until Tuesday.

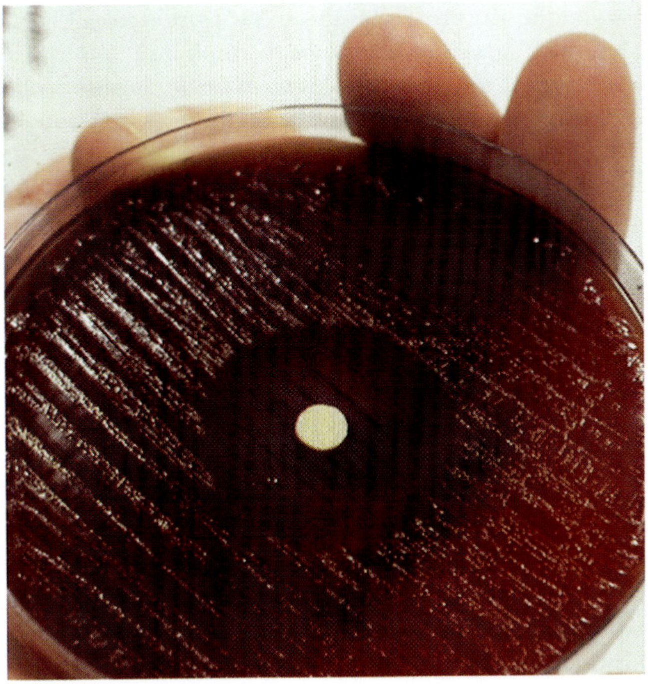

In-vitro.

a. Inhibition of *Helicobacter pylori* growth by disc containing bismuth citrate (De-Nol).

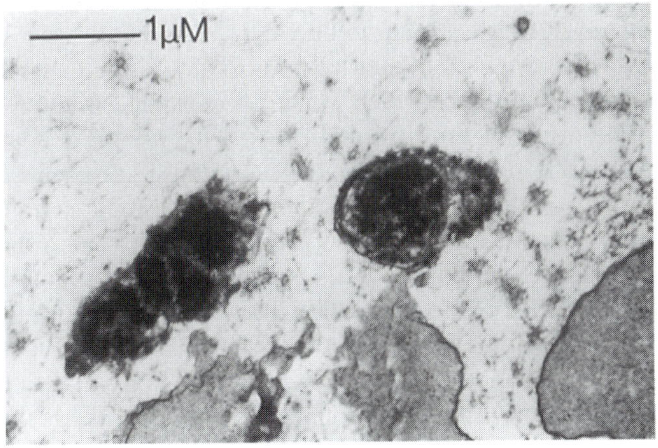

In-vivo.

b. Electron micrograph of *Helicobacter pylori* in the gastric mucosa 30 minutes after treatment with bismuth.

Figure 5. Bismuth effect on *Helicobacter pylori in-vitro* and *in-vivo.*

When examined on the fourth day, a clear zone of bacterial inhibition was present around each of the discs (as shown in Figure 5). I had discovered that *Helicobacter* was exquisitely sensitive to bismuth.

It was probably the most exciting day of my life when I saw that bismuth had killed the *Helicobacter*. It all fitted too perfectly to be a coincidence. Everyone had forgotten that, in the days before penicillin, bismuth was an important antibacterial therapeutic agent. I think that was the first time it crossed my mind that we might win the Nobel Prize. My keen intern, Vinod Ganju, of Indian descent, then agreed to take some bismuth and undergo a gastroscopy. As shown in Figure 5, the numerous *Helicobacter* organisms could be seen practically exploding with dense bismuth precipitates all around them.

The experiments described above enabled me to solve a major clinical dilemma. To test my theory, I wanted to perform a double-blind clinical study in which antibiotics were compared with standard acid-lowering ulcer treatment such as cimetidine.

But antibiotics were so experimental that it would be unethical to ask people with potentially fatal ulcer disease to try out something which might not work. Ulcer patients always had to have an active therapy to heal their ulcers. However, bismuth was already proven as an ulcer healing agent, and it did not lower stomach acid. Therefore, in order to suit our placebo-controlled study design, I could boost it with antibiotics and have a control group with placebo in various ways.

After telephone advice from Walter (Pete) Peterson in Dallas, and various primitive power calculations, I designed a four arm study using H2 blocker alone vs. H2 blocker with antibiotic vs. bismuth alone vs. bismuth with antibiotic. My grant application to the National Health and Medical Research Council (NHMRC) reached the interview stage but it was clear that the panel members were rather skeptical. They opened the interview by stating that they refused to separately fund my plan to develop a serological test for ulcers. It just seemed too far fetched. Somewhat annoyed, I replied that I had already developed a serological test which worked very well. Grudgingly perhaps, the NHMRC did agree to fund my clinical trial for one year. The title was "The effect of antibiotics on duodenal ulcer relapse." It seemed very unlikely to them that I could recruit 100 patients from a single center therefore I was instructed to provide a satisfactory progress report 12 months hence, in order to obtain the remaining two years of funding. The budget of about 50,000 Australian dollars included my salary, but no computer upon which I could save my data. During 1984, before the NHMRC funding came through, I received some extra support from Smith Kline and French, Pfizer and Abbott, each for about 12,000 Australian dollars.

DRINKING HELICOBACTER: THE ATTEMPT TO FULFILL KOCH'S POSTULATES

Disbelievers

In the months after my failed pig experiments, things went badly. I could see that bismuth was healing gastritis, albeit temporarily, but I had no convincing data to prove the bacteria were indeed pathogens. I could also see that several years might go by before we could discover a cure for the infection. The extreme skepticism of my colleagues led me to believe that I might never be funded to perform the crucial trial of antibiotics.

I found the response to my presentations very illogical and rather irritating. One day, after I presented my histology data showing the healing of gastritis with bismuth, the senior hospital pathologist stated "Dr Marshall these changes seem very subtle." Actually the changes were quite dramatic, and this was the first time anyone in the world had been able to heal gastritis! I bit my tongue to stop myself from saying "are you crazy?" Others suggested again that these commensal bacteria merely infected people who already had ulcers. But quite clearly I had presented data from patients with gastritis who did not have ulcers.

I realized then that the medical understanding of ulcer disease was akin to a religion. No amount of logical reasoning could budge what people knew in their hearts to be true. Ulcers were caused by stress, bad diet, smoking, alcohol and susceptible genes. A bacterial cause was preposterous.

At about that time also, I began to realize that there was some urgency to this work. I had admitted a young man with diffuse gastric bleeding. Basically, there was oozing of blood from his whole stomach and he was receiving daily blood transfusions. He had not taken any aspirin, had not consumed alcohol, and had no coagulation disorder. I wondered if *Helicobacter* could be involved. At endoscopy I encouraged Ian Hislop to obtain a few gastric biopsies to search for the bacteria. The specimens were full of pus cells but, unfortunately I could not find any *Helicobacter.* A colleague commented "Well Barry, you can't expect *Helicobacter* to cause everything." The patient continued to bleed, the transfusions continued, and therapy continued with higher doses of acid blockers, antacids and anticholinergics. Maybe they even tried iced water gastric lavage. A few days later I tested his serum with my prototype antibody test. He gave a strong reaction, positive at 1:5012 dilution. The patient had been transferred to the surgical ward and was still bleeding. The political situation at Fremantle was becoming rather delicate. Clearly I was obsessed with the gastric bacteria. I detected a certain coolness amongst my more senior colleagues. The patient was being managed by the surgeons now, perhaps they would like to try a course of amoxicillin; it seemed such a simple thing to do. I discussed my findings with the registrars in charge of the patient. Two days later, as the bleeding continued, the patient underwent total gastrectomy. I was too upset to go back to see him and to this day have never followed up his case further.

HELICOBACTER IS PRESENT IN HEALTHY PEOPLE

By mid 1984, by using a somewhat crude serological test (by today's standards) I had discovered that 43% of "healthy blood-donors" in the port of Fremantle had *Helicobacter.* These people were apparently quite well which meant that it was not necessarily fatal to have *Helicobacter.* By then also, I had treated several patients with bismuth and metronidazole, with a 100% cure rate for my first 4 cases. It seemed that I had a cure. Patients were having a fantastic clinical response. Maybe it was safe for me to try swallowing the bacteria to see what really happened.

ETHICS?

Although I am not one to stew over such things, the implications of my experiment did pass through my mind, albeit fleetingly. The only person in the world at that time who could make an informed consent about the risk of swallowing the *Helicobacter* was me. Therefore I had to be my own guinea pig. If I submitted an ethics committee application and had it rejected I probably would have performed the experiment anyway and then I would not have been able to publish it. Perhaps I would have been sacked and my medical career would have been over. So I decided to do it anyway using the "don't ask don't tell" strategy. I had some unofficial support from my senior colleagues because the experiment required endoscopies by Ian Hislop and assistance from the chief of microbiology.

I remember proposing a hypothetical experiment with my microbiology boss, David McGechie, who laughingly declined to "take the bug." While thinking about the project I asked my gastroenterology chief, Ian Hislop, to carry out an endoscopy on me in order to obtain some healthy control tissue. I did not explain that this was to be the baseline sample but I suspect he knew.

That same day, I biopsied yet another patient who was positive for *Helicobacter pylori* but who did not have an ulcer. In-vitro experiments revealed that his organism was sensitive to all our antibiotics, so I treated him for 14 days with my new therapy and arranged a follow-up endoscopy two weeks after that. His biopsy was now negative. It was now or never.

DRINKING HELICOBACTER

On the morning of the experiment, I omitted my breakfast but took 400 mg of cimetidine, believing that the infection might be easier if my stomach acid level was lowered. Two hours later, Neil Noakes scraped a heavily inoculated 4 day culture plate of *Helicobacter* and dispersed the bacteria in alkaline peptone water (a kind of meat broth used to keep bacteria alive). I fasted until 10 am when Neil handed me a 200 ml beaker about one quarter full of the cloudy brown liquid. I drank it down in one gulp then fasted for the rest of the day. A few stomach gurgles occurred, was it the bacteria or was I just hungry?

For the next three days I had no symptoms and continued to work normally. On the third day I felt over-full after a modest evening meal of Chinese noodles; making me sip water during the evening to help my digestion. On about the fifth to the eighth day I awoke very nauseated, just as the dawn was breaking, and ran into the bathroom to vomit in the toilet. On each day I vomited mostly clear slimy liquid, without any acid present. I recall thinking that it was rather strange, but in my sleepy state never thought to catch any of the material and merely flushed it away. During those days I spent many hours performing serological testing on hundreds of serum samples so I had been working extra hours each day and also over the weekend. I felt very tired and lethargic but assumed it was merely the many hours of sitting at the bench. I was also sleeping poorly, feeling clammy at night. My wife had told me I had "a putrid breath." Unbeknown to me, my colleagues at Fremantle Hospital had also noticed my halitosis that week, but were more polite. If I felt depressed it may have been the illness, or just loneliness as my laboratory staff found other duties far away from my bench!

After 10 days I asked Ian Hislop to endoscope me again which he did at the end of a rather busy endoscopy afternoon. I was very uncomfortable during the endoscopy and rather weak afterwards as I had been fasting all day. I recall that it was someone's birthday and I was able to finish off the chocolate cake after my test. The endoscopy had been a preliminary study, done just to confirm or refute the presence of a bacterial infection. To my joy, spiral bacteria were present on the Gram stain of the first biopsy. The next day Ross Glancy showed me a pathology specimen teeming with *Helicobacter* and pus cells. The experiment had succeeded – *Helicobacter* was a proven pathogen.

MY WIFE'S REACTION

I had not discussed the experiment with Adrienne until the evening of the day I swallowed the culture and she had been observing my deteriorating condition without saying too much. While I was sure she would not have approved of the experiment ahead of time, once I began she accepted it was an important milestone. Like me she felt the need to fast track the research. However, she had been in a car accident two weeks before and had two cracked ribs and a moderate whiplash. Now she was caring for 4 children and a husband who was becoming worse by the day. I don't recall wondering if I might transmit the bacteria to her or the children, rather selfish of me I suppose. I think the reality was that we were both quite young, barely past 30 years old and like most young people had a strong belief in our own invincibility.

Although I had discussed the proposed self experiment in general terms a few months earlier, and Adrienne had not been radically opposed to it, there was probably another reason for not telling my wife that the time had come. I could see that the outcome might make a very large difference to our lives. I had submitted a grant application for funding in 1985, but it was quite likely that my application would fail. In addition, if nothing happened in my experiment, if the bacteria did not take, if gastritis did not develop, then my whole hypothesis could be wrong. At the very least, the disease was far more complicated than I had supposed and it would be extremely hard to convince the skeptics that we had found something important. If that occurred, my future jobs might be in clinical medicine and I would be off interviewing for placement in a private practice, perhaps in a remote area where my eccentric ulcer theories were less well known.

On the other hand, a successful infection with *Helicobacter* would point towards a career in clinical research, more exciting but likely to be financially insecure. I chose not to raise the issue until the family settled down a little. (Nevertheless, a few months later I did endoscope and biopsy Adrienne just in case she had picked up the *Helicobacter* infection.)

SPONTANEOUS CURE

The two biopsies taken on day ten were not enough to really define the pathology, so I scheduled another endoscopy four days later. By then my vomiting had stopped so I assumed I had entered the asymptomatic phase. At the next endoscopy the stomach seemed normal and, surprisingly, we could not find *Helicobacter* in any of the eight samples which were taken. Cultures, histology and electron micrographs were all bacteria-free, with the appearance of healing gastritis being the only abnormality. I had apparently eradicated the *Helicobacter* myself, without antibiotics. Serum samples taken at the time and a few months later were negative for *Helicobacter* antibody.

Whatever happened to cause the *Helicobacter* to disappear continues to be a mystery to this day. Many reports say that I took antibiotics to eradicate the infection on the instructions of my wife. This was not really the case. I decided to terminate the experiment and treat myself with the antibiotic, tinidazole, but I did not take the tablets until after my endoscopy on the 14th day. In retrospect, tinidazole as a single agent would not have worked in any case. In a subsequent trial, 23 out of 24 patients merely developed antibiotic resistant bacteria which were then rather difficult to eradicate.

SYNTHESIS

As I wrote the paper in the subsequent months, I reflected on the achlorhydric vomiting and the halitosis. I recalled some passages from William Osler's 1910 textbook of medicine which describe a similar illness in children. I then re-read papers I had discovered which described epidemics of gastritis in laboratory volunteers. There too achlorhydria had been observed. Suddenly the whole process became clear. The reason why ulcer patients could not recall an acute infection with *Helicobacter* was because it mostly occurred when they were tiny children, aged 2–3 years. This transient vomiting illness then settled into a lifelong asymptomatic phase, sometimes punctuated by clinical ulcer disease in adulthood. Because the bacteria were not affected by any of the usual ulcer therapies, ulcer disease became a lifelong problem, with a relapsing type of pattern. You could be infected by family members, even your mother, so it seemed to be hereditary. It was spread by the faecal-oral route so individuals in lower socio-economic groups were more likely to catch it. Varying epidemiologic patterns could explain many of the differences in ulcer incidence around the world. I could see there was plenty of interesting research ahead. As Robin and I had just had our main paper published in *The Lancet* (7), the editor of *the Medical Journal of Australia* wrote to me requesting a paper about the bacteria so I sent him two papers. After detailed review by the late Professor Doug Piper of UNSW in Sydney, and substantial revision, the papers were accepted and published in April 1985 (8)(9). There is no doubt that the editor was sticking his neck out very far to publish the self experiment with its enclosed hypothesis. However, it was very timely because yet another epidemic of achlorhydric gastritis had been published in

Figure 6. Ulcer Tales. A Comic describing the Self Ingestion of *Helicobacter.*
Notes: This page shows Neil Noakes saying "Dr Marshall you're crazy" and I am saying "There is no other way."

the preceding month in *the British Medical Journal,* again without the authors being able to detect a pathogen (10). *The Lancet* editors, seeking to claim the *Helicobacter* highground for ever more, then editorialized my paper in their journal, giving it far more notoriety than it might otherwise have had. I kept my head down for a few months, held my breath and waited for the sparks to fly. But nothing happened. The gastroenterology community might have been too stunned. Pete Peterson, in a telephone conversation to me during 1984, said "Wow. You're a real cowboy Barry." Coming from a Texan I thought that was a supreme complement.

Many years later the experiment was immortalized in a comic created by the Abbott company, makers of the most important antibiotic for *Helicobacter*, clarithromycin. By combining that drug with amoxicillin and an acid blocker called omeprazole, it became rather easy to treat *Helicobacter* with an 85% cure rate. A panel of the Abbott comic is shown in Figure 6. In it, my colleague Neil Noakes says "you're crazy" whereas, while drinking the brew I say "There's no other way!" In 1984, both statements were true enough.

TREATMENT FOR HELICOBACTER

Towards the end of 1984 things began to accelerate at Fremantle Hospital. By pre-screening potential endoscopy patients with a serological test, I dramatically increased the concentration of *Helicobacter*-positive patients in my clinic. By combining bismuth with metronidazole, I was able to eradicate *Helicobacter* in most patients, in just two weeks. When that failed I could use bismuth with amoxicillin to cure half the remainder. For people facing stomach surgery for ulcers, all these antibiotics seemed rather trivial.

The local GP's noticed the remarkable results and kept sending more patients. Very soon it seemed that this was a miracle cure for people with stomach problems. I had a treatment which cured most of them in two weeks. Not only that, patients who had carried a diagnosis of "functional" or psychosomatic gastric symptoms were so pleased to have a real diagnosis – *Helicobacter* gastritis.

In July 1984, I was called by Dr. Larry Altman, medical writer for *the New York Times*. I did not know at the time that he was almost finished writing a book on self experimentation. After the interview he published a major article about *Helicobacter* which he later said was rather difficult to get past his editor as the medical community in USA was extremely skeptical at the time. However, his article triggered a further interview between a journalist from a tabloid newspaper, *The Star* and Dr. Warren. The outcome was that people from all walks of life, in many countries, suddenly became aware of gastric bacteria and the possibility that ulcers could be cured with antibiotics. For years after that, I spent many hours per week sending out treatment advice by mail.

DIAGNOSTIC TESTS

In 1984 Dutch investigators reported that *Helicobacter* produced massive quantities of urease enzyme. I realized that this allowed it to produce ammonia and thereby survive in the acidic stomach. In search of a rapid diagnostic method for gastroenterologists, I discovered that the urease was also detectable in gastric biopsies. With a little trial and error, I built a rapid urease test which enabled me to make the diagnosis of *Helicobacter* in a few minutes, without laborious laboratory testing. This test, which I called the CLOtest (for "**C**ampylobacter **L**ike **O**rganism test"), became the world's first commercial diagnostic test for *Helicobacter pylori.* It is still marketed by the Kimberly Clark Corporation.

My other studies with Simon Langton in the Biochemistry department at Fremantle Hospital revealed that urea was absent from the gastric juice of patients with *Helicobacter.* The bacteria had presumably destroyed all the urea which would have been converted to ammonia and CO_2. CO_2 would be expired in the breath. This understanding eventually matured into an idea for a breath test in which a carbon radioisotope tracer (C14) was used to detect *Helicobacter.* In the simplified final version of this test, a patient swallows a capsule containing labeled urea, and merely blows up a 2 liter balloon ten minutes later. The presence of C14-labeled CO_2 in the breath indicates that *Helicobacter* is present. The test has an important role in confirming that *Helicobacter* has been eradicated after treatment. Once it became available, non invasive testing allowed many investigators to develop very effective treatments for *Helicobacter.* The urea breath test is still a major diagnostic test, commonly used in many countries. Based on either the C13 stable isotope method described initially by Graham and co-workers from Baylor Hospital in Houston (11), or on the C14 method described by myself and Ivor Surveyor at Royal Perth Hospital (12).

PROOF: A PLACEBO CONTROLLED DOUBLE-BLIND TRIAL

In 1985 I moved back to Royal Perth Hospital because I believed that it would be easier there to recruit the 100 patients necessary for a clinical trial. I was under pressure as I had only been funded for one year. The study was designed to see if the new treatment (antibiotics) was the same as the standard of care (cimetidine) and cimetidine plus placebo was therefore the control group. The most active group was bismuth plus tinidazole, a combination which we knew eradicated nearly all *Helicobacter* infections in two weeks. I knew that most people continued to believe that ulcers were psychosomatic so it was important to ensure that patients were completely unaware of which treatment they received. But clearly, smart patients could figure out that they were taking bismuth, because it caused black faeces. Therefore, it was necessary to add a group who took only bismuth with placebo tinidazole. I knew that the *Helicobacter* cure rate with this therapy was not more than 30% so in the final analysis most of the patients so treated would be expected to have an ulcer recurrence. Finally, to be fair to the makers of cimetidine, it seemed appropriate to include a cimetidine plus tinidazole group. If it worked, this would be a convincing test of the hypothesis, since it avoided many of the blinding issues present with bismuth treatment. The four treatment groups were thus:

1. cimetidine plus placebo,
2. cimetidine plus tinidazole,
3. bismuth plus placebo,
4. bismuth plus tinidazole.

The study design meant that all groups contained a well known ulcer healing treatment, either cimetidine or bismuth for 8 weeks. In addition, for each of these therapies, an antibiotic or placebo would be given during the first 14 days. The treatment groups therefore contained increasingly strong anti-*Helicobacter* action.

After 8 weeks of treatment, patients would stop all medication for two weeks and then undergo endoscopy to see if the ulcer was healed. After that, patients with healed ulcers would be re-examined at three, six and 12 months to see if the ulcer had recurred. If the ulcer was seen to have returned, then the patient was removed from the study and was called a "relapse." In addition, patients were removed from the study if they became unwell and needed to restart ulcer treatment. In this case they were asked to have a final unscheduled endoscopy to document the presence or absence of a visible ulcer, but were removed from the study in any case, regardless of the findings.

Whenever an endoscopy was performed, biopsy tests were taken to see if *Helicobacter* was still present, or had been eradicated. The results of these biopsies were only made available if patients had already been removed from the study according to the other rules. In addition, in order to maintain stringent blinding, I was never allowed to speak with patients in the study in case they mentioned side effects and thereby indicated what medication they were taking.

As part of the study I included a psychometric test called the Jung Scale, which approximately measures things such as sleep patterns, optimism, well-being etc. I wanted to see if the so-called ulcer personality had anything to do with the ulcer disease. Since I could not talk to patients, I hired my brother and sister-in-law, Andrew and Kim, to interview patients before they came to each endoscopy.

STUDY PROGRESS

With the aid of the local television news, hundreds of patients applied to take part in the study. I could then include only the most severe cases with already proven ulcers. Patients who were cigarette smokers were known to have rapid ulcer recurrence so I included these as preference. This meant that the results of my study would be known rather soon. If patients were not cured they would only last a few weeks without treatment.

Within four months I had screened about three hundred patients and selected 100 of these who had both a duodenal ulcer and the bacteria. Seven patients were found to have the ulcer but no bacteria. One of these turned out to be a lymphoma so our special testing for *Helicobacter* resulted in an early diagnosis and a lucky cure for the patient (with chemotherapy and radiotherapy).

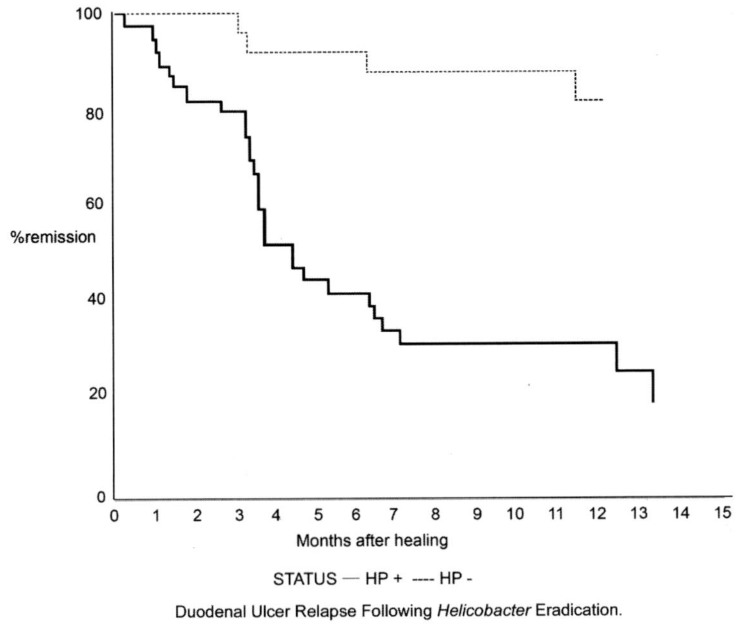

Figure 7. Duodenal Ulcer Relapse Following *Helicobacter* Eradication.

By the time of my NHMRC grant review interview in 1985, I had recruited the necessary 100 patients and quite of few of the early recruits had already experienced healing and then recurrence of their ulcer. By deduction I could tell that we already had a significant result. I knew this because all the patients who had relapsed were still infected with *Helicobacter*. Since the active treatment would have eradicated the bacteria in at least half the patients, I could assume that patients without bacteria were not relapsing. I knew that I had chosen the worst ulcer patients I could find for the study, so for them to last more than two months without symptoms was unusual.

In the final result, the study showed that ulcer healing was more common and recurrence less common if *Helicobacter* was eradicated. Since the aim of our study was to cure patients, success only occurred if a patient healed his ulcer and remained healed, with no further treatment, for twelve months. Using these criteria, a dramatic difference was seen between patients who still had bacteria and those in whom the bacteria had been eradicated, as shown in Figure 7. Simply stated, in order to cure 10 patients with the old treatment, one would have to treat 104 patients. But to cure 10 patients with the new treatment one would only need to treat 14 patients. Variations of our study have since been published hundreds of times with the same or better results.

Two other results are worthy of note. Firstly, in our study, cigarette smoking made no difference to the result, provided that the *Helicobacter* had been eradicated. Of all the ulcer risk factors, smoking was known to be the most adverse, causing almost all doctors to insist that ulcer patients discard their cigarettes. Other studies have confirmed that so-called risk factors for peptic ulcer are all rather inconsequential compared to the risk of persistent *Helicobacter*. When ulcers occur without *Helicobacter*, NSAID's type drugs are usually implicated.

Secondly, the Jung Scales showed that the patients' mental status improved when I eradicated their *Helicobacter*. When patients were in remission from their ulcer, *Helicobacter* eradication correlated with significantly lower Jung scores. Thus it seemed likely that the "ulcer personality" merely reflected a diminished state of health related to chronic infection of the stomach. Perhaps low grade gastrointestinal symptoms persisted even when the ulcer was not present. The psychosomatic aspects of peptic ulcer have since been largely ignored but have been studied as part of treatment for non-ulcer dyspepsia. In general, they have been found to be irrelevant. The idea that high acid levels are caused by "stress" was based on erroneous models of peptic ulcer in rats and in monkeys where the most important factor, *Helicobacter*, remained uncontrolled. It has subsequently been shown that *Helicobacter* gastritis, by interfering with local endocrine negative feedback in the stomach, can lead to excessive acid secretion.

ACCEPTANCE

Even more convincing studies of antibiotic use were published by Rauws and Tytgat (Amsterdam) in 1990 (13), Graham (Houston, Texas) in 1991 (14) and Hentschel (Vienna) in 1993 (15). For the extreme skeptics, Hentschel's was the most convincing study because he did not use bismuth. Therefore the blindedness of the study could not be questioned. In 1994 the National Institute of Health convened a consensus conference in Washington which was attended by an experienced international faculty of gastroenterologists, physicians, and infectious disease experts. It was concluded that in all cases of peptic ulcer, the essential first step was the identification and eradication of *Helicobacter pylori*.

POSTSCRIPT

The history of the discovery of *Helicobacter pylori* has been described in more detail in the book *Helicobacter* Pioneers, by Blackwell (16). The genome of *Helicobacter* was sequenced by Tomb *et al.* in 1997 (17). Presently, more than half the people in the world are still infected by *Helicobacter* and it is believed that about 800,000 persons die from related stomach cancer each year. Much clinical and basic research continues, primarily focused on the aetiology and prevention of gastric cancer. In developed countries, gastric surgery is now a rarity and ulcer disease is cured by general practitioners at its first presentation. In Australia, the cure of peptic ulcer has reduced the number of upper endoscopies by about 75%, freeing up clinical resources for colonoscopic diagnosis and treatment. This should have a flow on effect causing a reduction in colon cancer as well.

REFERENCES

1. Ito, S. Anatomic structure of the gastric mucosa. In Code CF (Ed). *Alimentary Canal.* Washington: American Physiological Society, 1967; 705–41.
2. Bizzozero, G. Sulle ghiandole tubulari del tubo gastro-enterico e sui rapporti del loro epitelio coll'epitelio di rivestimento della mucosa. *Atti della Reale Accademia delle Scienze di Torino* 1892; **28:** 233–51.
3. Lee, A.; Gordon, J.; Dubos, R. Enumeration of the oxygen sensitive bacteria usually present in the intestine of healthy mice. *Nature* 1968; **220**: 1137–1139.
4. Magnus, H. A. Gastritis. In: Jones F. A. (ed.) Modern Trends in Gastroenterology. London: Butterworth, 1952; 323–51.
5. Warren, J. R.; Marshall, B. Unidentified curved bacilli on gastric epithelium in active chronic gastritis. *Lancet* 1983; 1(8336): 1273–5.
6. Martin, D. F.; Hollanders, D.; May, S. J.; Ravenscroft, M. M.; Tweedle, D. E.; Miller, J. P. Difference in relapse rates of duodenal ulcer after healing with cimetidine or tripotassium dicitrato bismuthate. *Lancet* 1981 Jan 3;1(8210): 7–10.
7. Marshall, B. J.; Warren, J. R. Unidentified curved bacilli in the stomach of patients with gastritis and peptic ulceration. *Lancet* 1984; 1(8390): 1311–1315.
8. Marshall, B. J.; McGechie, D. B.; Rogers, P. A.; Glancy, R. J. Pyloric campylobacter infection and gastroduodenal disease. *Med J Aust* 1985; 142: 439–44.
9. Marshall, B. J.; Armstrong, J. A.; McGechie, D. B.; Glancy, R. J. Attempt to fulfill Koch's postulates for pyloric Campylobacter. *Med J Aust* 1985; 142: 436–439.
10. Gledhill, T.; Leicester, R. J.; Addis, B.; Lightfoot, N.; Barnard, J.; Viney, N.; Darkin, D.; Hunt, R. H. Epidemic hypochlorhydria. *Br Med J.* 1985;290(6479): 1383–6.
11. Graham, D. Y.; Klein, P. D.; Evans, D. J. Jr.; Evans, D. G.; Alpert, L. C.; Opekun, A. R.; Boutton, T. W. *Campylobacter pylori* detected noninvasively by the ^{13}C-urea breath test. Lancet 1987; 1: 1174–1177.
12. Marshall, B. J.; Surveyor, I. Carbon-14 breath test for the diagnosis of *Campylobacter pylori* associated gastritis. *J Nuc Med* 1988; 29(1): 11–16.
13 Rauws, E. A. J.; Tytgat, G. N. J. Cure of duodenal ulcer associated with eradication of *Helicobacter pylori. Lancet* 1990; 335: 1233–35.
14 Graham, D. Y.; Lew, G. M.; Evans, D. G.; Evans ,D. J. Jr.; Klein, P. D. Effect of triple therapy (antibiotics plus bismuth) on duodenal ulcer healing. A randomized controlled trial. *Ann Intern Med.* 1991;115: 266–9.
15 Hentschel, E.; Brandstatter, G.; Dragosics, B.; Hirschl, A. M.; Nemec, H.; Schetze K.; Taufer, M.; Wurzer, H. Effect of ranitidine and amoxicillin plus metronidazole on the eradication of *Helicobacter pylori* and the recurrence of duodenal ulcer. *N Engl J Med* 1993; 328: 308–12.
16. Marshall, B. J. (ed.) Helicobacter Pioneers: Firsthand Accounts from the Scientists Who Discovered Helicobacters, 1893–1983. Melbourne: Blackwell Science Asia, 2002; 15–24.
17. Tomb, J. F.; White, O.; Kerlavage, A. R.; Clayton, R. A.; Sutton, G. G.; Fleischmann, R. D.; Ketchum, K. A.; Klenk, H. P.; Gill, S.; Dougherty, B. A.; Nelson, K.; Quackenbush, J.; Zhou, L.; Kirkness, E. F.; Peterson, S.; Loftus, B.; Richardson, D.; Dodson, R.; Khalak, H. G.; Glodek, A.; McKenney, K.; Fitzegerald, L. M.; Lee, N.; Adams, M. D.; Hickey, E. K.; Berg, D. E.; Gocayne, J. D.; Utterback, T. R.; Peterson, J. D.; Kelley, J. M.; Cotton, M. D.; Weidman, J. M.; Fujii, C.; Bowman, C.; Watthey, L.; Wallin, E.; Hayes, W. S.; Borodovsky, M.; Karp, P. D.; Smith, H. O.; Fraser, C. M.; Venter, J. C. The complete genome sequence of the gastric pathogen *Helicobacter pylori. Nature* 1997; 388, 539–47.

Portrait photo of Barry J. Marshall by photographer C. Northcott.

Robin Warren

J. ROBIN WARREN

ADELAIDE

I was born on the 11th of June 1937, in North Adelaide, South Australia, the first child of middle-class parents. I am a fifth generation South Australian. South Australia is a charming state, a little overfull of the fact that, along with Victoria, it was settled by free settlers from England, who took part in a land investment program. There were no convicts. Elsewhere in Australia, the major cities began as prisons for Britain's unwanted, when America became closed to them after the War of Independence. The original settlers arrived in 1836–7 and the beautiful city of Adelaide was chosen as the main settlement. It was imaginatively designed, with the main city square mile crossed by broad streets and surrounded by parkland. Adelaide rapidly expanded during the Victorian era, and by the end of the 19th century it was a bustling city of about 100,000 people.

The Warrens migrated from Aberdeen in 1840. Their eldest son, my great grandfather John Campbell Warren, was a member of the local government, Captain of the Light Cavalry, and patriarch of a family of 16 children. He owned a large estate in the Adelaide Hills. At the turn of the century, he sent his sons to outback South Australia (Anna Creek) and Western Australia (Katanning) to open up huge areas for cattle and wheat. My father, Roger Warren, studied viticulture and became one of Australia's leading winemakers.

My mother's ancestors migrated from England to Adelaide with the first settlers in 1836–7. My grandfather, Sydney Verco, belonged to a dynasty of doctors. The Verco family still make up many of Adelaide's doctors. He died young, leaving my grandmother, Alice, with no income and four children to feed and educate. Somehow with the help of the extended family, she managed to send all the children to private schools. She scratched and saved to get enough money to send her son, Luke, through medical school. My mother, Helen, had desperately wanted to be a doctor, but could not be similarly financially supported. She eventually trained as a nurse instead. I cannot remember my mother ever pressuring me to study medicine, but somehow this always seemed to be my aim. We were all very proud of my uncle Luke Verco, who was a captain in the Army Medical Corps during the Second World War. I still have clear memories of him in his uniform. After the war, he became a country general practitioner, and my favourite uncle.

My parents married during the depression. Life was not easy. When I was born, we lived at the seaside suburb of Brighton, which I only remember from photographs taken by my father. In 1939, we moved to my grandmother's old home in the southern suburb of Unley. My earliest memories were of riding my little tricycle to a local private school, which would have roughly coincided with Japan's entry into World War II. For me, the war was rather unreal. We knew it was something to do with the Japanese up north, who were bombing Northern Australia. I recall attending an exhibition of Japanese submarines caught in Sydney Harbour. There were books about our brave soldiers fighting Germans in North Africa. For children in quiet little Adelaide these were all faraway events, almost an adventure story.

Other things that we saw, without really understanding them, were the almost total absence of cars, and the impossibility of getting electrical goods for the home. One car on our street used a 'gas producer', which burnt coke and produced carbon monoxide to fuel the motor. We used to watch these cars, with rusty black burners attached to the back, in some wonderment. A car near our home had a gas bag on top of the roof, as big as the car itself, filled with coal gas. Petrol was almost unavailable, but this had little effect on me. Food and clothing were rationed, but my mother seemed to be able to keep the family well fed and clothed, even if our idea of luxury was 'bread and dripping', made with fat and meat juices from the roasting dish, smeared on bread. When the war ended, we were able to buy a refrigerator. Previously, food was kept cool in an ice chest.

During and after the war, I attended the local public primary school, Westbourne Park School. Education in those days was much more by rote than today, particularly learning how to spell columns of words and memorize multiplication tables. I suspect the balance now has tilted too far the other way. With the onset of calculators and then computers, arithmetic was no longer done in the head, although I never regret the ability. Rote learning was actually only a small part of our education – certainly not all of it, as the media often seems to suggest today. My children seemed to be expected to pick up spelling as they went along, reading and studying literature, but with the onset of increasing home entertainment they seem to read less than we did. We used to listen to the radio, and I made myself a crystal radio set, which enabled me to hear the radio in bed. My mother made sure that I could get any books I wanted, and I was an avid reader of all the usual boy's adventure stories, both new and classic. I also used to read books about science.

As I grew older, I obtained my first bicycle, which I used to ride to school and visit my friends. In those days, the Adelaide suburbs were remarkably safe. You hardly needed to lock the house to go out (although mother always did). There seemed to be no worry about children being attacked. There were still very few cars on the roads, and the idea that it was unsafe for children to ride alone was a generation away. I also used to amuse myself riding in the Adelaide foothills, sometimes picking wild blackberries (which my mother made into beautiful jam) or catching occasional 'yabbies' (the local wild freshwater lobsters). I watched my father take family photographs with his old Voitlander camera, and I finally persuaded him to buy me a Kodak box camera for my 10th birthday. I soon obtained developing dishes and printing paper, and my lifelong hobby of photography was definitely started.

I was never good at sports at school, although I enjoyed a game of cricket (played at a pretty low level), but I did become more adventurous with my bicycle and camera, touring around the Adelaide Hills and taking landscape photographs. I was a bit of a loner, riding on my own and doing whatever I wanted, not having to worry about how fast or slow companions were. This did somewhat hinder my social skills, but no doubt made my eventual profession in pathology much easier.

As I entered my teenage years, I began my secondary education. This was at the oldest school in Adelaide, St Peter's College, the same school Lord Florey attended. At least two previous generations of Warrens went there. I found my father's name carved into a desk and dated 1917. Life there was quite different from the local primary school. It was much more competitive. The best students were in the top level, with four levels in each grade (year). They had an arrangement similar to English league football, with the bottom students in each term exams being relegated to the class below. This happened to me once, to my extreme annoyance, and I vowed to return to level one and stay there. Most students who later attended university were in the top class. The curriculum there included English literature, two foreign languages (usually Latin and French) mathematics (my favourite subject), physics and chemistry. All were required for matriculation to the science oriented university schools such as medicine or engineering.

Sport was an important part of the school curriculum. Unfortunately, I was never much good at it, and did not really enjoy it. However, I continued my weekend cycling and photography. I attended the school's army cadet unit, where I learned army skills and, presumably, instant obedience. I particularly enjoyed the rifle shooting. Target shooting, which I learnt in the school cadets, became the primary sporting activity of my adult life.

I matriculated from the school in 1954, gaining a Commonwealth scholarship. These scholarships were the initial attempt by the Commonwealth government to provide free tertiary education for the masses. I still think it was an excellent idea. There were just about enough scholarships to cover anyone who wanted to enter the major University schools, such as medicine or engineering, so most people attending the University were provided with free education. The situation is much more complex now, with the government paying the fees, but requiring them to be repaid by the students when they start earning money.

About this time I had my first experience of girls. I only had brothers, and somehow my mother did not count. Girls had always been very distant and strange people, especially the snooty-looking ones who walked past our front gate on the way to the local catholic convent school. (I married a catholic girl from a different catholic school, who thoroughly agreed with my description above!) Just before matriculation, the students were expected to attend dan-

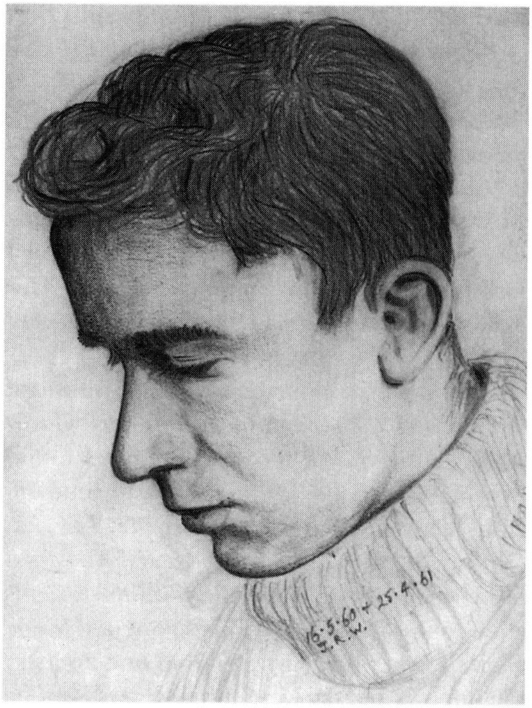

Figure 1. A self portrait, drawn while I was working for my final exams in Adelaide, 1960.

cing classes and then invite girls to the annual school ball. At first this seemed a strange affair, trying to learn ballroom dancing with a lot of girls who seemed just as shy as me. However, we soon got the idea, and we began to have a lot of pleasure, with groups of boys and girls developing personal relationships. There was none of the 'going steady' culture, which seems to dominate that period in America, and certainly a generation later with my children. For us, it was all very easygoing and pleasant, nothing serious.

Also, during my matriculation year, an event occurred that was to mark most of my life. One morning, my mother found me unconscious on the back lawn. I recovered without apparent ill-effect, but some time later it recurred. I was soon diagnosed as suffering grand mal epilepsy. I was placed on phenobarbital, and then a variety of other drugs to try and control the seizures, but control remained imperfect. I was unable to obtain a driving licence, a major event for young people at that age. There were apparently numerous comments to my mother at the time from people, both professional, friends and relatives, who should have known better. Many apparently suggested keeping me at home, no University, definitely no medical school, and so on. Thank goodness, none of this filtered through to me at the time; it made life difficult and embarrassing enough as it was at that rather sensitive teenage period. Mother had enough sense to give me a free rein to do what I liked, and none of this was ever mentioned. It was only years later that I came to appreciate just how much my mother had gone through to support my independence and personal maturation. My cycling in the hills continued, never interrupted by epileptic attacks or by admonitions that it was 'too dangerous.' Apparently, mother was worried sick, but she never said a word about it.

I obtained entry to the medical school of the Adelaide University in 1955, and the next stage of my life began. The first year was a wonderful entry to the university environment. Much of the work was a repetition of the final year at school. There was far more freedom. For the first time, we learnt about responsibility; we could work or not, as we liked. The only difference was that at the end of the year, a pass or fail was on the student's own shoulders. Help was always there, if we wanted it, but some of my colleagues revelled in the ability to do nothing. Luckily for me, I was enjoying the work too much to miss it. I particularly enjoyed botany and zoology, new subjects for me. I remember dissecting a frog and setting up its skeleton – my specimen showing a marked absence of any imagination, just bones glued to a piece of cardboard!

Both before and after starting university, I always read widely, including numerous scientific books and medical history books. Astronomy was a particular interest of mine at the time. I remember reading Fred Hoyle's books about the universe. I read the Oxford Junior Encyclopaedia, all 12 volumes. I probably should have spent as much time reading textbooks! Unfortunately, while I found textbooks fascinating to read as my curiosity and interest dictated, when I had to learn them for exams, they tended to lose their charm.

Medicine began in earnest with the preclinical years two and three. I was at the medical school almost full-time. The medical school was set apart from the main university buildings and the direct association with the university during first year was largely lost. Most of our time was spent in the anatomy department dissecting a cadaver or learning bone and joint structure and attachments, then the inner organs and the brain. I have never regretted the chance to learn anatomy (and, in later years, pathology) in such detail, which today's students do not have time for. While I do not remember all the anatomy, it soon comes back with a glance at Gray's textbook.

We also learned physiology, biochemistry, pharmacology, embryology and histology. I illustrated my notes for the latter two subjects with full colour sketches from the practical classes, using purple and pink pencils for haematoxylin and eosin. In those days there was only a very simple range of available drugs and, though we did not realise at the time, pharmacy and pharmacology were very simple in comparison to what the students need to know today.

The next adventure began with the clinical years four, five and six. In those days, the Royal Adelaide Hospital was the only general teaching hospital in South Australia. In both its scattered buildings and in its function, it was like a living museum. Some of the best wards were "temporary" buildings erected for World War I. One was named Verco Ward, after one of my mother's uncles. Some of the more Victorian wards were like an old English moving picture; a huge barn-like structure, with beds along each wall and Sister at the high table at one end. Nobody argued with Sister, especially not the students, who were assigned patients to study and follow-up and were treated as the bottom of the social ladder (by everyone but the nurses – who wanted future doctors to marry – and the patients, who generally seemed to appreciate our attention). This three-year term of surgery, medicine and obstetrics and gynaecology, passed very quickly. There always seemed to be something new and interesting around the corner.

After medical school, life became very busy. We were called 'Junior Resident Medical Officers' – the equivalent of today's interns. We actually were residents, living at the hospital residents' quarters. A second government hospital opened, and I obtained a position there. At that stage in life, I was still very much an innocent, with very little it exposure to the outside world of finance and employment. All graduating medical students were given Junior Resident positions. These entailed about 100 to 120 hours per week working, with very little payment (I received £17–10 per week). Nevertheless, I was young and fit, and the constant work was actually very enjoyable. I bought a Leica M3 camera, and started to turn my hobby to professional subjects, photographing interesting clinical lesions. I do not think I have ever worked in a branch of medicine that I did not enjoy.

This period changed my life around again. The new hospital also provided a second obstetrics facility for the medical school. The obstetrics students lived in the resident's quarters. Some of the students were young women. Their dormitory was near mine. We soon became good friends, and then I found myself very attached to one Winifred Williams. In fact, before we knew what had happened, I was spending all my spare time at her home. I remember one magic day, we wandered through the Botanic Gardens together, and it started to rain. Next thing we were arm in arm in the glasshouse, but we did not notice the flowers. Soon we were engaged and, a year later, married. For me, that was the biggest and best decision of my life, and so easy to do at the time.

I remained a little innocent about the big bad world outside. I did not realise how easy it was for us, with automatic positions provided – even if it was almost slave labour. Law students for instance, had to find clerking positions with private firms to complete their Articles of Clerkship, before they could practice at all. Other students had to find employment for themselves. I learned a valuable lesson at the end of my resident year. I assumed it was good manners to only apply for the second year position that I wanted, registrar in psychiatry. I did not get the post and found myself, to my surprise, unemployed. Luckily there were still a few positions available, and the one which appealed to me most was Registrar in Clinical Pathology at the Institute of Medical and Veterinary Science, attached to the Royal Adelaide Hospital.

In practice, 'Clinical Pathology' meant mainly laboratory haematology, which I thoroughly enjoyed. We had a good deal of freedom and responsibility. Although the usual work entailed reporting on blood smears and bone marrow, we had a wide range of other tasks, including examining faeces for parasites, examining urine and testing skin and nails for fungus. I put my drawing skills to work again, with detailed sketches of the various ova, amoebae and other organisms we saw. It really was an excellent all-round position to give one an overall feeling for pathology.

By the end of that year, Win graduated MB BS, and we had our first baby on the way. My pay was just about double what I earned in the intern year and, in pathology, my time of work was largely during the day. Life was settling into an enjoyable routine, if less adventurous. I had learnt about employment by them, and I applied for every position advertised at the end of the year. My first choice was Temporary Lecturer in Pathology at the Adelaide University, and I obtained the position. The work there consisted largely of morbid anatomy and histopathology, under the guidance of Professor Jim Robertson, which completed my overview of pathology and convinced me to go for membership of the (then) new College of Pathologists of Australia. Our baby boy, John, was born and we thought he was so wonderful, we had to start another. Of course, the baby soon started to crawl around, still wonderful, but a constant handful. By the time we realised our mistake, it was too late and our second baby was on the way. Win was still trying to fit in her intern year (it took her four years in bits and pieces to finish).

MELBOURNE

Again I applied for every position possible, both in Australia and overseas. As it turned out, I was offered the position of Clinical Pathology Registrar at the Royal Melbourne Hospital. We moved to Melbourne, for one of the happiest periods of our lives. The pathology community in Melbourne was much bigger and more active than in Adelaide, and Sydney was only a short distance up the coast. Everything seemed to be at our fingertips. The work was similar to that at the Medical and Veterinary Institute two years before. A couple of years of clinical pathology under the tutelage of Dr David Cowling and Dr Bertha Ungar enabled me to pass the college exams in haematology and microbiology.

After this, I became Registrar in Pathology, for training in morbid anatomy and histopathology. All hospital deaths received a post-mortem examination, for which we were responsible each morning. After morning tea, we examined the day's biopsy slides and presented them in a Grand Round, using a primitive old slide projector that was lit with two arcing rods of carbon. It really was quite an entertaining and educational show, with most of the surgeons present. Dr Doug Hicks, the head of department, would bark out questions to the resident and comments or answers to other questions. We had to work hard and fast, but Dr Hicks was an excellent teacher, and everybody learned from the show.

After four years in Melbourne, I finished my College membership and was a fully-fledged pathologist. Our second son, David, was born during our first year in Melbourne. Win somehow managed to fit in her intern year between babies. This was complicated by the unexpected arrival of twin sons, Patrick and Andrew, two years later. We had four sons under 3½ years old! After this we decided that the Catholic Church methods of contraception needed assistance! Win was now able to practice, if only part time. This was very difficult, in an era when "women stayed home". Part-time positions required a lot of talking, with promises to leave if she was not suitable (she was never asked to leave).

I was trying to obtain a position as pathologist at Port Moresby in Papua New Guinea, which was then in the process of obtaining independence from Australia. I thought I could obtain experience with some of the more exotic and unusual diseases. However, this had to be done through the Department of Foreign Affairs, and government red tape was extraordinarily slow. I was working in my room one afternoon, when a thickset man with a strong Germanic accent walked in and said, "You are working with me next year." and walked out. There was no chance for explanation or argument. I discovered he was Professor Rolf ten Seldam, the Professor of Pathology at the University of Western Australia and the Royal Perth Hospital. He was apparently a man no one argued with, or at least not twice. So I gave the Department of Foreign Affairs one day to settle the position in Papua and when they failed to reply, I accepted the position in Perth, Western Australia.

PERTH

We arrived in Perth in January 1968. In Melbourne, we had no trouble renting a flat for the family. In Perth, there was nothing to rent, especially with four children. We had little money and no experience of buying property. Luckily, our first real estate agent was a good choice. He found us a good house to buy and a rental property in the interim. What I had not realised was that he told the owners that I was 'the new doctor at the Royal Perth' – quite true, but the owner thought he meant the new medical superintendent, who had received a good deal of newspaper publicity, and not just the junior pathologist! Life quickly settled down.

Pathology at the Royal Perth Hospital was totally different from Melbourne. We went along at a much more leisurely pace, but with a much more flexible timetable. We did less work, but it seemed to take longer to do it. Perth was a small and isolated community. It has doubled in size over the last 30 years, with a population now of over one million. These days, most pathology training in Australia is in a single specialty. In Perth, this had always been the case, perhaps because of a lack of positions for training, other than haematology or morbid anatomy. Trainees were encouraged to keep the same position. In contrast, the eastern states, with most of Australia's population in one quarter of the continent, provided numerous training positions in all branches of pathology. General training was considered to be the norm, to give a broad base in all branches of pathology, before specialising in one area, as I did. I think this helped me later when dealing with *Helicobacter*, but it made me something of an oddity to my colleagues in Perth, who had all specialised in their particular branch of pathology from the start, giving them an in-depth knowledge on a narrow base.

In 1970, our last child, Rebecca, was born, our first daughter. Since Win only had sisters and I only had brothers, our knowledge of the difference between boys and girls was strangely superficial. Somewhat to our surprise, with no pushing from us, our daughter was quite different from the boys. Where they enjoyed running around in the dirt and pulling things to pieces, she would sit inside playing with dolls. After this, Win managed to get more work, and became quite an experienced general practitioner. She decided to specialise in psychiatry, and was accepted into the college training scheme, which took most of her time into the early 1980s.

HELICOBACTER

During the 1970s, I wrote up a few interesting cases and developed an interest in the new gastric biopsies that were becoming frequent. I also attempted to develop improved bacterial stains for use with histological sections, as I describe more completely in my Nobel lecture. Then, in 1979, on my 42nd birthday, I noticed bacteria growing on the surface of a gastric biopsy. From then on, my spare time was largely centred on the study of these bacteria. Over the next two years, I collected numerous examples and showed that they were usually related to chronic gastritis, usually with the active change described by Richard Whitehead in 1972. I attempted, with some difficulty, to obtain a negative control series, by collecting cases reported as normal gastric biopsies. This was more difficult than I expected, because all gastric biopsies were coded the same, wherever they came from in the stomach. Almost all so-called normal biopsies were from the corpus. Normal biopsies from the gastric antrum were very rare, but I eventually found 20 examples, and none showed the bacteria. With this material available, I began to prepare a paper for publication.

In 1981, I met Barry Marshall, and we agreed to undertake a more complete clinico-pathological study. He could cover the clinical aspects and provide improved biopsies, specimens for culture, clinical history and endoscopy findings. This resulted in our papers of 1983 and 1984, linking the infection to duodenal ulcer and culturing a new organism. My letter to the Lancet in 1983 was a summary of the paper that I was preparing when I first met Barry Marshall. After this Barry and I continued our association, but he moved to the Fremantle hospital. I was involved with the pathology related to several studies: Professor Goodwin, improving the accuracy of culture to diagnose the infection; Dr Ivor Surveyor, producing the breath test for diagnosis; and doctors Marshall and Morris, attempting to fulfil Koch's postulates to demonstrate that the bacteria was a pathogen.

Winifred, who was much more literate than me, used to read our papers. She was able to point out clichés or excessive jargon, and suggest ways of making the work more widely readable. Before I met Barry, Win was the only person to accept my work and encourage me. Considering that much of this work was done after hours or at home, thereby stealing her husband, she had every right to be annoyed. Particularly as she was a doctor, and knew the stan-

Figure 2. Barry and I working in the pathology department, Royal Perth Hospital, 1984.

dard teaching that nothing grows in the stomach, and therefore that I was trying to prove the 'impossible.' As a psychiatrist, she could have suggested I was mad. But she stood beside me and helped me when no one else would.

My last major work was the pathology for a large study by Barry Marshall *et al.* to show the effect of eradicating the bacteria on the relapse rate of duodenal ulcer. The study extended over seven years. It clearly showed that, after successful treatment of the infection, recurrence of peptic ulcer was rare; otherwise, it was usual. For me, the study provided a wealth of material to study the associated pathology. It soon became clear that active gastritis was very closely related to the infection. Treatment of the infection produced a very rapid and complete resolution of the active changes in the surface epithelium. Other changes, including lymphoid infiltration of the stroma and some epithelial changes, disappeared more slowly or not at all. The active changes varied considerably. The classic severe changes described by Whitehead were present in about 10 to 20% of cases. At the other end of the spectrum, some biopsies showed only occasional single intraepithelial polymorphs. These epithelial changes were almost absolutely related to the infection. With experience, I found the same features in the mucosa from the corpus, usually much more mild, superficial and focal than the antrum. Finally, the duodenal ulcers seemed to be distal pyloric ulcers rather than true duodenal ulcers. They appeared to arise in the distal pyloric mucosa, or perhaps the gastroduodenal junction.

By 1990, our findings began to be recognised by the medical community. We started to receive increasing numbers of honours and requests for atten-

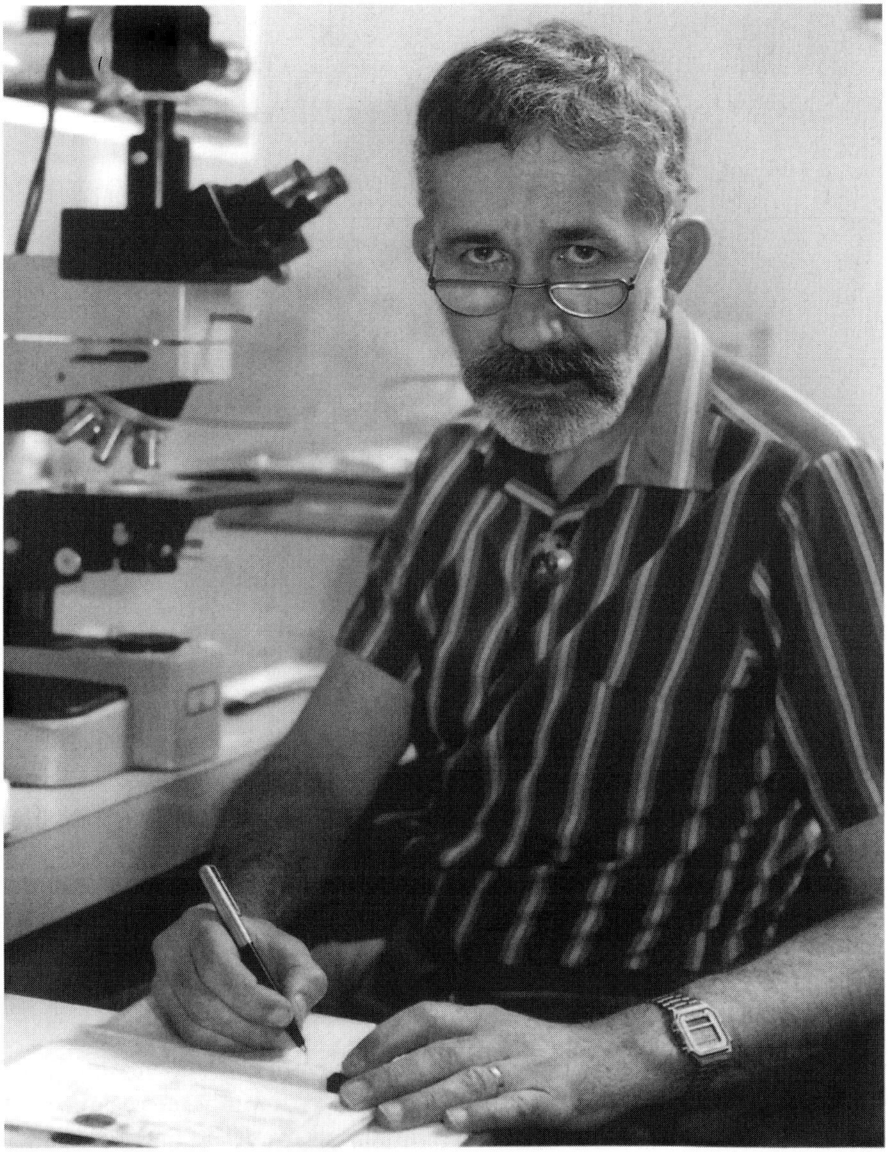

Figure 3. Working in my room at RPH, 1986.

dances at meetings and lectures. It had been an interesting decade. After our initial publications in 1983–84, a wealth of further studies appeared, most of them apparently just repeating our work, with similar results. No one proved we were wrong. Yet in spite of this, no one but patients and local general practitioners appeared to believe our findings. Many patients demanded treatment, and some GPs were very keen to treat them. Otherwise, it seemed that only our wives stood beside us.

In 1996, I was invited to Japan for a lecture tour. The following year, a

Figure 4. The new generation – with my new granddaughter in 1997.

three-month tour of Germany and adjacent countries was arranged by Manfred Stolte and Hansjörg Meyer. This provided us with some real recognition for our work, and it seemed the fighting was over. The tour was a wonderful working holiday for Win and me. However, soon after our return, Win experienced difficulty eating, and investigation showed duodenal obstruction due to an inoperable pancreatic carcinoma. Win gradually deteriorated and died four months later. After spending this time caring for her, I decided the time had come to retire.

At first, I spent most of my time trying to return to my hobby, photography. I intended to print my own pictures, using today's improved digital technology. The early results were interesting, although more complicated than I had expected. Digital pictures only provide a narrow band of information. Only 256 tones are available for each colour. Outside this range is either total black or total white, with no information. This is quite unlike photographic film, which shows a flattening of the curve at each end of the tonal range, with an almost continuous variation still present in light and dark areas. I had to put this project aside to try and digitise all my old publications, microphotographs and other works, since I was receiving many requests for them. Now I have to put that aside, because the Nobel Prize has brought a stream of requests for my presence at meetings and presentations.

HONOURS, AWARDS, PRIZES FOR MEDICINE.

Sixth International Workshop Campylobacter, Helicobacter and related organisms, 1991: Guest of honour.

Warren Alpert Foundation Prize 1994,
Harvard Medical School, jointly with Dr. B. J. Marshall, "for research that has led to improved understanding and treatment of a specific disease: identifying *Helicobacter pylori* as a cause of peptic ulceration."

Australian Medical Association – Western Australian Branch: 1995 Award.

Royal College of Pathologists of Australasia: Distinguished Fellows Award 1995,
"for Distinguished Service to the Science and Practice of Pathology."

The First Western Pacific Helicobacter Congress, February 1996: Inaugural Award,
"In recognition of his contribution to the advancement of medical science through the co-discovery of the gastric pathogen *Helicobacter pylori*."

The medal of the University of Hiroshima, September 1996.

The University of Adelaide Alumni Association: Distinguished Alumni Award, 24 October 1996,
"in recognition of your contribution to the healing of peptic ulcers, to the relief of human suffering and to huge world wide economic savings."

The Paul Ehrlich and Ludwig Darmstaedter Award 1997,
Paul Ehrlich Foundation, Johann Wolfgang Goethe-Universität, Frankfurt am Main, Germany, jointly with Dr. B. J. Marshall, "for your discovery of *Helicobacter pylori* as cause for peptic ulcer."

Guest speaker at the Centenary Meeting of the German Society of Pathology, Berlin, May 1997.

Honorary degree of Doctor of Medicine, University of Western Australia September 1997.

Guest speaker at the World Helicobacter meeting, Lisbon 1997.

Faulding Florey medal, 1998,
centenary of the birth of Lord Florey.

Australian Institute of Political Science, Cavalcade of Australian Scientists of the 20th Century,
for "excellence in intellectual endeavour and contribution to international scientific research." Presented at the Tall Poppy Dinner, Melbourne, May 18, 2000.

Nobel Laureate in Physiology or Medicine for 2005,
for "the discovery of the bacterium *Helicobacter pylori* and its role in gastritis and peptic ulcer disease."

HELICOBACTER – THE EASE AND DIFFICULTY OF A NEW DISCOVERY

Nobel Lecture, December 8, 2005

by

J. Robin Warren

Perth, WA 6000, Australia.

PREFACE

This is the story of my discovery of *Helicobacter*. At various times I have been asked: did I steal the discovery; did I find it by accident; did it follow some brilliant research work; or was it serendipity. My answer to most of these is a definite "No." Obviously, as with any new discovery, there is an element of luck, but I think my main luck was in finding something so important. I think the best term is serendipity; I was in the right place at the right time and I had the right interests and skills to do more than just pass it by. First, let us examine this.

Before 1970, well-fixed specimens of gastric mucosa were rarely seen in clinical practice. Biopsies, taken with the rigid gastroscope or the suction method, were very uncommon. Gastrectomy specimens are clamped at each end, with the contents inside. They fix slowly from the outside. Meanwhile the mucosa autolyzes and any organisms disappear. Autopsy specimens are even worse. Most surgical specimens were taken to remove tumours or ulcers and pathology descriptions centred on this rather than the fine histology of the mucosa. If they described gastritis at all, pathologists gave it names such as 'superficial' or 'atrophic,' which showed little real relationship to the histology.

Since the early days of medical bacteriology, over one hundred years ago, it was taught that bacteria do not grow in the stomach. When I was a student, this was taken as so obvious as to barely rate a mention. It was a "known fact," like "everyone knows that the earth is flat." Known facts can be dangerous; to quote Sherlock Holmes (Conan Doyle, *The Boscombe Valley Mystery*) "There is nothing more deceptive than an obvious fact." As my knowledge of medicine and then pathology increased, I found that there are often exceptions to "known facts." In the stomach, organisms, usually yeast or fungus, often grow in the necrotic debris in ulcers or tumours. Unusual infections sometimes do involve the gastric wall. Once I saw tuberculosis. Bacteria, floating above the mucus layer on the epithelium, are often seen in gastric biopsies. They appear to be mixed varieties, probably just passing through, dead, or contaminants; they are relatively sparse in cultures.

The introduction of the flexible endoscope changed all this. It enabled gastroenterologists to biopsy many of their patients. Small biopsies, placed

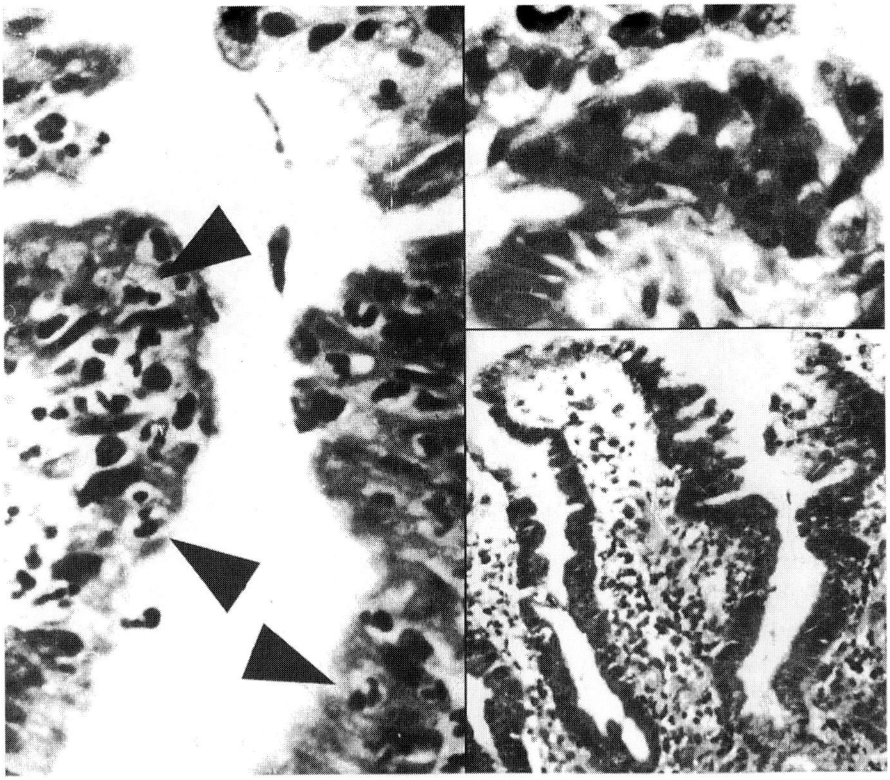

Figure 1. Whitehead's illustration of active change shows gross distortion of the superficial epithelium (above) and intra-epithelial polymorphs in the neck of a gland (arrowheads). (Whitehead R. Mucosal Biopsy of the Gastrointestinal Tract, 1st edition, figures 15, 16, 17, pages 20–22. © 1973 Elsevier Inc., reprinted with permission.)

immediately into formalin, fixed well. Instead of rare, these became some of our most frequent biopsies. Whitehead accurately described them in 1972. He described 'active' changes, which become important in my story. His pictures of this (figure 1) show intraepithelial polymorph infiltration in the necks of the gastric glands and a remarkable distortion of the foveolar (surface) epithelium. These features proved to be quite common and easy to diagnose. They were remarkably consistent in appearance, although often much more focal or mild than in the original illustrations (figure 2 and 3). The changes were superficial, usually involving only the epithelium.

Whitehead devised a classification based on the features he actually saw and described. Most of these features are mentioned in the diagnosis. This allows any associations between histology and other clinical features to be noted. I was very impressed with Whitehead's work. I simplified his classification for my own use (table), and the pathology of the stomach suddenly seemed to make sense. The diagnosis describes in one short line the features actually seen.

Microbiological stains are excellent for staining bacteria in smears, espe-

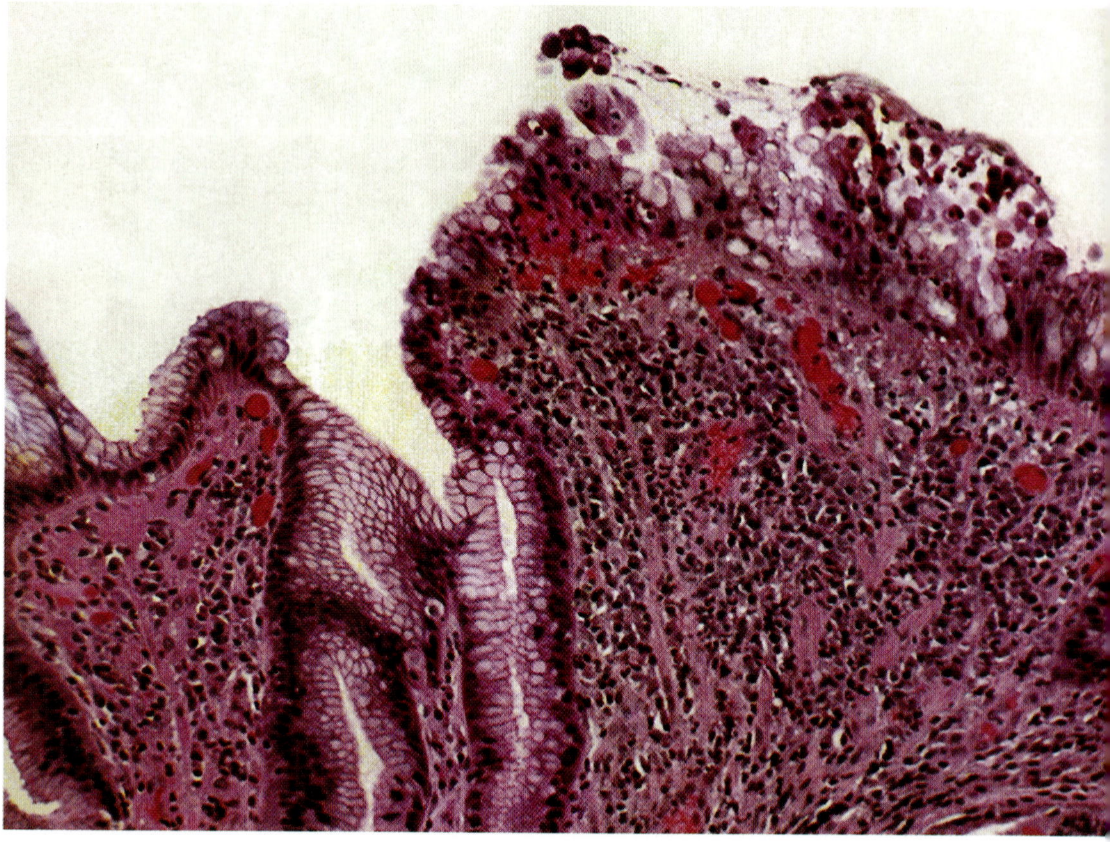

Figure 2. The surface (foveolar) epithelium to the right shows a focus of gross epithelial irregularity, of the type described by Whitehead. Elsewhere the epithelium shows only mild non-specific changes. In many biopsies the changes are often much milder than shown here (H&E x100).

cially from a clean culture. However, histology shows a complex mass of tissue structures that also stain. To see bacteria, it is necessary to contrast them with the tissue. Gram positive organisms and acid fast organisms contrast with tissue sections. Warthin-Starry silver stain of tissue shows spirochaetes (in

Table. My simplification of Whitehead's Classification of Gastritis

Pathology	Description
Severity	Mild, Moderate, Severe
'Active'	Active (if present)
Type of Inflammation	Acute, Chronic etc.
Other features present	Atrophy, Metaplasia, Dysplasia, Reduced mucus secretion

Using this table, the diagnosis may be written as a single line. In the following example, replace the headings (in brackets) with the appropriate descriptive terms.

Diagnosis: (Severity) – (Active?) – (Type) – gastritis – (with any other features).

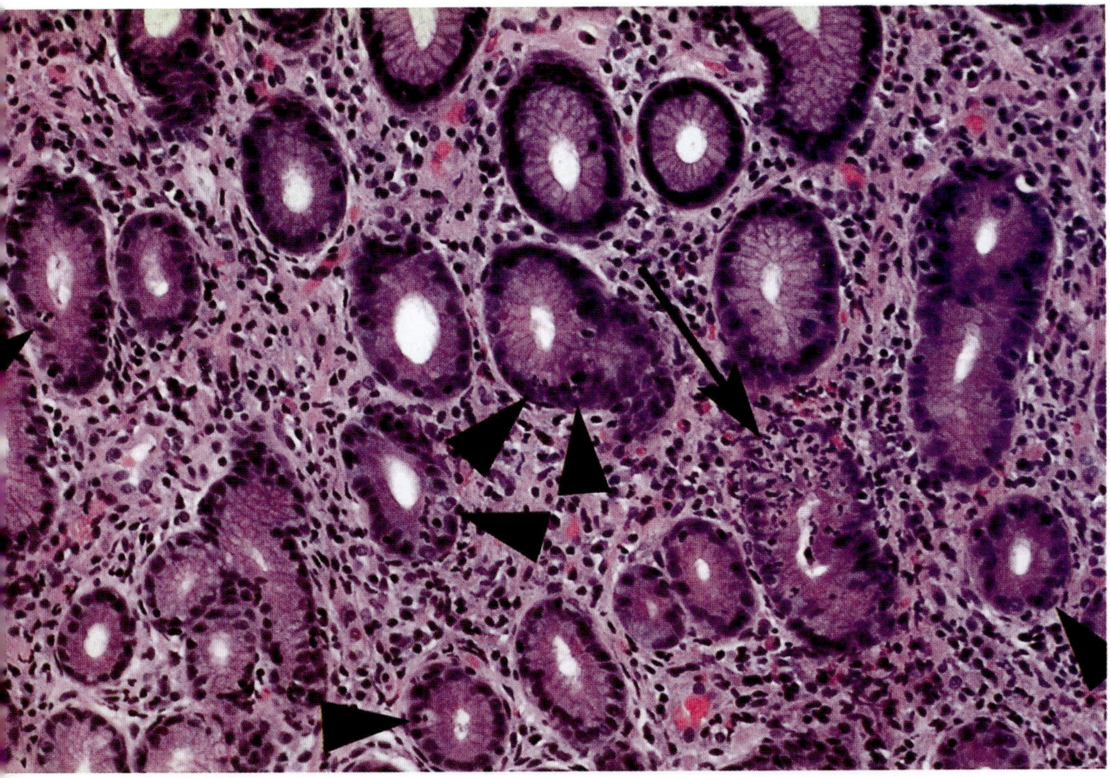

Figure 3. The section is cut obliquely through the necks of the gastric glands. This shows numerous gland necks in transverse section, lined by foveolar type epithelium. Glands are visible in the lower area, lined by smaller mucus-secreting cells. Polymorphonuclear leucocytes infiltrate the epithelium of the neck of one gland (arrow). There are also individual PMN's in other gland necks (arrow heads). Sometimes a few of these is all that is found, and the infiltration is often focal, as shown here (H&E x100).

syphilitic chancres) and bipolar Donovan bodies (the Gram negative bacilli in granuloma inguinale). I was interested in microbiological stains. After seeing several cases of granuloma inguinale in which the bacteria were clearly visible with the silver stain, I was experimenting with this stain on other Gram negative organisms, with variable success.

Thus, I was a young pathologist when high quality gastric biopsies became frequent. By 1979, I had a particular interest in gastric pathology, based on Whitehead's work and, in particular, his description of active gastritis. I was interested in bacterial stains, especially the use of silver stains for Gram negative bacilli. In addition, electron microscopy had recently started in our department. I found this interesting, giving another dimension to histology. Finally, I was interested in drawing specimens, and also in photography, both of which helped me to discern detail.

DISCOVERY: THE EASY PART

My adventure with *Helicobacter* began in June 1979. A routine biopsy showed severe active chronic gastritis (figure 4). The epithelium showed gross cob-

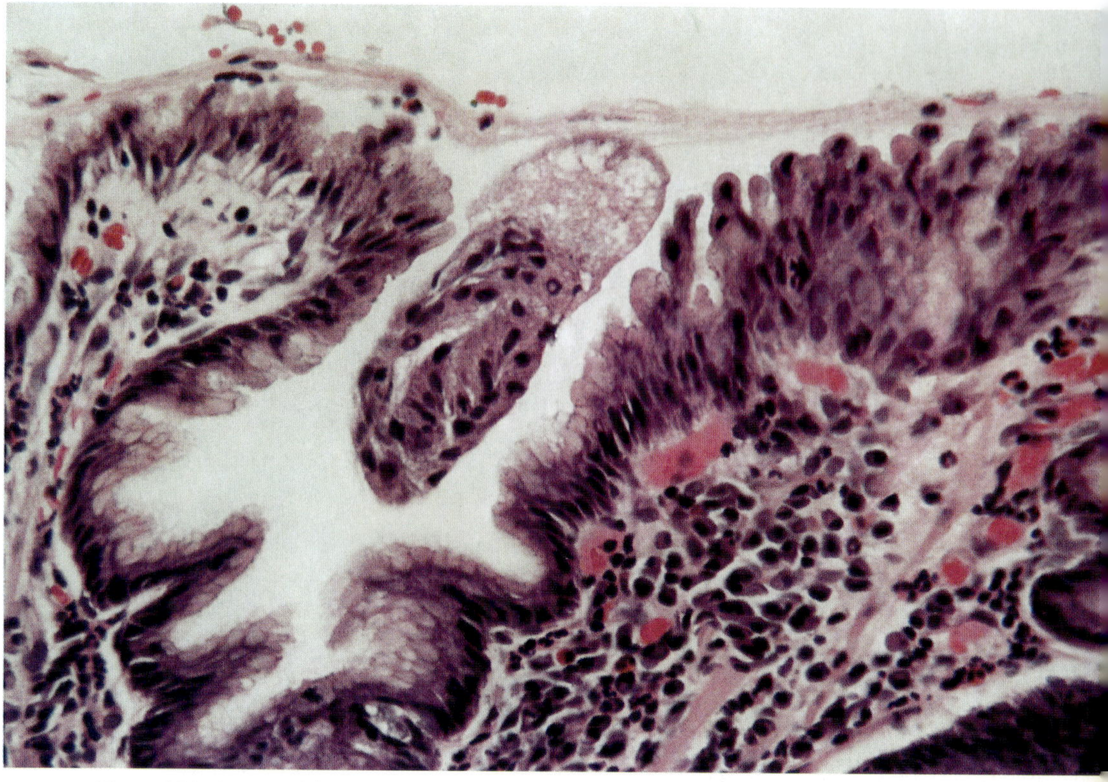

Figure 4. My first case. The epithelium shows gross cobblestone change, most marked to the right, resembling Whitehead's 'active' change. A thin blue line on the surface shows bacteria at high power (H&E x100).

blestone change, very similar to Whitehead's description. Nuclei were out of alignment. Mucus secretion showed a marked patchy reduction. Focal intraepithelial polymorphonuclear leucocytes were present (figure 5). There were numerous lymphocytes and plasma cells in the stroma. A thin blue line was visible on the surface, which on high power I thought consisted of numerous bacteria. My colleagues could not see them, so I stained them with the Warthin-Starry silver stain and numerous bacteria were easily visible at low power. At high power (figure 6), they were obviously small curved and spiral bacilli, closely applied to the epithelial surface and often arranged in palisades.

I took tissue from the wax block used for standard histology and obtained the electron microscopy. The images were of good quality and showed the bacteria well (figure 7). There were small curved bacilli closely applied to the surface. Some were attached to microvilli. The top of the cells bulged out. Mucus secretion was reduced. Bacteria were infiltrating between the bulging tops of the cells. They were not obviously penetrating past the cell junctions; however they may do so, because occasional bacterial fragments were present in the superficial stroma.

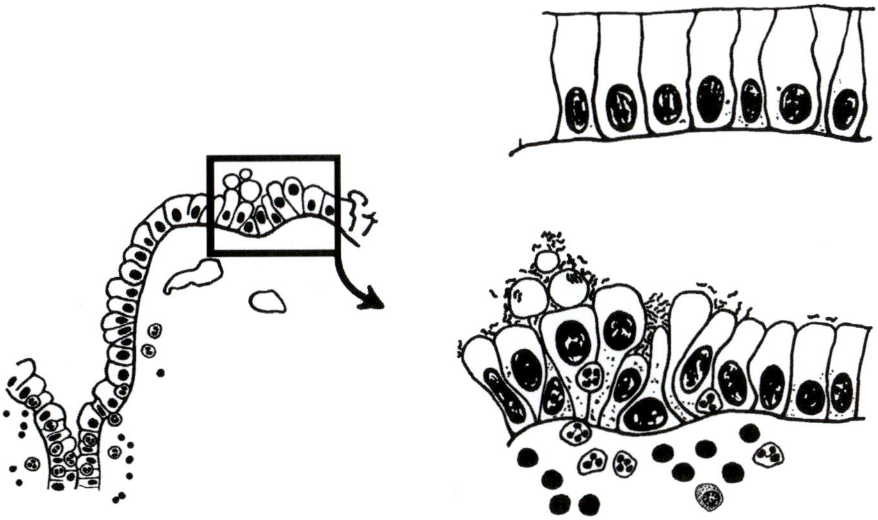

Figure 5. Diagram from my first case shows active changes in the infected epithelium (below). Normal (above) shows a flat surface and well aligned basal nuclei.

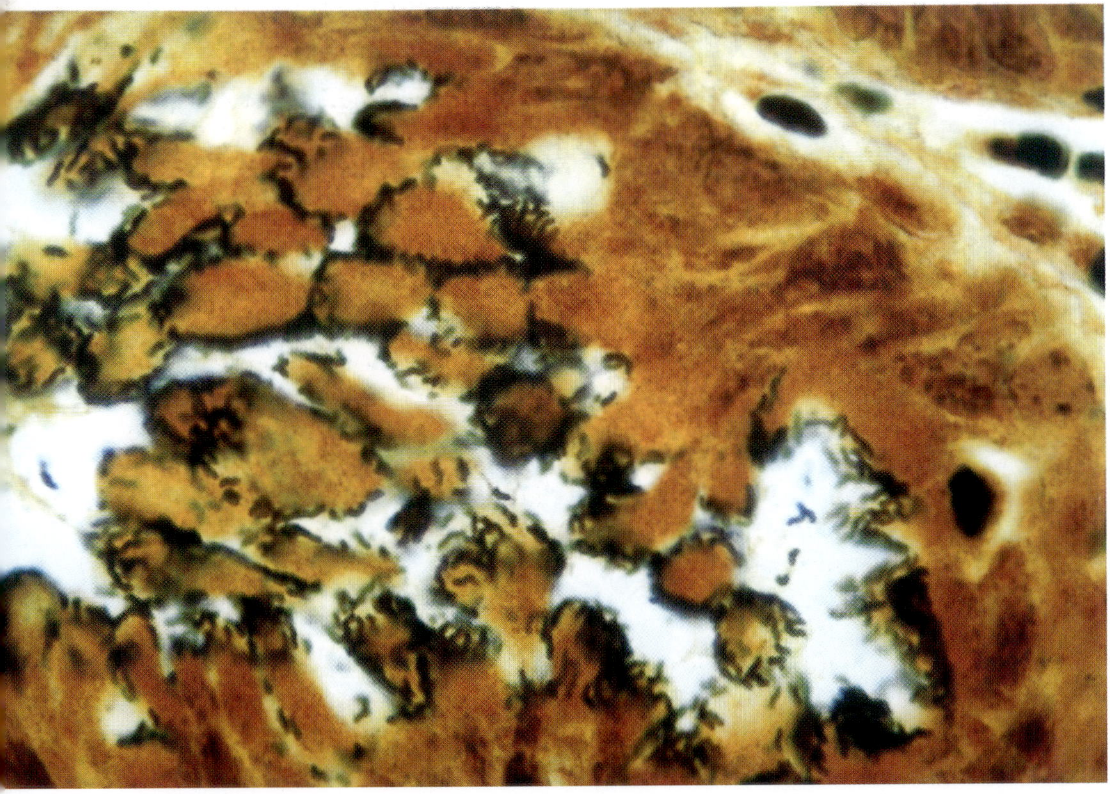

Figure 6. My first case. High power view with the silver stain shows numerous curved bacilli on the distorted epithelium (Warthin Starry x 1000).

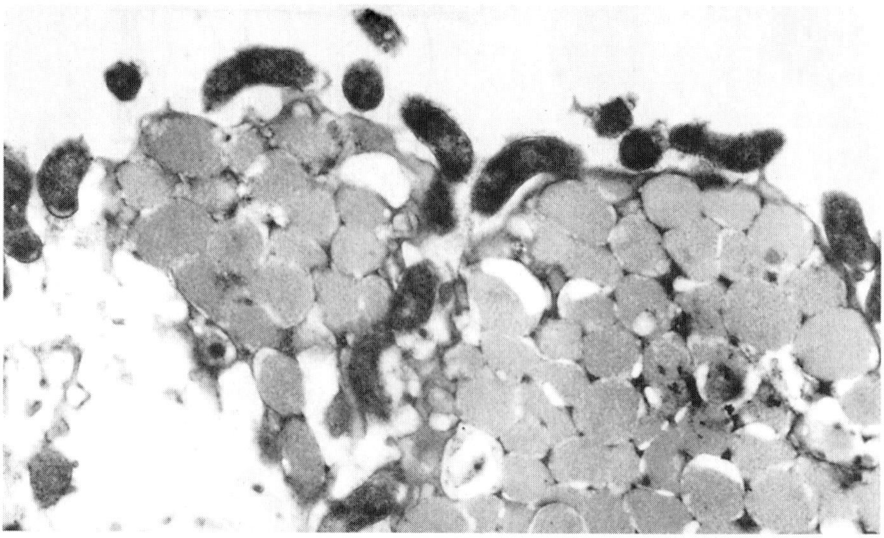

Figure 7. My first case. High power electron microscopy shows the top of two epithelial cells bulging out, with small curved bacilli closely applied to the surface. Few microvilli are seen.

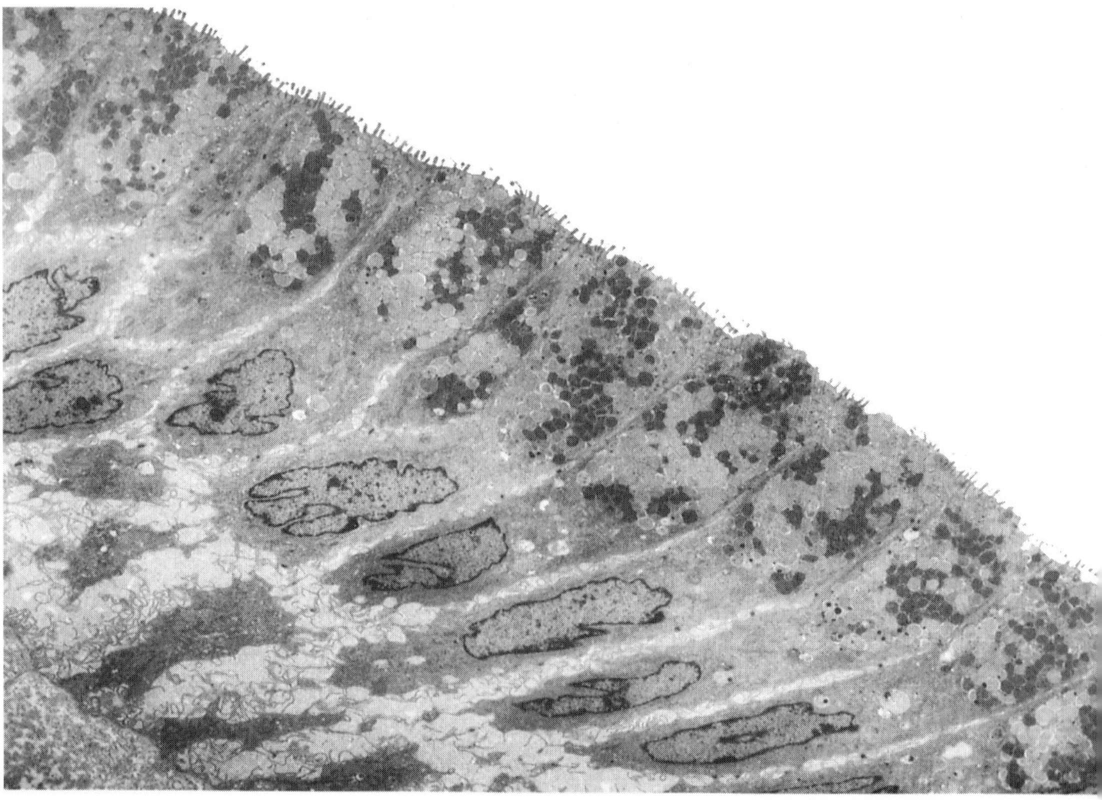

Figure 8. Electron microscopy, low power, of normal foveolar epithelium shows a flat surface with numerous tiny microvilli just visible.

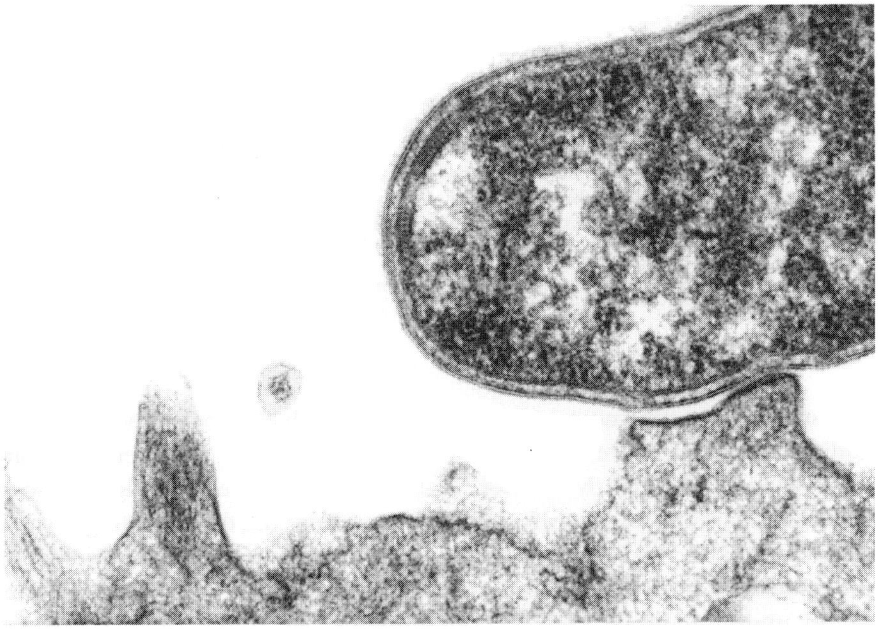

Figure 9. Very high power electron microscopy shows how the bacteria attach to the surface microvilli and flatten them. Bundles of filaments are visible within the microvilli to the left.

Electron microscopy demonstrates the normal anatomy of the columnar (foveolar) epithelium and the mechanism of the active change. The normal epithelium shows a flat surface, but there are numerous tiny microvilli (figure 8). The microvilli contain bundles of filaments that attach to the top of them. These filaments normally extend through the cells and attach to the cell base, giving the cells a rigid structure. This fixes their shape and also maintains their internal architecture, with basal nuclei and superficial mucus secretion. The normal columnar epithelium can be scraped from the mucosal surface, smeared onto a glass slide and still retain its columnar structure on cytological examination. *Helicobacter pylori* attach to the microvilli (figure 9) and often flatten and destroy the microvilli. The filaments become detached and the cells loose their structure. They behave in an amoeboid fashion, with nuclei floating through the cytoplasm and the surface bulging out.

My colleagues finally believed the bacteria were there. However, they doubted their importance, and challenged me to find any more cases. I thought they were worthy of further study (figure 10) so I continued to search and, to my surprise, I found them in quite a significant number of biopsies. The number increased with experience. Many cases showed only mild pathology, but the basic changes were still present. Eventually I was finding them in about a third of the gastric biopsies.

Another interesting feature gradually became apparent as my experience increased. I found the bacteria were easily visible on many surgical specimens. They were only seen along the cut edge of the specimens, where a nar-

Conclusion:

There is chronic gastritis with a small erosion. The quality of the surface mucus appears slightly more dense than normal in many areas, and it contains numerous bacteria in close contact with the surface epithelium. These bacteria have the morphology of Campylobacter. They appear to be actively growing and not a contaminant. I am not sure of the significance of these unusual findings, but further investigation of the patient's eating habits, gastro-intestinal function and microbiology may be worthwhile.

Figure 10. My original conclusion when I first reported the bacteria.

row strip of mucosa came into rapid contact with the formalin fixative. In addition, they were often mixed with a variable number of spherical organisms, particularly slightly further (2–3 mm) from the cut edge. It soon became apparent that the spherical organisms were the degenerating form of *Helicobacter*. This strip of 'mixed' organisms, only seen along the cut edge of the specimen, probably helps explain the absence of past reports. They would undoubtedly be seen as contaminants. We found these specimens a very useful source of positive control specimens when performing the bacterial stain.

DIFFICULTIES

I was unable to convince the clinicians of the importance of the organisms. Generally, they did not believe they were there at all. 'Everybody knows the stomach is sterile'. Gastritis was not considered to be of much significance anyway. Most thought that if the bacteria were there, they were just secondary to the gastritis. The histology suggested the opposite to me, but it was hard to prove. Another common question was 'If they are there, why has not anyone described them before?' At that stage I did not know why I had not seen them, let alone no one else.

It has become apparent over the years that gastric bacteria have been described many times over the last 100 years (ref 4). However, these descriptions were not generally known. Most of them were either veterinary biopsies or from research animals, which provided well fixed specimens without regard for 'patient' well-being. Most descriptions were looked on as peculiarities, of no particular importance, even by their authors. The apparent absence of any previous report was given to me as one of the main reasons why they could not be there at all.

I worked in a laboratory, without patient contact. Although the tissue quality was far better than it had been before the flexible endoscopes, most gastric biopsies were taken from visible lesions such as ulcers, to diagnose or exclude carcinoma. As a result, the histology often showed the effects of the nearby lesion. I needed biopsies from apparently intact antral mucosa, to show the effects of the bacteria without the competing effects of other lesions. The idea of taking gastric biopsies for culture was considered ludicrous. The patient's well-being was the prime consideration.

Acute inflammation in the stroma is not specific for *Helicobacter* infection, and is often due to nearby ulceration. As might be expected with a surface infection, only superficial polymorphs within the epithelium are closely associated with the infection. Flattening of the foveolar epithelium is often due to the healing edge of an ulcer, particularly when associated with grossly reduced mucus secretion, gland atrophy or stromal fibrosis and polymorph infiltration. *Helicobacter* is often rare in such areas, even when it is plentiful on nearby intact mucosa.

After two years I had collected many cases and was almost ready to publish my findings. Then Barry Marshall, the new gastroenterology registrar, came to my room and asked to see my work. He had been told to find a research project, and since he did not like the one suggested, his superiors sent him to me. He was the first person to show any interest in my work, so I showed him. He did not seem impressed at first, but he agreed to send me a series of biopsies from apparently normal gastric antrum, to see if the same findings were present. He soon became more enthusiastic, and I finally had a clinical collaborator.

SUCCESS

In 1982, we obtained biopsies for culture and histology from 100 consecutive outpatients referred for gastroscopy. Most of them complained of peptic symptoms or pain, so this could not be investigated. They all completed a detailed clinical protocol that listed every symptom Barry could think of.

The results were totally unexpected. First, the bacteria were not related to any significant symptoms, only bad breath and burping. The gastroscopy reports were surprising. They showed that the gastric infection was most closely related to duodenal ulcer. Most gastric ulcers were associated with the infection, but every patient with a duodenal ulcer was infected. "Gastritis," as observed on gastroscopy, was not related to either the histology or the bacteria.

At first, no bacteria were cultured. Finally, plates incubated for five days over the Easter holiday showed a culture of a new type of bacteria, not described previously. The microbiology technicians had previously treated our research culture plates as routine cultures and discarded negative plates at 48 hours. After this, the plates were allowed to mature, and several more cultures were obtained. The bacteria showed many features of *Campylobacter*, but they were unusual and were eventually considered to be a new genus, now termed *Helicobacter*.

I sent a letter to the Lancet in 1983, a summary of the work I had done before I met Barry (ref 1). Barry sent an accompanying letter describing our joint work. He also presented our findings at the Brussels Campylobacter conference. Martin Skirrow, who chaired the conference, was very impressed with our work.

We sent our definitive paper to *the Lancet* in 1984 (ref 2). Although the editors wanted to publish, they were unable to find any reviewers who believed our findings. Our contact with Skirrow became crucial here. We told him of our trouble, and he had our work repeated in his laboratory, with similar results. He informed *the Lancet* and shortly afterwards they published our paper, unaltered.

I continued as a clinical pathologist, with an interest in *Helicobacter*. The subject rapidly expanded throughout medicine over the next decade. The original methods for diagnosis and treatment were all suggested by Barry. I was involved with the pathology from: two attempts to fulfil Koch's postulates; the development of the breath test for diagnosis; improved methods of culture; studies of duodenal ulcer.

Helicobacter patients show considerable variation. I was involved with these early examples.

- Barry gave himself a severe active gastritis, to the disgust of his wife, in an attempt to fulfil Koch's postulates.
- Morris, in New Zealand, gave himself chronic gastritis and took years to cure it.
- My wife developed arthritis and as soon as she took NSAIDs she developed severe epigastric pain. Stopping the NSAIDs reversed this. And again. I sent her to Barry, who found *Helicobacter*, treated it and she was able to take the NSAIDs. Do not take it for granted that NSAIDs are the only guilty party.
- Most patients are symptomless. This was actually one of our major difficulties. I was an example. After she was treated, my wife complained that I had bad breath. I was positive for *H pylori* and after treatment marital bliss returned.

ACTIVE GASTRITIS

In 1986, we undertook a double blind trial to find the effect of treatment of *Helicobacter pylori* infection on ulcer relapse (ref 3). All patients received treatment for their ulcers. They received antibacterial therapy or placebo for *Helicobacter* infection. All were examined by multiple gastroscopies and biopsies for 12 months and again after 7 years. This provided me with excellent material for the study of the pathology related to *Helicobacter* and, also, the pathology of duodenal ulcers.

I quantified the grade of gastritis on a 0–36 scale by giving a value 0–9 for each of the main four features seen with active gastritis: intraepithelial polymorphs; typical epithelial distortion; reduced mucinogenesis in the foveolar epithelium; increased stromal lymphoid cells (a non-specific change seen with all chronic inflammation). This gave easily obtainable and remarkably consistent grades of gastritis for each case. From these results I made a histogram to show the grades of inflammation before and after eradication of *H pylori* (figure 11).

The grade of gastritis when *Helicobacter pylori* was present was usually above 20. This includes all patients in the study, including pre-treatment biopsies of

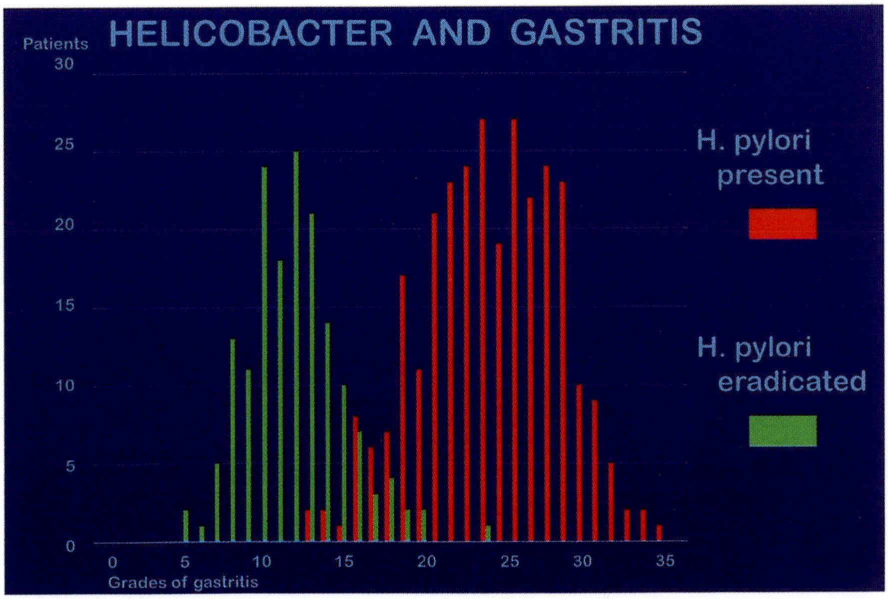

Figure 11. Histogram, comparing gastritis before and after eradication of *H pylori*. The normal range is (0–14), in the absence of pre-existing disease or infection.

those in whom the bacteria were later eradicated. Biopsies were taken 2 weeks after treatment. After successful eradication of *H pylori*, the active changes disappeared very quickly, and the grades in the histogram for these patients were mainly below 20. The true normal range is 0–14, but our cases show treated active gastritis, many biopsies taken only 2 weeks after treatment, not random normal samples. The stromal lymphoid cell infiltration disappeared more slowly, over about twelve months or more.

The absolute difference between the two groups is very impressive. There is some overlap, but the difference in the gastric pathology with and without *Helicobacter pylori* is incontrovertible (figure 11). One interesting feature was the consistency of the results over time. Repeated biopsies from each patient showed remarkably constant histological features throughout the 7 years of the study, as long as the bacteria remained. The active changes vanished as soon as the bacteria were eradicated, within weeks. This strongly suggests the bacteria caused these changes. 'Active' changes are almost never seen in the absence of *H pylori*. Other changes remained longer, particularly structural damage such as scarring, and epithelial changes such as gland atrophy, metaplasia and dysplasia.

DUODENAL ULCER

We were surprised to find duodenal ulcer so closely related to *Helicobacter*. However, further investigation shows that most duodenal ulcers can be viewed as distal pyloric ulcers. They are in the duodenal cap and the pyloric

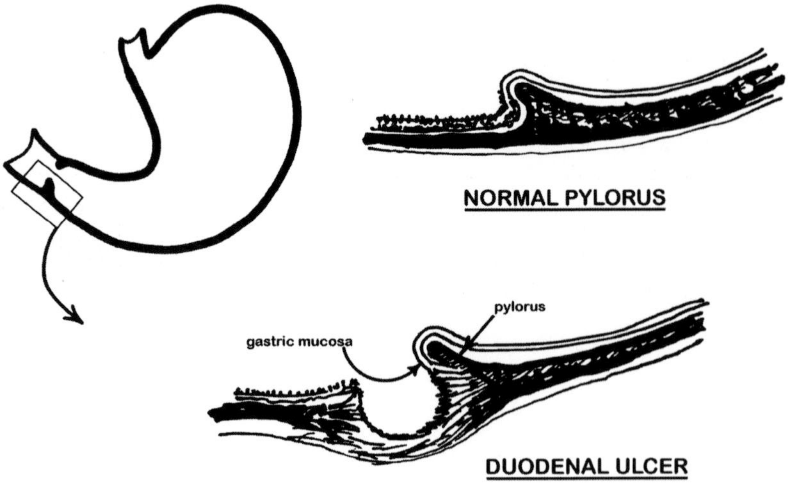

Figure 12. Diagram of the pylorus; the gastric mucosa normally extends into the proximal duodenum, and forms the proximal border of most duodenal ulcers.

mucosa normally extends through the pylorus (figure 12). Biopsies from the proximal border of all duodenal ulcers in this study showed either gastric mucosa or scarred mucosa, consistent with a gastric origin and with no apparent Brunner's glands, as seen in duodenal mucosa.

The pyloric mucosa is very mobile and easily moves some distance through the pylorus. When the stomach contracts, a mixture of food fragments and corrosive gastric juice squirts through the pylorus. Perhaps it is not surprising that ulcers are so common here, especially when the epithelium is damaged by infection and active inflammation.

CONCLUSION

Now, the importance of *Helicobacter* is generally recognised, particularly with regard to duodenal ulcer. As a pathologist, I am disappointed that active gastritis is not considered worthy of treatment. I see it in all infected stomachs, although often mild. Unfortunately, it does not cause many symptoms and nobody is interested. In conclusion, we now know that *Helicobacter* had been seen and largely ignored for over 100 years. I saw them 25 years ago and linked them with active gastritis. Barry Marshall and I cultured the bacteria and linked them to duodenal ulcer. In various different ways over the next few years we proved these relationships.

REFERENCES

1. Warren, J. R. Unidentified curved bacilli on gastric epithelium in active chronic gastritis. (letter) *Lancet 1983; i: 1273.*
2. Marshall, B. J.; Warren, J. R. Unidentified curved bacilli in the stomach of patients with gastritis and peptic ulceration. *Lancet 1984; i: 1311–1315.*
3. Marshall, B. J.; Goodwin, C. S.; Warren, J. R.; Murray, R.; Blincow, E. D.; Blackbourn, S. J.; Phillips, M.; Waters, T. E.; Sanderson, C. R. Prospective double blind trial of duodenal ulcer relapse after eradication of *Campylobacter pylori. Lancet 1988; ii: 1437–1442.*
4. Marshall, B. J. Ed. Helicobacter Pioneers. Firsthand accounts from the scientists who discovered helicobacters, 1892–1982. *Blackwell Publishing, Melbourne 2002.*

Portrait photo of J. Robin Warren by photographer U. Montan.